INFERENTIAL PROBLEMS IN SURVEY SAMPLING

INFERENTIAL PROBLEMS IN SURVEY SAMPLING

PARIMAL MUKHOPADHYAY
Professor
Computer Science Unit,
Indian Statistical Institute
Kolkata, West Bengal

Publishing Globally

NEW AGE INTERNATIONAL (P) LIMITED, PUBLISHERS
New Delhi • Bangalore • Chennai • Cochin • Guwahati
Hyderabad • Kolkata • Lucknow • Mumbai
Visit us at **www.newagepublishers.com**

Published by New Age International (P) Ltd., Publishers
First Edition: 1996
Second Edition: 2016

BRANCHES

- **Bangalore** 37/10, 8th Cross (Near Hanuman Temple), Azad Nagar, Chamarajpet Bangalore-560 018, Tel.: (080) 26756823, Telefax: 26756820 **E-mail: bangalore@newagepublishers.com**
- **Chennai** 26, Damodaran Street, T. Nagar, Chennai-600 017 Tel.: (044) 24353401, Telefax: 24351463 **E-mail: chennai@newagepublishers.com**
- **Cochin** CC-39/1016, Carrier Station Road, Ernakulam South, Cochin-682 016 Tel.: (0484) 2377303, Telefax: 4051303 **E-mail: cochin@newagepublishers.com**
- **Guwahati** Hemsen Complex, Mohd. Shah Road, Paltan Bazar, Near Starline Hotel Guwahati-781 008 Tel.: (0361) 2513881, Telefax: 2543669 **E-mail: guwahati@newagepublishers.com**
- **Hyderabad** 105, 1st Floor, Madhiray Kaveri Tower, 3-2-19, Azam Jahi Road, Near Kumar Theater Nimboliadda, Kachiguda, Hyderabad-500 027, Tel.: (040) 24652456 Telefax: 24652457 **E-mail: hyderabad@newagepublishers.com**
- **Kolkata** RDB Chambers (Formerly Lotus Cinema) 106A, 1st Floor, S N Banerjee Road Kolkata-700 014 Tel.: (033) 22273773, Telefax: 22275247 **E-mail: kolkata@newagepublishers.com**
- **Lucknow** 16-A, Jopling Road, Lucknow-226 001, Tel.: (0522) 2209578, 4045297 Telefax: 2204098, **E-mail:lucknow@newagepublishers.com**
- **Mumbai** 142C, Victor House, Ground Floor, N.M. Joshi Marg, Lower Parel Mumbai-400 013, Tel.: (022) 24927869, Telefax: 24915415 **E-mail: mumbai@newagepublishers.com**
- **New Delhi** 22, Golden House, Daryaganj, New Delhi-110 002 Tel.: (011) 23262368, 23262370, Telefax: 43551305 **E-mail: sales@newagepublishers.com**

ISBN: 978-81-224-3855-0

₹ 799.00

C-15-02-8264

Printed in India at Glorious Printers, New Delhi.
Typeset at Excellent Graphics, Delhi.

PUBLISHING GLOBALLY
NEW AGE INTERNATIONAL (P) LIMITED, PUBLISHERS
7/30 A, Daryaganj, New Delhi - 110002
Visit us at **www.newagepublishers.com**

Dedicated to the memory of my father,
Late Jnan Sagar Mukhopadhyay

Preface to the Second Edition

After the good response of the earlier editions it is a pleasure for us to present the second edition. In this edition the style pattern of the book has been modified to give a new look. Necessary corrections are incorporated in response to the suggestions made by many teachers and students and all attempts are made to make the book error free. We would like to acknowledge the valuable advice and suggestions of many teachers and students who used the earlier edition. We would be grateful to our esteem readers for suggestions in the interest of the subject.

Publisher

Preface to the First Edition

During the last few decades theory of Survey Sampling has undergone rapid growth, often involving speculations about the nature of the finite population at hand. The purpose of this book is to make a comprehensive review of the overall development of the theory in the fixed population approach and in the setup which essentially needs superpopulation model for its thriving.

Chapters 1 through 3 are based on the model of fixed population. Chapter 1 introduces preliminary concepts, sampling designs, estimators, sampling strategies, etc. Chapter 2 provides an overview of various unequal probability without replacement sampling designs and estimation of variance of estimators based on these designs. Chapter 3 addresses some inferential problems, e.g. uniformly minimum variance unbiased estimation, admissibility, sufficiency and minimax estimation in survey sampling. The remaining chapters have developed essentially from the assumption of some superpopulation models depicting the survey population. Chapter 4 considers model-dependent optimal strategies in the prediction theoretic approach; Chapter 5 deals with the robustness of these strategies under specific model failures; Chapter 6, the class of strategies which combine randomisation both due to sampling design and superpopulation models; Chapters 7 and 8 consider some asymptotic properties of these strategies and their robustness. The concept of near-optimal sampling strategies under a class of priors are considered in Chapter 9. The concept of ϕ-optimality in survey sampling and its role in reconciliating different optimality criteria are discussed in the next chapter. Lastly, we consider various replication and resampling techniques useful in variance estimation in complex surveys and model-based estimation of conditional variance. In the treatment of topics we have tried to systematically arrange the main results, supplied short proofs, examples, explanatory notes and remarks and indicated research areas. Our hope is that the treatise will be helpful to scholars who choose to pursue research in this field. Practitioners may find some parts of the book very useful, specifically, Chapters 2 and 11 (variance estimation—different replication and resampling procedures).

The draft text was written in sections at the Indian Statistical Institute, Calcutta and during visits to the Moi University, Kenya, University of Windsor, Canada and University of Western Australia. I am indebted to the authorities of these institutions for the necessary support. A few words are due to my wife Manju and sons Pabak and Pralay, who were source of strength and encouraged me in my endeavour.

Parimal Mukhopadhyay

Contents

CHAPTER

1 A Model for a Survey Population

1.1 INTRODUCTION

The basic purpose of a sample survey is to draw inference about the characteristic of a finite population through observing only a sample of units and devising suitable estimators of population parameters with desired degrees of precision within limitations of resources. Different statistical techniques have been developed to cater to these needs during the last few decades.

The interest in putting the art of sampling from a finite population on a strong mathematical basis, known as the 'Theory of Sampling from a Finite Population', also known as 'Survey Sampling' was aroused towards the early part of this century mainly due to the classical work of Neyman (1934). Its development has travelled a long since then passing through mazes of various postulates and assumptions. In this chapter, we examine a model for a fixed finite population which formed the basis of the initial exposition of the theory. The set-up established in this chapter would be fundamental to our discussion mainly up to Chapter 3.

1.2 THE BASIC MODEL

We assume that we have a finite population of distinct units, with a known population size and a variable of interest taking real values on these units. The primary task of the survey statistician is to estimate some characteristics of the population, *e.g.*, population total, mean, variance by suitably choosing a subset (sample) of the population and observing the values of this variable only on units in this selected subset. To formulate the problem more precisely let us consider a few definitions.

Definition 1.1: A finite (survey) population $\mathscr{P}$ is a collection of a known number N of identifiable units labelled 1, ..., N; $\mathscr{P} = \{1, ..., i, ..., N\}$, where i stands for the physical unit labelled i.

The above definition excludes from its coverage the populations of the following types: batches of industrial products of the same specification coming out from a production process as the units are not distinguishable individually, population of fishes in a lake as the population size is unknown. Collection of households in an area, industrial units in an urban complex and agricultural fields in a village are examples of survey populations.

Let y be a study variable having value y_i on $i = 1, \ldots, N$. Associated with $\mathscr{P}$ we have a vector of real numbers $\mathbf{y} = (y_1, \ldots, y_N)$ which constitutes the parameter for the model of a survey population. $\mathbf{y} \in R^N$, the parameter space. One is often interested in estimating a parameter function θ, *e.g.*, population total, $Y = \left(\sum_{i=1}^{N} y_i\right)$, population mean, $\bar{Y}$ (= Y/N), population variance $S^2 = N - 1^{-1} \sum_{k=1}^{N} (y_k - \bar{Y})^2$ by choosing a sample (a part of the population, defined below) from $\mathscr{P}$ and observing the values of y only on the units in the sample.

Definition 1.2: A sample is a part of a population.

(*i*) Consider a with-replacement (wr) sampling. A sample may be a sequence,

$$S = \{i_1, \ldots, i_n\},\ 1 \le i_t \le N, \tag{1.2.1}$$

i_t not necessarily $\neq i_{t'}$, for $t \neq t'$ (= 1, ..., N), i_t denoting the result of the t-th draw.

(*ii*) For a without-replacement (wor) sampling, a sample may be a sequence,

$$S = \{i_1, \ldots, i_n\},\ 1 \le i_t \le N, i_t \neq i_{t'},\ \text{for } t \neq t' (= 1, \ldots, N) \tag{1.2.2}$$

since repetition of units in S is not possible.

Arranging the units in the sample in increasing (decreasing) order of magnitudes of labels, a sample may also be written as a set s. In a wr-sampling of n draws, therefore, a sample written as a set is

$$s = (j_1, \ldots, j_2, \ldots, j_{\nu(S)}),\ 1 \le j_1 < \ldots < j_{\nu(S)} \le N \tag{1.2.3}$$

where $\nu(S)$ is the number of distinct units in S. In wor-sampling, a sample of n draws, written as a set is

$$s = (j_1, j_2, \ldots, j_n),\ 1 \le j_1 < \ldots < j_n \le N. \tag{1.2.4}$$

Clearly, information on order of selection and repetition of units in the sample is not available from s.

Definition 1.3: Number of distinct units in a sample is the effective sample size. Number of draws n, in a sample, is its nominal sample size. In (1.2.3), $\nu(S)$ is the effective sample size, $1 \le \nu(S) \le n$. For a wor-sample of n draws, $\nu(S) = \nu(s) = n$.

Definition 1.4: The collection of all possible samples is the sample space, $\mathscr{S}$. Thus, $\mathscr{S} = \{S\}$ or $\{s\}$ according as we are interested in S or s.

If the samples (s) of all possible sizes are considered in a wor-sampling, there are 2^N samples in $\mathscr{S}$.

In a simple random sampling wr (srswr) of n draws, $\mathscr{S}$ consists of N^n sequences S. In a simple random sampling wor (srswor) of n draws, $\mathscr{S}$ consists of $(N)_n$ sequences S or $\binom{N}{n}$ sets s, where $(a)_b = a(a-1) \ldots (a-b+1)$ for non-negative integers a, b $(a \geq b)$.

Definition 1.5: Let A be the minimal σ-field over $\mathscr{S}$ and p be a probability measure defined over A such that $p(s)$ [or $p(S)$] denotes the probability of selecting s[or S], satisfying

$$p(s)\ [p(S)] \geq 0 \tag{1.2.5}$$

$$\sum_{\mathscr{S}} p(s) \left[\sum_{\mathscr{S}} p(S) \right] = 1$$

One of the main tasks of the statistician is to find a suitable $p(s)$ or $p(S)$. The collection $(\mathscr{S}, p)$ is called a sampling design, s.d., often denoted as $D(\mathscr{S}, p)$ or simply p. The triplet $(\mathscr{S}, A, p)$ is the probability space for the model of the finite population.

The expected effective sample size of a s.d. p is

$$E\{\nu(S)\} = \sum_{\mathscr{S}} \nu(S)p(S) = \sum_{\mu=1}^{N} \mu P[\nu(S) = \mu] = \nu \tag{1.2.6}$$

We shall denote the class of sampling designs p for which the effective sample size $\nu(S)$ is a constant ν $\forall$ $S \in \mathscr{S}$ as ρ_ν, *i.e.*,

$$\rho_\nu = \{p : p(s) > 0 \quad \Rightarrow \quad \nu(S) = \nu\} \tag{1.2.7}$$

ρ_ν is the class of fixed effective size-ν, FES (ν) or simply FS (ν) designs. A s.d. p is said to be non-informative if $p(S)$ [or $p(s)$] does not depend on y-values. In this treatise we consider non-informative designs only.

Informative designs have been considered by Basu (1969), Zacks (1969), Liao and Sedransk (1975), Stenger (1977), Bethlehem and Schuerhoff (1984) among others.

Henceforward, we shall write the results for samples, s only, when they are valid for both S and s. When the results differ, we will write them explicitly. The quantities

$$\pi_i = \sum_{s \ni i} p(s), \quad \pi_{ij} = \sum_{s \ni ij} p(s) \tag{1.2.8}$$

$$\pi_{i_1 \ldots i_k} = \sum_{s \ni i_1 \ldots i_k} p(s)$$

are, respectively, the first order, second order, ..., k^{th} order inclusion probability of units in the sample in a s.d. p.

We shall often write Σ to denote $\sum_{i=1}^{N}$, Σ' to denote $\sum_{i \neq i'=1}^{N}$, when these are clear otherwise.

Lemma 1.1: For any s.d. p,

(*i*) $\pi_i + \pi_j - 1 \leq \pi_{ij} \leq \min(\pi_i, \pi_j)$ (*ii*) $\Sigma\pi_i = v$

(*iii*) $\Sigma'\pi_{ij} = v(v-1) + V(v(S))$.

If $p \in \rho_v$,

(*iv*) $\Sigma_{j(\neq i)} \pi_{ij} = (v-1)\pi_i$ (*v*) $\Sigma' \pi_{ij} = v(v-1)$.

Proof: (*i*) $\pi_i + \pi_j - \pi_{ij} = P$ (at least one of i and j is selected in the sample) ≤ 1; hence the left hand side; the right hand side follows immediately.

(*ii*) For each i, define a indicator random variable I_{iS} over each S such that

$$I_{iS} = 1(0) \text{ if } i \in (\notin)S. \ \Sigma I_{iS} = v(S).$$

$$E(I_{iS}) = \pi_i,\ V(I_{iS}) = \pi_i(1-\pi_i),\ E(I_{iS}I_{jS}) = \pi_{ij}.$$

$$\sum \pi_i = \sum \sum_{S \in \mathcal{S}} p(S)\, I_{iS} = \sum_{S \in \mathcal{S}} \left[\sum I_{iS}\, p(S)\right]$$

$$= \sum_{S \in \mathcal{S}} v(S)\, p(S) = v$$

(alternatively, $\sum \pi_i = E\left(\sum I_{iS}\right) = E(v)(S)) = v)$.

(*iii*)
$$\sum{}' \pi_{ij} = E\left(\sum{}' I_{iS}\, I_{jS}\right)$$

$$= E\left\{\left(\sum_i I_{iS}\right)^2 - \sum_i I_{iS}^2\right\} = E(v(S))^2 - v$$

$$= V\{v(\text{S})\} + v(v-1)$$

(*iv*)
$$\sum_{j(\neq i)} \pi_{ij} = \sum_{j(\neq i)} E(I_{iS}I_{jS}) = E\{I_{iS}(v - I_{iS})\} = (v-1)\pi_i$$

(*v*) Follows from (*iv*) and (*ii*). Results (*ii*), (*iii*) and (*iv*), (*v*) are respectively, due to, Godambe (1955), Hanurav (1962*b*), Yates and Grundy (1953).

Further, for any s.d. p,

$$\theta(1-\theta) \leq V\{v(S)\} \leq (N-v)(v-1) \tag{1.2.9}$$

where $v = [v] + \theta$, $0 \leq \theta < 1$, θ being the fractional part of v.

The lower bound in (1.2.9) is attained by a s.d. for which

$$P[v(S) = [v]] = 1 - \theta \text{ and } P[v(S) = [v]+1] = \theta.$$

Mukhopadhyay (1975) gives a s.d. with fixed nominal sample size $n\ (> v)[p(S) > 0 \Rightarrow n(S) = n\ \forall\ S]$ such that $V\{v(S)\} = \theta\,(1 - \theta/(n - [v]))$, which is very close to the lower bound in (1.2.9).

Lemma 1.2: (Mukhopadhyay, 1972). For any s.d. $p \in \rho_v$,

$$\Sigma''\pi_{i_1 \ldots i_t i_{t+1} \ldots i_r} = (v - t)_{(r-t)}\ \pi_{i_1 \ldots i_t}, \quad t < r < v \tag{1.2.10}$$

where Σ'' denotes summation over $i_{t+1} \neq \ldots \neq i_r (\neq i_1 \neq \ldots \neq i_t) = 1, \ldots, N$.

Proof: For each $S \in \mathscr{S}$ define two indicator functions, $Q(S), Q'(S)$ such that

$$Q(S) = 1(0) \text{ if } (i_1, \ldots, i_t) \in (\notin) S$$

$$Q'(S) = 1(0) \text{ if } (i_{t+1}, \ldots, i_r) \in (\notin) S.$$

Now, $$\sum{}''\ \pi_{i_1,\ldots,i_t i_{t+}\ldots i_r} = \sum{}''\ \sum_{S \in \mathscr{S}} p(S)\,Q(S)\,Q'(S)$$

$$= \sum_{S \in \mathscr{S}} \left[\sum{}''\ Q'(S)\right] p(S)\,Q(S) \tag{1.2.11}$$

$\Sigma''\ Q'(S)$ is the number of ways in which $(r - t)$ units taken out of $(v - t)$ units in a S containing $(i_1, \ldots, i_t)$ can be permuted among themselves and this number is $(v - t)_{(r-t)}$. Hence, (1.2.11) is

$$(v - t)_{(r-t)} \sum_{S \in \mathscr{S}} p(S)\,Q(S)$$

$$= (v - t)_{(r-t)}\ \pi_{i_1 \ldots i_t}.$$

We shall denote by

$p_r(i_r)$ = probability of selecting i_r at the rth draw,

$p_r(i_r \mid i_1, \ldots, i_{r-1})$ = conditional probability of drawing i_r at the rth draw given that $i_1, \ldots, i_{r-1}$ were drawn at the first, ..., $(r - 1)^{\text{th}}$ draw respectively,

$p(i_1, \ldots, i_r)$ = the joint probability that $(i_1, \ldots, i_r)$ are selected at the first, ..., r^{th} draw respectively.

Definition 1.6: A sampling scheme (s.s.) gives the conditional probability of drawing a unit at any particular draw given the results of the earlier draws.

A s.s., therefore, specifies the conditional probabilities $p_r(i_r \mid i_1 \ldots i_{r-1})$. The following theorem shows that any sampling design can be attained through a draw-by-draw mechanism.

Theorem 1.1: (Hanurav, 1962*a*). For any given sampling design, there exists at least one sampling scheme which realises this design.

Proof: Let $n_0 = \max_S \nu(S)$. Extend any S with $\nu(S) < n_0$ to a sample S' adding 0's to the last $(n_0 - \nu(S))$ places. Occurrence of a 0 at a place in S' means no unit is drawn at the corresponding draw. Let

$$A_k = \{S : d_1 = i_1, d_2 = i_2, ..., d_k - i_k\}$$

d_r denoting the result of the rth draw. Then, for the sampling scheme,

$$p_1(i) = \sum_{A_1} p(S),\ p_2(i_2 \mid i_1) = \sum_{A_2} p(S) \Big/ \sum_{A_1} p(S),$$

$$p_r(i_r \mid i_1, ..., i_{r-1}) = \sum_{A_r} p(S) \Big/ \sum_{A_{r-1}} p(S), r = 2, ..., n$$

the s.d. p is realised.

Example 1.1: (Mukhopadhyay, 1972). Consider a wor-design $p \in \rho_n$. For the following scheme, p is realised:

$$p_1(i_1) = \pi_{i_1} / n$$

$$p_r(i_r \mid i_1, ..., i_{r-1}) = \pi_{i_1 ... i_r} / (n - r + 1)\, \pi_{i_1 ... i_{r-1}} (0)$$

if $$\pi_{i_1 ... i_{r-1}} > 0 \text{ (otherwise)}, \quad r = 2, ..., n.$$

Suppose the values $x_1, ..., x_N$ of a closely related auxiliary variable x on units 1, 2 ..., N, respectively, are available. The quantity $P_i = x_i / X$, $X = \Sigma x_k$ is called the size-measure of unit $i(= 1, ..., N)$ and is often used in selection of samples.

1.3 DIFFERENT TYPES OF SAMPLING DESIGNS

The following types of sampling designs are generally used:

(*a*) Simple random sampling with replacement (srswr)

(*b*) Simple random sampling without replacement (srswor)

(*c*) *Probability proportional to size with replacement* (*ppswr*): A unit i is selected with probability p_i at the rth draw and a unit once selected, is returned to the population before the next draw (r = 1, 2, ...).

(*d*) *Probability proportional to size without replacement* (*ppswor*): A unit i is selected at the rth draw with probability proportional to its normed measure of size and a unit once selected is removed from the population. Here,

$$p_1(i_1) = p_{i_1}$$

$$p_2(i_2 \mid i_1) = p_{i_2}/(1-p_{i_1}) \tag{1.3.1}$$

$$\ldots \quad \ldots \quad \ldots$$

$$p_r(i_r \mid i_1 \ldots i_{r-1}) = p_{i_r}/(1-p_{i_1} - \ldots p_{i_{r-1}}),\ r = 2, \ldots, n.$$

For $n = 2$, for this scheme,

$$\pi_i = p_i \left[1 + A - \frac{p_i}{1-p_i}\right],$$

$$\pi_{ij} = p_i p_j \left(\frac{1}{1-p_i} + \frac{1}{1-p_j}\right), \text{ where } A = \sum p_k/(1-p_k).$$

The scheme is also known as 'successive sampling'. The sampling design arising out of (1.3.1) may be attained also by an inverse sampling procedure where units are drawn by ppswr, until for the first time n distinct units occur. The n distinct units each taken only once constitutes the sample.

(*e*) *Rejective sampling*: n draws are made with ppswr; if all the units turn out to be distinct, the selected sequence constitutes the sample; otherwise the whole selection is rejected and fresh draws are made.

(*f*) *Unequal probability wor (upwor) sampling*: This is a generalisation of ppswor sampling scheme in which the selection is made at the rth draw with probability $\{p_i^{(r)}\}$ (for which $p_i^{(r)}$ has to be suitably normed) and a unit once selected is removed from the population ($i = 1, \ldots, N;\ r = 1, \ldots, n$). The quantities $\{p_i^{(r)}\}$ are generally some functions of p_i and the p-values of the units already selected.

(*g*) *Generalised rejective sampling*: Draws are made wr and with probability $\{p_i^{(r)}\}$ at the rth draw. If all the units turn out distinct, the selection is taken as a sample otherwise the whole selection is rejected and fresh draws are made.

(*h*) Systematic sampling with equal probability.

(*i*) Systematic sampling with varying probability.

(*j*) *Sampling from groups*: The population is divided into L groups either at random or following some suitable procedure and a sample of size n_h is drawn from the hth group by using any of the above-mentioned sampling design such that the desired sample size, $n = \sum_{h=1}^{L} n_h$ is attained.

Based on the above methods, there are many uni-stage or multi-stage sampling procedures.

As is evident, samples for all the sampling designs may be selected by a whole-sample procedure or mass-drawal procedure in which a s is selected with probability $p(s)$.

Sometimes, as will be evident from Section 1.4, a $FS(n)$ s.d. with π_i proportional to p_i is preferred. Such a design is called a πps design or IPPS (inclusion probability proportional to size) design. Since $\pi_i \le 1$, it is required that $x_i \le X/n \forall i$ for such a design. Various sampling designs suggested in the literature have been overviewed in Chapter 2.

1.4 THE ESTIMATORS

After the sample is selected, the statistician collects the following data from the field.

Definition 1.7: Data collected through a sequence sample S is

$$d' = \{(k, y_k), k \in S\} \tag{1.4.1}$$

For the set sample data are

$$d = \{(k, y_k), k \in s\} \tag{1.4.2}$$

It will be seen (Section 3.3) that data given in (1.4.2) are sufficient for making inference about θ, whether the sample is a sequence (S) or a set (s). Data are said to be unlabelled if after the collection of the data its label part is ignored. Unlabelled data may be represented by a sequence or a set of the observed values without any reference to the labels.

It may not be possible, however, to collect the data from the sampled units correctly and completely. If the information is collected from a human population, the respondent may not be 'at home' during the time of collection of the data or may refuse to answer or may give incorrect information *e.g.*, in stating income, age, etc. The investigators in the field may also fail to register correct information due to their own lapses.

We assume throughout that our data are free from such types of errors due to non-response and errors of measurement and it is possible to collect the information correctly and completely.

Definition 1.8: An estimator $e = e(s, \mathbf{y})$ or $e(S, \mathbf{y})$ is a function defined on $\mathscr{S} \times R^N$ such that for a given $(s, \mathbf{y})$ or $(S, \mathbf{y})$, its value depends on $\mathbf{y}$ only through those i for which $i \in s$ (or S).

Clearly, the value of e in a sample survey does not depend on the units not included in the sample.

e is unbiased for Y wrt a sampling design p if

$$E_p\,(e(s, \mathbf{y})) = Y \ \forall \ y \in R^N \tag{1.4.3}$$

i.e.,

$$\sum_{s \in \mathscr{S}} e(s, \mathbf{y})\, p(s) = Y \ \forall \ \mathbf{y} \in R^N$$

where, E_p denotes expectation wrt s.d. p. We shall often omit the suffix p when it is clear otherwise.

The mean square error (mse) of e around Y wrt a s.d. p is

$$M(e) = E(e - Y)^2 = \sum_{s \in \mathscr{S}} (e(s, \mathbf{y}) - Y)^2\, p(s),$$

$$= E(e - E(e))^2 + \{E(e) - Y\}^2$$

$$= V(e) + (B(e))^2 \quad (1.4.4)$$

where $B(e)$ denotes the design-bias of e. If e is unbiased for Y, $B(e)$ vanishes and (1.4.4) gives the variance of e.

Definition 1.9: A combination (p, e) is called a sampling strategy, often denoted as $H(p, e)$.

$H(p, e)$ is unbiased for Y if (1.4.3) holds and then its variance is

$$V\{H(p, e)\} = E(e - Y)^2.$$

A sampling strategy $H(p, e)$ is said to be better than another strategy $H'(p', e')$, in the sense of variance, written as $H \succ H'$, if

$$V\{H(p, e)\} \leq V\{H(p', e')\} \;\forall\; \mathbf{y} \in R^N \quad (1.4.5)$$

with strict inequality holding for at least one $\mathbf{y}$.

If the s.d. p is kept fixed, an unbiased estimator e is said to be better than another unbiased estimator e' in the sense of variance, if

$$V_p(e) \leq V_p(e') \;\forall\; \mathbf{y} \in R^N \quad (1.4.6)$$

with strict inequality holding for at least one $\mathbf{y}$.

Unless otherwise stated we shall throughout compare the strategies (estimators or designs) in the sense of variance.

A strategy H^* will be said to be uniformly minimum variance unbiased (UMVU) strategy in a class of unbiased strategies $\mathcal{H} = \{H\}$ if it is better than any other strategy $H \in \mathcal{H}$, i.e., if

$$V(H^*) \leq V(H) \;\forall\; \mathbf{y} \in R^N \quad (1.4.7)$$

with strict inequality holding for at least one $\mathbf{y}$, holds for all $H \not\equiv H^* \in \mathcal{H}$.

For a fixed p, an estimator e^* is said to be UMVU-estimator in a class of unbiased estimators $\eta = \{e\}$ if it is better than any other estimator e in η *i.e.*, if

$$V_p(e^*) \leq V_p(e) \;\forall\; \mathbf{y} \in R^N \quad (1.4.8)$$

with strict inequality holding for at least one $\mathbf{y}$, holds for all $e \in \eta$.

We now consider different estimators for $\bar{Y}$, first considering the cases when the s.d. is either srswr or srswor, based on n draws.

(1) *mean per unit estimator:*

(*a*) *srswr*: (*i*) Sample mean, $\bar{y}S = \frac{1}{n}\sum_{1}^{n} y_i$, (when the summation is over $i = 1, ..., n$; z_i will denote the value of a variable z on the unit selected at the ith draw);

(*ii*) mean of the distinct units, $\overline{y}'_S = \frac{1}{v(S)} \sum_{i \in s} y_i$

where s is the set corresponding to the sequence S. It follows that $\overline{y}'_S \succ \overline{y}S$ (Section 3.3)

(*b*) *srswor:* Sample mean, $\overline{y}_s = \frac{1}{n} \sum_{i \in s} y_i$

(2) *ratio estimator:*

(*a*) srswr: $\hat{\overline{Y}}_{P(S)} = (\overline{y}_s / \overline{x}_s)\, \overline{X},\ \overline{X} = X / N$

(*b*) srswor: $\hat{\overline{Y}}_R = (\overline{y}_s / \overline{x}_s) \overline{X}$

(3) *product estimator:*

(*a*) srswr: $\hat{\overline{Y}}_{P(S)} = \overline{y}_S \overline{x}_S / \overline{X}$

(*b*) srswor: $\hat{\overline{Y}}_P = \overline{y}_S \overline{x}_S / \overline{X}$

(4) *difference estimator:*

(*a*) srswr: $\hat{\overline{Y}}_{D(S)} = \overline{y}_S + d(\overline{X} - \overline{x}_S)$

(*b*) srswor: $\hat{\overline{Y}}_D = [\overline{y}_s + d(\overline{X} - \overline{x}_s)]$, where d is a constant.

(5) *regression estimator:*

(*a*) srswr: $\hat{\overline{Y}}_{lr(S)} = \overline{y}_S + b'(\overline{X} - \overline{x}_S)$

(*b*) srswor: $\hat{\overline{Y}}_{lr} = \overline{y}_s + b(\overline{X} - \overline{x}_s)$

where, $$b' = \sum_{i=1}^{n} (x_i - \overline{x}_S)(y_i - \overline{y}_S) \Big/ \sum_{i=1}^{n} (x_i - \overline{x}_S)^2$$

$$b = \sum_{s} (x_i - \overline{x}_s)(y_i - \overline{y}_s) \Big/ \sum_{s} (x_i - \overline{x}_s)^2$$

(6) *mean of the ratio estimator:*

(*a*) srswr: $\hat{\overline{Y}}_{MR(S)} = \overline{X}_{\overline{r}S}$ where $\overline{r}_S = \frac{1}{n} \sum_{i=1}^{n} r_i,\ r_i = y_i / x_i$

(*b*) srswor: $\hat{\overline{Y}}_{MR} = \overline{X}_{\overline{r}}$ where $\overline{r} = \frac{1}{n} \sum_{i \in s} r_i$

Estimators stated at (2), (3), (5) and (6) are biased estimators of $\overline{Y}$ in srs.

Different modifications of ratio estimator, regression estimator, product estimator and estimators obtained by taking convex combinations of these estimators have been considered in the literature for srs. For srswor, the unbiased ratio estimator of Hartley and Ross (1954) is

$$\hat{\hat{Y}}_{HR} = \bar{r} + \frac{n(N-1)}{N(n-1)} \frac{\bar{y} - \bar{r}\,\bar{x}}{\bar{X}}.$$

Other modifications of ratio estimator are due to Quenouille (1956), who uses the jackknife technique, Mickey (1959), Tin (1963) and Pascual (1961), among others. Unbiased regression estimators have been developed following Mickey and Williams (1963) (see Rao, 1969 for a review of ratio and regression estimators). Mixture of ratio and regression estimators have been considered by Vos (1980), Upadhyay *et al.* (1985), among others.

In ppswr-sampling the estimators

(*i*) $\hat{Y}_{pps} = \frac{1}{n}\sum_{i=1}^{n} y_i / p_i$

(*ii*) $\hat{\hat{Y}}'_{pps} = \frac{1}{n}\sum_{i \in s} \{y_i / p_i\}\, c_{(i)}$ – ($c_{(i)}$'s given in Example 3.2), where s is the set corresponding to the sequence S, have been considered for Y. $\hat{Y}_{pps}$ is the Hansen-Hurwitz estimator for Y and will be denoted by $\hat{Y}_{HH}$. The corresponding strategy is known as Hansen-Hurwitz strategy.

$$V(\hat{Y}_{pps}) = \frac{1}{n}\sum \left(\frac{Y_i}{P_i} - Y\right)^2 P_i$$

$$= \frac{1}{2^n}\sum{}' P_i P_j \left(\frac{Y_i}{P_i} - \frac{Y_i}{P_j}\right)^2 = V_{pps} \text{ (say)}.$$

An unbiased variance estimator is

$$v(\hat{Y}_{pps}) = \frac{1}{n(n-1)}\sum_{1}^{n} \left(\frac{y_i}{p_i} - \hat{Y}_{pps}\right)^2 = v_{pps} \text{ (say)}, \qquad (1.4.9)$$

It follows that $\hat{Y}'_{pps} \succ \hat{Y}_{pps}$ (Section 3.3).

We now consider estimators applicable to any sampling design. For any s.d. p, consider a non-homogeneous linear estimator,

$$e'_L = b_{os} + \sum_{i \in s} b_{si} y_i \qquad (1.4.10)$$

where the constant b_{os} may depend only on s and b_{si} on (s, i) ($b_{si} = 0$ if $s \not\ni i$). e'_L is unbiased *iff*

$$\sum_{s \in \mathcal{S}} b_{os}\, p(s) = 0 \tag{1.4.11.1}$$

$$\sum_{i} y_i \sum_{s \ni i} b_{si}\, p(s) = Y \ \forall\ \mathbf{y} \in R^N \tag{1.4.11.2}$$

which imply

$$b_{os} = 0\ \forall\ s : p(s) > 0 \tag{1.4.12.1}$$

$$\sum_{s \ni i} b_{si}\, p(s) = 1,\ i = 1, \ldots, N \tag{1.4.12.2}$$

Hence, a linear unbiased estimator must be a homogeneous function,

$$e_L = \sum_{i \in s} b_{si} y_i \tag{1.4.13}$$

subject to the condition (1.4.12.2).

Letting $b_{si} = b_i \forall\ s \ni i$, (1.4.12.2) implies $b_i = \pi_i^{-1}$ when e_L reduces to the Horvitz-Thompson estimator (HTE),

$$e_{HT} = \sum_{s} y_i / \pi_i \tag{1.4.14}$$

which is unbiased for Y for any s.d.

$$V\{e_{HT}) = \sum \frac{Y_i^2}{\pi_i}(1 - \pi_i) + \sum{}' \frac{Y_i Y_j}{\pi_i \pi_j}(\pi_{ij} - \pi_i \pi_j)$$

$$= V_{HT}\ \text{(say)} \tag{1.4.14.1}$$

If $p \in \rho_n$, (1.4.14.1) can be shown to be equal to

$$\frac{1}{2}\sum{}' \left(\frac{Y_i}{\pi_i} - \frac{Y_j}{\pi_j}\right)^2 (\pi_i \pi_j - \pi_{ij}) = V_{YG} \tag{1.4.14.2}$$

The form V_{HT} is due to Horvitz and Thompson (1952) and v_{YG} is due to Yates and Grundy (1953). Liu (1984) showed that the condition $p \in \rho_n$ is also a necessary condition for V_{HT} to be expressed in the form V_{YG}.

Two unbiased estimators of $V(e_{HT})$, due to Horvitz and Thompson (1952) and Sen (1953), Yates and Grundy (1953) respectively, are

$$v_{HT} = \sum_{s} \frac{y_i^2}{\pi_i^2}(1 - \pi_i) + \sum_{s}{}' \frac{y_i y_j}{\pi_i \pi_j} \frac{(\pi_{ij} - \pi_i \pi_j)}{\pi_{ij}} \tag{1.4.15}$$

$$v_{YG} = \frac{1}{2}\sum_s{}'\left(\frac{y_i}{\pi_i} - \frac{y_j}{\pi_j}\right)^2 \frac{(\pi_i\pi_j - \pi_{ij})}{\pi_{ij}} \tag{1.4.15.1}$$

the later being unbiased *iff* p is a FS(n)-design with $\pi_{ij} > 0 \ \forall\ i \neq j$. Here and subsequently, we shall denote $\sum\sum_{i \neq j \in s} (\cdot)$ as $\sum_s{}'$. v_{HT} is unbiased for any p with $\pi_{ij} > 0$. Both v_{HT} and v_{YG} can take negative values for some samples and this leads to the difficulty in interpreting the reliability of these estimates.

It is evident from (1.4.14.2) that sufficient conditions for v_{YG} to be non-negative are

$$\pi_{ij} \leq \pi_i\ \pi_j,\ i \neq j = 1, \ldots, N \tag{1.4.15.2}$$

Sen (1953) suggested a biased non-negative variance estimator v' such that $v'(s) = v_{HT(s)}$ (0) if $v_{HT(s)} > 0$ (otherwise). Sampling designs for which (1.4.15.2) holds and the conditions for non-negativity of variance-estimators are discussed in the next chapter.

Ramakrishnan (1971) suggests a modified version of v_{YG} which is applicable to designs of varying sizes also. He (1975) also suggested a generalisation of v_{YG}. Assuming that $p \in \rho_n$ and $\pi_i\ \pi_j - \pi_{ij}$ $(i \neq j = 1, \ldots, N)$ is approximately a constant, Jessen (1969) proposed,

$$v_{JA} = \frac{n - \sum \pi_i^2}{2n(n-1)} \sum_s{}'\left(\frac{y_i}{\pi_i} - \frac{y_j}{\pi_j}\right)^2 \tag{1.4.16}$$

as an estimator of v_{YG}. The estimator v_{JA} is a biased estimator.

It may be noted that if $p \in \rho_n$ and $\pi_i \propto y_i$ $(\forall_i)$, $V_{HT} = 0$ (this is often called the 'ratio estimator' property of HTE). In practice, y-values will be unknown and instead, values of some auxiliary variable x closely related to the main variable y will be known. In such cases it may be assumed that x is approximately proportional to y.

One may, therefore, reasonably expect that V_{HT} based on a πps design will be small provided the auxiliary variable x has been suitably chosen. Hanif and Brewer (1979) suggest a generalisation of HTE.

Letting again $b_{si} = b_s\ \forall\ i \in s$ in (1.4.13), one requires,

$$b_s = \frac{1}{M_1 p(s)}\ \forall\ s : p(s) > 0 \tag{1.4.17}$$

for e_L to be unbiased, where $M_i = \binom{N-i}{n-i}$, $i = 1, 2 \ldots$

Hence, the estimator

$$e'_R = \left(\sum_{i \in s} y_i\right) \Big/ M_1 p(s) \tag{1.4.18}$$

is unbiased for Y.

$$V(e'_R) = \sum y_i^2 \left\{ \frac{1}{M_1^2} \sum_{s \ni i} \frac{1}{p(s)} - 1 \right\}$$

$$+ \sum{}' y_i\, y_j \left\{ \frac{1}{M_1^2} \sum_{s \ni ij} \frac{1}{p(s)} - 1 \right\} \qquad (1.4.19)$$

$$v(e'_R) = \frac{\sum_s y_i^2}{M_1 p(s)} (M_1^{-1} - 1) + \frac{\sum_s{}' y_i y_j}{p(s)} (M_1^{-2} - M_2) \qquad (1.4.20)$$

(1.4.18) is a ratio type estimator of Y. The case of special interest is when $p(s) \propto \sum_s p_i$, which is known in the literature as Midzuno (1950,1952)-Lahiri (1951)-Sen (1952) sampling scheme (discussed in Section 2.6). In this case, (1.4.18) reduces to the ordinary ratio estimator

$$\hat{Y}_R = \left(\sum_s y_i \right) \Big/ \left(\sum_s p_i \right).$$

Des Raj's estimator: Suppose a sample $S = \{i_1, ..., i_n)$ is selected by ppswor-sampling procedure. Des Raj (1956) considered n ordered estimators $\{t_r, r = 1, ..., n\}$, each unbiased for Y, where t_r is based on the results of the first r ordered observations.

$$t_1 = y_{i_1} / p_{i_1}$$

$$t_r = y_{i_1} + ... + y_{i_{r-1}} + \frac{y_{i_r}}{p_{i_r}} (1 - p_{i_1} - ... - p_{i_{r-1}}),\ r = 2, ..., n.$$

$(E\,(t_r \mid d_1 = i_1, ..., d_{r-1} = i_{r-1}) = Y$ and hence,
$E\,(t_r) = Y\ (r = 2, ..., n)$ Raj proposed,

$$e_D = \frac{1}{n} \sum_{r=1}^{n} t_r \qquad (1.4.21)$$

as an unbiased estimator of Y. The estimators $\{t_r, t_{r'}\}$ $(r \neq r')$ are uncorrelated and hence, an unbiased estimator of $V(e_D)$ is

$$V\,(e_D) = \frac{1}{n(n-1)} \sum_{r=1}^{n} (t_r - e_D)^2 = v_D \text{ (say)} \qquad (1.4.22)$$

which is always non-negative. Raj's strategy is uniformly better than the Hansen-Hurwitz strategy, both being based on same $\{n, p_i, i = 1, ..., N\}$.

For $n = 2$, and $s = \{1, 2\}$

$$e_D = \frac{1}{2} \left\{ \frac{y_1}{p_1} (1 + p_1) + \frac{y_2}{p_2} (1 - p_1) \right\} \qquad (1.4.23)$$

$$V(e_D) = \frac{1}{2}\sum{}' P_i P_j (2 - P_i P_j)\left(\frac{y_i}{P_i} - \frac{y_j}{P_j}\right)^2 \tag{1.4.24}$$

$$v(e_D) = \frac{1}{4}(1 - p_1)^2\left(\frac{y_1}{p_1} - \frac{y_2}{p_2}\right)^2 \tag{1.4.25}$$

Pathak (1967) considered expression for $V(e_D)$ and gave asymptotic expressions for its upper bounds. Mukhopadhyay (1977) improved upon these bounds. [For the ppswor sampling scheme, Das (1951) suggested an unbiased estimator based on the order of selection of the units. However, its variance estimator is not always non-negative].

Murthy's (1957) estimator: Murthy (1957) and Das (1962) showed that an estimator t_S which is based on an ordered sample S drawn by a wor-sampling procedure can always be improved upon by taking the weighted averages of t_{S_i} with weight proportional to $p(S_i)$, where all the sequences S_i's correspond to the same set s, $i = 1, \ldots, n!$

The improved estimator,

$$t_s = \sum_{\substack{S_i: \\ r(S_i) = s}} t_{S_i} p(S_i)/p(s) \tag{1.4.26}$$

where, $r(\cdot)$ is a function defined on each S_i, reducing a sequence to its set.

Using this unordering (symmetrisation) technique over e_D, Murthy (1957) suggested

$$e_M = \left\{\sum_{\substack{s: \\ r(s) = s}} e_D(S)_p(s)\right\} \Bigg/ p(s) \tag{1.4.27}$$

as an estimator of Y. It is shown

$$e_M = \frac{1}{p(s)}\sum_{i \in s} y_i P(s/i), \tag{1.4.28}$$

where $P(s/i)$ is the conditional probability of getting the unordered sample s, given that i was selected at the first draw.

An unbiased variance estimator of (1.4.28) is

$$v(e_M) = \frac{1}{2\{p(s)\}^2}\sum_s{}'\{p(s)\,P(s/ij) - P(s/i)\,P(s/j)\}$$

$$p_i p_j \left(\frac{y_i}{p_i} - \frac{y_j}{p_j}\right)^2 = v_M \text{ (say)}, \tag{1.4.29}$$

where $P(s/ij)$ is the conditional probability of getting the sample s given that i and j were selected in the first two draws.

For $n = 2$, $s = (1, 2)$,

$$e_M = \frac{1}{2 - p_1 - p_2}\left\{(1 - p_2)\frac{y_1}{p_1} + (1 - p_1)\frac{y_2}{p_2}\right\} \tag{1.4.30}$$

$$V(e_M) = \frac{1}{2}\sum{}' P_i P_j \frac{(1 - P_i - P_j)}{(2 - P_i - P_j)}\left(\frac{y_i}{p_i} - \frac{y_j}{p_j}\right)^2 \tag{1.4.31}$$

$$v_M = \frac{(1 - p_1)(1 - p_2)(1 - p_1 - p_2)}{(2 - p_1 - p_2)^2}\left(\frac{y_1}{p_1} - \frac{y_2}{p_2}\right)^2 \tag{1.4.32}$$

Murthy (1957) considered an improved estimator v'_D of $V(e_D)$, by unordering v_D in (1.4.22)

$$v'_D = \left\{\sum_{\substack{s:r(S)\\ = s}} v(e_D(S))p(S)\right\} \Big/ p(s).$$

For $n = 2$, $s = (1, 2)$

$$v'_D = \frac{1}{4}(1 - p_1)(1 - p_2)\left(\frac{y_1}{p_1} - \frac{y_2}{p_2}\right)^2 \tag{1.4.33}$$

For ppswor samples of size 2, Murthy (1963) proposed

$$e'_M = \frac{1}{2 - p_i - p_j}\left[\frac{y_i}{\pi_i - p_i}(1 - p_i) + \frac{y_j}{\pi_j - p_j}(1 - p_j)\right],$$

for $s = (i, j)$, $i < j$, $\pi_i = p_i + p_i \sum_{j(\neq i)} p_j/(1 - p_j)$ as an unbiased estimator of Y.

Andreatta and Kaufman (1986) considered ppswor but with $p_i = y_i/Y$ ('the y-values are both what are observed and what determine the sample design') and Murthy's estimator under this modified ppswor-procedure. They gave an integral representation to $p(s)$, showed that v_M is non-negative and gave an upper bound to $V(e_M)$. They gave asymptotic expressions for $p(s)$, $p(s/i)/p(s)$ and considered Monte Carlo experiments for studying the use of approximations to Murthy's estimator suggested by these asymptotic expressions.

Mukhopadhyay (1977) considered generalisation of Des Raj Procedure and Murthy Procedure to any upwor-sampling design.

Consider now some generalised estimators of Y:

(*a*) *Generalised difference estimator:* Basu (1971) considered an unbiased estimator,

$$e_{GD}(\mathbf{a}) = \sum_s \frac{y_i - a_i}{\pi_i} + A,\ A = \sum a_i \tag{1.4.34}$$

where $(a_1, ..., a_N)$ is a set of known quantities.

(*b*) Generalised regression estimator,

$$e_{GR} = \sum_s \frac{y_i}{\pi_i} - b\left(\sum_s \frac{x_i}{\pi_i} - X\right), \tag{1.4.35}$$

where b is the sample regression coefficient of y on x, was first considered by Cassel, Sarndal and Wretman (1976).

(*c*) Hajek (1959) considered a generalised ratio estimator,

$$e_{Ha} = X\left(\sum_s \frac{y_i}{\pi_i}\right) \Bigg/ \left(\sum_s \frac{x_i}{\pi_i}\right).$$

The estimators e_{GR}, e_{Ha} are not unbiased for Y. Besides these, specific estimators have been suggested for specific procedures. An example is Rao-Hartley-Cochran procedure (1962) a review of which may be seen in Section 2.

Denoting the strategy (πps, e_{HT}) as H_1 and Hansen-Hurwitz strategy as H_2, both based on the same $\{n, P_i, i = 1, ..., N\}$, Narain (1951) and Des Raj (1966) (respectively) showed that the conditions

$$\pi_{ij} \leq \frac{2(n-1)\,\pi_i \pi_j}{n} \tag{1.4.36}$$

$$\pi_{ij} > \frac{(n-1)\,\pi_i \pi_j}{n} \tag{1.4.36.1}$$

are respectively, necessary conditions and sufficient conditions for $H_1 \succ H_2$. Hence, for $n = 2$, non-negativity of v_{YG} is a necessary condition for $H_1 \succ H_2$. Gablar (1984) gave a sufficient condition for $H_1 \succ H_2$ (vide Exercise 9, Chapter 2). Note that only the designs with $\pi_i > o(\forall_i)$ admit an unbiased estimator of population total.

❑ ❑ ❑

CHAPTER

2 Varying Probability Without Replacement Sampling Designs

2.1 INTRODUCTION

In this chapter, we make a brief review of some of the unequal probability without replacement (upwor) sampling designs for estimating a population total, mean. Their numerousity and paucity of space restricts consideration of some selected designs only in some details. In Section 2.3 is considered relation between a srs and a t-design. Lastly, we discuss estimation of variance of different estimators based on upwor-sampling designs, with special reference to the non-negativity.

2.2 UPWOR-SAMPLING DESIGNS

2.2.1 Desirable Properties of a s.d.

Starting from the work of Madow (1949), many upwor-sampling designs have been suggested. Horvitz and Thompson (1952) first derived an unbiased estimator for any upwor-sampling design. Some desirable properties of a sampling design to base e_{HT} are (Hanurav, 1967):

(*i*) $\pi_i = nx_i/X$, $nx_i \leq X$, $i = 1, ..., N$

(*ii*) $v(s) = n \forall s : p(s) > 0$

(*iii*) $\pi_{ij} > 0 \ \forall \ (i, j)$

(*iv*) $\pi_{ij} \leq \pi_i\pi_j \ \forall \ (i, j)$

(*v*) $\pi_{ij}/\pi_i\pi_j > \beta$, where β is not too close to zero

(*vi*) π_{ij}'s should be easily computable.

Throughout, x will denote a closely related auxiliary variable with value x_i on i. The properties (*iii*) and (*iv*) are required for ensuring the existence of an unbiased variance estimator $v(e_{HT})$ and the non-negativity of $v_{YG}(e_{HT})$, respectively. The property (*v*) is desirable to make the values of $v_{YG}(e_{HT})$ expectedly stable over different samples.

As the values of $V(e_{HT})$ are uniquely determined in terms of π_i and π_{ij} values, and since y-values are not exactly proportional to x, one should also pay heed to the problem of choosing π_{ij}'s in more specific manner, apart from choosing the π_i's as above. Thus, if the values of inclusion probabilities be suitably chosen with a view to controlling $V(e_{HT})$, the problem becomes one of choosing a sample from a finite population so that these values of π_i's and π_{ij}'s are attainable.

Apart from sampling strategies consisting of (approximately) πps designs and corresponding e_{HT}'s, some special procedures like Rao-Hartley-Cochran strategy, Midzuno-Lahiri-Sen strategy have been suggested. The following is a list of different sampling designs, which, however, is not claimed to be exhaustive. The designs are listed by the names of their authors.

(1) Yates-Grundy (1953) (draw-by-draw) scheme, (2) Narain (1951), Brewer Undy (1962) scheme, (3) Brewer (1963) scheme, (4) Fellegi (1963), (5) Choudhry (1979), (6) Durbin's (1967) ungrouped procedure, (7) Durbin's (1967) grouped procedure, (8) Dodds and Fryer (1971) procedure, (9) Systematic sampling with varying probability due to Goodman and Kish (1950), Hartley and Rao (1962), (10) Grundy's (1954) systematic sampling, (11) Modified systematic sampling due to Singh and Singh (1978), (12) Rao-Hartley-Cochran (1962) procedure, (13) Chikkagoudar's (1967) generalisation of RHC-procedure, (14) Fuller (1971) procedure, (15) Yates-Grundy (1953), Durbin (1953) rejective procedure, (16) Rao (1963), Carroll-Hartley (1964) rejective procedure, (17) Rao (1965) rejective procedure, (18) Kumar and Agarwal (1982) rejective design, (19) Poisson sampling and modified Poisson sampling, (20) Hájek (1964) rejective procedure, (21) Sampford (1967) design, (22) Hanurav (1962) schemes I, II, III, (23) Hanurav (1967) schemes A, B, A′, A″ (with Mukhopadhyay (1974) generalisation of scheme A′), (24) Vijayan (1968) scheme, (25) Sengupta (1981) procedure, (26) Fuller (1970) procedure, (27) Brewer (1975) procedure, (28) Stevens (1958) procedure, (29) Agarwal *et al.*(1984) procedure, (30) Chikkagoudar (1967) generalisation of (28), (31) Midzuno (1950, 1952); Lahiri (1951)-Sen (1952) procedure, (32) Sankaranarayanan (1969) procedure, (33) Deshpande (1978) procedure, (34) Deshpande (1982) procedure, (35) Deshpande-Ajgaonkar (1987) generalisation of procedure (31), (36) Choudhry and Singh (1979) procedure, (37) Gupta, Nigam, Kumar (1982) procedure, (38) Singh and Srivastava procedure, (39) Das and Mahanty procedure, (40) Singh (1978) procedure, (41) Chromy (1981) procedure, (42) Chao (1982) procedure, (43) Samiuddin and Asad (1981) procedure, (44) Srivastava and Singh (1981) procedure, (45) Saxena, Singh and Srivastava (1986) procedure, (46) Sunter's (1977) list sequential sampling plan I, (47) Sunter's list sequential sampling plans II—IV, (48) Jessen (1969) procedure, (49) Nigam, Kumar, Gupta (1984) procedure, (50) Gabler (1987) procedure, (51) Chaudhuri (1971) procedure, (52) Mukhopadhyay (1972) procedure, (53) Sinha (1973) procedure, (54) Wright's (1990) procedure and (55) Schultz (1990) procedure.

The designs have been arranged more or less on the basis of their similarity in the method of construction.

Procedures (12, 13) use a special estimator other than e_{HT}. Similarly, procedures [(31)-(35)] give unbiased estimation for ratio estimator. The procedure (38) was developed with regression estimator in view. However, e_{HT} remains unbiased for all these procedures. The remaining procedures were developed with HTE in view.

The procedures 1, 2, 3, 6, 7, 8, 14, 17, 23, 34 are applicable for $n = 2$ only. Procedure 18 is applicable for $n \geq 3$. The remaining procedures are applicable for any n. For designs 19, and scheme I of (22), scheme II of (47) sample sizes vary.

Procedures 1, 2, 15, 20 are inexact in the sense, π_i's are not exactly proportional to p_i values. For other procedures applicable mainly to HTE, $\pi_i = np_i$, subject to the satisfaction of some conditions by x_i values.

The procedures (52) and (53) attempt to obtain sampling designs realising the given sets of inclusion probabilities. This problem was also considered by Herzel (1986).

Besides the above schemes, there are controlled sampling designs suited for selecting some desired samples with higher probability [Vide, *e.g.*, Sukhatme and Avadhani (1965), Avadhani and Sukhatme (1967, 1968, 1973)]. Section 2.4 discusses this idea to some extent while considering relation between srs and *t*-design. Inverse sampling has been considered by Pathak (1964), Chikkagoudar (1969), Dahiya *et al.* (1977), among others.

Different upwor-designs have been reviewed in Mukhopadhyay (1991). Excellent reviews on sampling designs are available due to Dalleneus (1962), Solomon and Zacks (1970), Vos (1974), Brewer and Hanif (1983), Chaudhuri and Vos (1988).

We shall consider here in some details procedures (1)–(4), (6), (7), (8), (12), (21), (31)–(33), (35), (37), (38), (42), (46), (50), (52), (53). Some general properties of some of the other schemes have been listed in the form of exercises appended to this chapter.

2.2.2 Procedures (1)–(4), (6), (7)

These are all πps designs developed for $n = 2$ [For procedure (4), any n]. In scheme (1), the first unit is selected with probability q_i (a revised size-measure) and the second unit with probability proportional to the remaining values of q_i. The solution for q_i is obtained through an iterative procedure, which converges very slowly, specially as

$$\max_i p_i \to \frac{1}{2}.$$

Narain solved the same problem by solving a quadratic equation in $\{q_i/(1-q_i)\}$. He chose the negative sign in the solution of the quadratic equation, which was found by Brewer and Undy to give incorrect results. Brewer and Undy gave a numerical procedure for solving out d_i. For *YG*-scheme, $v_{YG}(e_{HT}) \geq 0$ (Sen, 1953) and $V(e_{HT}) \leq V(e_{HH})$, both being based on same $\{n, P_i, i = 1, \ldots N\}$. Sen (1953), Des Raj (1956) proved that $\pi_{ij} \leq \pi_i\pi_j$ for Narain-Brewer-Undy procedure.

Brewer (1963) chose $p_1(i) = q_i$ and $p_2(j\,|\,i) = p_j/(1-p_i)$ when q_i is found to be

$$q_i = \frac{2p_i\,(1-p_i)}{(1-2p_i)\,(1+B)}$$

where, $$B = \Sigma P_k/(1-2P_k) \tag{2.2.1}$$

For this scheme $V(e_{HT}) \leq V(e_{HH})$. Rao (1963) considered expression for $V(e_{HT})$ assuming $P_i = O(N^{-1})$ and N moderate or large. To $O(N)$, he showed that this scheme has smaller variance than each of procedures (15) and (9) and to $O(N)$, all these schemes have the same variance.

Fellegi (1963) attached with each draw r a set of non-negative quantities,

$$\{p_r(i)\}(i = 1,..., N)\left(\sum_i p_r(i) = 1\right), r = 1, ..., N$$

such that $$p_r(i_r \mid i_1, ..., i_{r-1}) = \frac{p_r(i_r)}{1 - p_r(i_1) - ... - p_r(i_{r-1})} \quad (2.2.2)$$

The $p_r(i)$'s are so chosen that the probability i is selected at the rth draw,

$$\Delta_r(i) = \sum_{i_1 \neq ... \neq i_{r-1}(\neq i)} p_r(i \mid i_1, ..., i_{r-1})$$
$$= p_i, \; r = 1, ..., n \quad (2.2.3)$$

so that $\pi_i = \sum_r \Delta_r(i) = np_i$. Fellegi showed that for $n = 2$, a unique solution exists if $p_i \leq 1/3 \; \forall \; i$. Brewer (1963) gave a geometrical proof of the existence and uniqueness of the solution when $p_i \leq 1/2 \; \forall \; i$. For $n = 2$, the scheme is equivalent to the symmetric mass draw method of Carroll and Hartley (1964) in that they share the same π_{ij} values. Here $\pi_{ij} \leq \pi_i \pi_j$.

For general n, $p_r(i)$'s may be found by successive approximations which converge very rapidly provided np_i values are not too close to unity. Fellegi's design is practically suitable if the units in the sample have to be rotated, as in the case of repeatitive surveys. For general n, it is not known if $v_{YG}(e_{HT}) \geq 0$ or whether $V(e_{HT}) \leq V(e_{HH})$.

In Durbin's ungrouped procedure, $p_1(i) = p_i$ and

$$p_2(j \mid i) \propto p_j \left\{\frac{1}{1 - 2p_i} + \frac{1}{1 - 2p_j}\right\} \quad (2.2.4)$$

so that π_{ij} is a symmetric function of (i, j). The scheme has the same π'_{ij} s as Brewer's (1963) and can sometimes be extended to $n > 2$.

In Durbin's grouped procedure, the units are arranged in groups such that each group (Gr) contains as few units as possible subject to the requirement that $\max_{i \in G_r} P_i \leq \sum_{k \in G_r} P_k/2$. Two units are selected with ppswr. If the units come from the different groups, they constitute the sample. If they come from the same group, we accept the first and reject the second, making a further selection from the same group by Durbin's ungrouped procedure with probability P_i replaced by $P'_i = P_i / \Sigma' P_k$, Σ' being overall units in the same group.

For this scheme, when i, j belong to two different groups, $\pi_i \pi_j / \pi_{ij} = 2$. Thus, the method is advantageous for estimation of $V(e_{HT})$. For estimation of $V(e_{HT})$ in stratified multistage sampling, one should calculate the contribution of variance from those strata in which sampled units come from different groups as if the sampling has been carried out wor. For strata in which the sampled units come from the same group, contribution to the variance may be calculated as if each stratum consisted only of the groups containing the sampled units.

2.2.3 Schemes due to Dodds and Fryer (1971)

The authors give three families of πps-schemes which include Brewer's (1963) scheme as a particular case. We require

$$p_1(i) + \sum_{j(\neq i)} p_1(j)\, p_2(i \mid j) = 2p_i \tag{2.2.5}$$

with $\Sigma_i p_1(i) = \Sigma_{i(\neq j)} P_2\,(i \mid j) = 1$. Brewer used $p_2(i \mid j) = \dfrac{p_i}{1-p_j}$, when solving from (2.2.5), $p_1(i) = \dfrac{2p_i\,(1-p_i)}{1-2p_i} \cdot \left\{1 + \sum \dfrac{p_k}{1-2p_k}\right\}$. Durbin (1967) considered $p_1(i) = p_i$ and found $p_2\,(j \mid i) \propto p_j \left(\dfrac{1}{1-2p_i} + \dfrac{1}{1-2p_j}\right)$. Here we take

$$p_2(i \mid j) = \frac{1}{\alpha - \beta p_j}\left\{p_i + \frac{(1-\beta)p_j}{N-1} + \frac{\alpha - 1}{N-1}\right\} \tag{2.2.6}$$

α, β being suitable constants; $\alpha = \beta = 1$ give Brewer's solution. (2.2.6) gives, from (2.2.5),

$$p_1(i) = \frac{2}{f(p_i)}\left\{p_i(1-A) - \frac{(1-\beta)B}{N-1} - \frac{(\alpha-1)A}{N-1}\right\} \tag{2.2.7}$$

where $f(p_i) = 1 - p_2(i \mid i)$, (from (2.2.6) when $j = i$); A, B are found from equations,

$$A\left\{1 + a_2 + \frac{\alpha - 1}{N-1} a_1\right\} + B\left\{\frac{1-\beta}{N-1} a_1\right\} = a_2 \tag{2.2.8}$$

$$A\left\{a_3 + \frac{\alpha-1}{N-1} a_2\right\} + B\left\{1 + \frac{1-\beta}{N-1} a_2\right\} = a_3$$

with

$$a_j = \sum_{i=1}^{N} \frac{p_i^{j-1}}{(\alpha - \beta p_i)\, f(p_i)} \tag{2.2.9}$$

(2.2.6) and (2.2.7) satisfy $\Sigma_i p_1(i) = \Sigma_{j(\neq i)} P_2(j \mid i) = 1$. To make $\{p_1(i)\}$, $\{p_2(i \mid j)\}$ non-negative some restrictions on α, β are required. The following obvious choices of α, β give three families of sampling schemes (however, there are many other choices of α, β). We denote $\min_i .\ p_i$ and $\max_i\ p_i$ by p_n and p_m respectively.

Scheme A: $\alpha = \beta = \dfrac{1}{\gamma}$, where γ is allowed to vary. If $p_n < \dfrac{1}{2(N-1)}$ and γ satisfies

$$1 \geq \gamma \geq \gamma_0 = \left[\frac{(1-p_m)(1-2(N-1)p_n)}{1-p_m-(N-1)p_n} < 1\right]$$

then
$$p_1(i) = \frac{2p_i - \dfrac{1-\gamma}{N-1} + \dfrac{\gamma(1-\gamma)}{N-1}(a_1 p_i - a_2)}{(1+\gamma a_2) f(p_i)}$$

and
$$p_2(j \mid i) = \frac{\gamma p_j}{1-p_i} + \frac{1-\gamma}{N-1}$$

If $p_n \geq \dfrac{1}{2(N-1)}$, then a sufficient condition for the validity of the system is $1 \geq \gamma \geq 0$. In particular when $\gamma = 0$, we get Midzuno type sampling scheme where first unit is drawn with probability $\dfrac{2p_i(N-1)-1}{N-2}$ and the remaining units by srswor.

If $p_n < \dfrac{1}{2(N-1)}$, we can group the smallest unit with another unit reducing the population size to $N-1$ and see if the new $p_n \geq \dfrac{1}{2(N-2)}$, when (for $\gamma = 0$) a Midzuno-type sampling as above can be applied. If the amalgamated unit is chosen we randomise out a unit with probability proportional to size.

Scheme B: $\alpha = k$, $\beta = 1$ and allow k to vary. If $p_n < \dfrac{1}{2(N-1)}$ and k is any number satisfying

$$\frac{1-(N-1)p_n(1+2p_m)}{1-2p_n(N-1)} \geq k \geq \max$$

$$\left[\frac{(N-1)p_m(1+2p_m)-1}{2(N-1)p_m-1}, 1-(N-1)p_n\right] = k_0 \text{ (say)}$$

then,
$$p_2(i \mid j) = \frac{p_i + \dfrac{k-1}{N-1}}{k-p_j},$$

$$p_1(i) = \frac{2\left\{p_i + (k-1)\dfrac{p_i a_1}{N-1} - (k-1)\dfrac{a_2}{N-1}\right\}}{f(p_i)\left\{1 + a_2 + \dfrac{(k-1)a_1}{N-1}\right\}}$$

give a πps design. If $p_n \geq \dfrac{1}{2(N-1)}$, a similar condition holds except that there should not be any upper bound on k.

Scheme C: $\alpha = 1$, $\beta = d$ and allow d to vary. If $p_n < \dfrac{p_m}{1+(N-2)p_m}$ and the number d is such that $\min\{d_2, d_0\} > d \geq \max\left\{d_1, -\dfrac{1}{N-2}\right\}$ where,

$$d_0 = \frac{(N-1) - Np_m}{(N-2)p_m} \quad (>1),$$

$$d_1 = \frac{p_m + Np_n p_m - Np_n}{p_m - p_n - (N-2)p_n p_m} \quad (<1),$$

$$d_2 = \left\{1 + \frac{(N-1)\,p_n}{p_m}\right\} \quad (>1),$$

then,
$$p_2\,(j \mid i) = \frac{p_j + \dfrac{1-d}{N-1}p_i}{1-dp_i},$$

$$p_1(i) = \frac{2\{(N-1)p_i + (1-d)p_i a_2 - (1-d)a_3\}}{f(p_i)\,[(N-1)\,(1+a_2) + (1-d)\,\{a_2(1+a_2) - a_1 a_3\}]}$$

give a πps design. If $p_n \geq \dfrac{p_m}{1+(N-2)p_m}$, then a sufficient condition is

$$-\frac{1}{N-2} \leq d < \min\,\{d_0, d_2\}.$$

For these schemes it is not known whether $v_{YG}(e_{HT}) \geq 0$ always. For different values of γ, k or d within the admissible range, these schemes give different values of $V(e_{HT})$ and one may thus, choose the values of parameters optimally. For Scheme B, five populations as listed below were considered for investigating into the dependence of the value of V_{HT} on k. For three of these populations, variance decreased steadily reaching a minimum value and then increased with increasing value of k; for one population the behaviour was just the reverse, while for the other the variance increased steadily.

Population	Range of k
Yates and Grundy (1953); $N = 4$	(0.8286, 1.150)
Cochran (1977); p. 152; $N = 10$	(0.8119, 1.1743)
Cochran (1977); p. 203; $N = 10$	(– 0.2688, –)
Cochran (1977); p. 268; $N = 4$	(0.4571, 1.8)
Cochran (1977); p. 325; $N = 10$	(– 0.1819, –)

2.2.4 Rao-Hartley-Cochran (RHC) Sampling Strategy (1962)

Rao, Hartley, Cochran (*RHC*) proposed a sampling strategy as follows: Divide the population into n groups $G_1,...,G_n$ of sizes $N_1,..., N_n$ respectively at random. From each of the n groups draw a sample i with probability proportional to p_i *i.e.*, with probability $\frac{p_i}{II_k}$ where $II_k = \Sigma t \in G_k\, p_t$ if $i \in G_k$. An unbiased estimator of population total is

$$e_{RHC} = \sum_{i=1}^{n} \frac{y_i}{p_i} II_k(i),$$

$G_{k\,(i)}$ denoting the group to which a sampled unit i belongs.

$$V(e_{RHC}) = \frac{n\left(\sum_{1}^{n} N_i^2 - N\right)}{N(N-1)} V(e_{HH}) \tag{2.2.10}$$

where,
$$V(e_{HH}) = \frac{1}{n}\left(\sum_{1}^{N} \frac{y_i^2}{P_i} - Y^2\right)$$

$V(e_{RHC})$ is minimum if,

$$N_i = \frac{N}{n} \forall i, \tag{2.2.11}$$

when (2.2.10) reduces to

$$\left(1 - \frac{n-1}{N-1}\right) V(e_{HH}) \tag{2.2.12}$$

If $N = nR + k$, $0 < k < n$, R a positive integer, one should choose $N_i = R + 1$, $i = 1, ..., k$, $N_{k+1} = ... = N_n = R$. Then (2.2.10) reduces to

$$\left[1-\frac{n-1}{N-1}+\frac{k(n-k)}{N(N-1)}\right]V(e_{HH}) \tag{2.2.13}$$

An unbiased estimator of $V(e_{RHC})$ is

$$V(e_{RHC}) = \frac{\left(\sum_1^n N_i^2 - N\right)}{N^2 - \sum_1^n N_i^2} \sum_1^n II_i \left(\frac{y_i}{p_i} - e_{RHC}\right)^2 \tag{2.2.14}$$

If $N_i = R \; \forall \; i$, this reduces to

$$\frac{N-n}{N(n-1)} \sum_1^n II_i \left(\frac{y_i}{p_i} - e_{RHC}\right)^2$$

Hartley and Rao (1962) obtained asymptotic expression for $V(e_{HT})$ under their scheme of systematic sampling with varying probabilities. If the *HR*-scheme is compared with the present scheme with $N_i = R \; \forall \; i$,

$$V(e_{HR}) < V(e_{RHC})$$

if,

$$\sum_{t=1}^N P_t^2 \left(\frac{y_t}{P_t} - Y\right)^2 > \frac{1}{N} \sum_{t=1}^N P_t \left(\frac{y_t}{P_t} - Y\right)^2 \tag{2.2.15}$$

If one assumes a superpopulation model (vide Section 3.7 for the concept of superpopulation in finite population sampling),

$$Y_t = YP_t + \in_t$$

where Y_t is a random variable whose one particular realisation is y_t, $\in_t$'s are random variables with

$$(\in_t) = 0, \; \in(\in_t^2) = aP_t^g, \; \in(\in_t \in_{t'}) = 0, \; t \neq t'$$

$\in$ denoting expectation wrt model, and if the model-average variance is taken as the criterion for comparison, then $e_{HR} \succ e_{RHC}$ if

$$\Sigma P_t^g > \Sigma P_t^{g-1}/N \tag{2.2.16}$$

which is so if $g > 1$. Thus for $g > 1$, $e_{HR} \succ e_{RHC}$, though it should be borne in mind that the comparison (2.2.16) is based on asymptotic expression for $V(e_{HR})$ valid for large N and for values of n relatively small compared to N and exact expression for $V(e_{RHC})$.

Expression (2.2.12) shows, under (2.2.11) *RHC* procedure is better than the ppswr-procedure. However, a ppswr-procedure involves a variable number of distinct units ($\leq n$) and as such two procedures should be compared when they involve the same expected cost. Singh and Kishore (1975) compared the *RHC* and ppswr-procedures when

(*i*) cost is proportional to the expected number of distinct units

(*ii*) cost is proportional to the total expected size of the sample.

Letting n' to be the number of draws in ppswr and n the sample size in *RHC, RHC* $\succ$ *PPSWR* if

$$(N + n' - 1) \sum_{j=1}^{N} Q_j^{n'} < n\,(N-1) \tag{2.2.17}$$

in case (*i*) and

$$n' < \frac{(N-1)\left(1-\sum_{1}^{N} P_j\, Q_j^{n}\right)}{\sum_{j=1}^{N} P_j(Q_j^{n'} + NP_j - 1)} \tag{2.2.18}$$

in case of (*ii*) where $Q_j = 1 - P_j$. Singh and Kishore did not consider number of draws in ppswr to be integer-valued and hence their comparison is not proper. Deshpande (1984) considered a modified ppswr scheme as follows. If m be the number of draws in ppswr, effective sample size is

$$E(m) = N - \sum_{i} (1 - P_i)^m$$

Noting $E(m)$ is a monotonically increasing function of m, find an integer r such that under ppswr,

$$E(r) < n \leq E(r + 1).$$

Let,
$$p = \frac{E(r+1) - n}{E(r+1) - E(r)}$$

Consider the modified ppswr-procedure where a sample of r draws is selected with probability p and of $(r+1)$ draws with probability $1-p$, when the effective sample size n is attained. The estimator considered is

$$e'_{HH} = \frac{1}{d}\sum_{1}^{d}\frac{y_i}{p_i}$$

where d is the number of draws and

$$V(e'_{HH}) = \frac{r+p}{r(r+1)}\sum_{1}^{N}\left(\frac{y_i}{p_i} - Y\right)^2 \tag{2.2.19}$$

Comparison with (2.2.10) shows $V(e_{RHC}) < V(e'_{HH})$ if

$$\frac{1}{n}\left[1 - \frac{n-1}{N-1} + \frac{k(n-k)}{N\,(N-1)}\right] \le \frac{r+p}{r(r+1)} \tag{2.2.20}$$

Examples have been given to show that (2.2.20) is realisable in practice.

The *RHC* sampling strategy is, however, not admissible. Berg (1974) has given a strategy with uniformly smaller variance. Gabler (1985) has shown there does not exist any (π_{ps}, e_{HT}) strategy which is uniformly better than *RHC* strategy, the two being equivalent iff $P_i = \frac{1}{N}\ \forall\, i$.

Singh and Lal (1978) considered a scheme of random group formation as follows: The first random group of size N_1 is selected in Midzuno's scheme of sampling and the remaining $(n - 1)$ random groups as in *RHC* procedure. Since e_{RHC} is unbiased for any scheme of random group formation it remains valid under the present scheme also. It has been shown the proposed method of random group formation will be more efficient than the usual *RHC* scheme in the sense of yielding lower variance if

$$N_1 \lesseqgtr \sum_{i=1}^{n}\frac{N_i^2}{N} \quad \text{according as } N\delta \ge 2, \text{ where}$$

$$\delta = \frac{1}{\sigma_z^2}\left[\sum_{1}^{N} y_j^2 + \left(\sum\frac{y_j^2}{P_j}\right)\left(\sum P_j^2\right) - 2Y\sum y_j P_j\right]$$

and

$$\sigma_z^2 = \sum\frac{y_j^2}{P_j} - Y^2$$

Recently Ohlsson (1986) generalised *RHC*-sampling procedure and estimator.

2.2.5 Sampford's (1967) Sampling Scheme

Define $\lambda_i = \frac{p_i}{1 - p_i^n}$, $S(m)$ = a set of $m(\le N)$ distinct units $i_i, \ldots, i_m$; $L_m = \sum_{s(m)}\lambda_{i_1}\ldots\lambda_{i_m}$;

$L_m(\bar{i})$ and $L_m(\overline{ij})$ are defined similarly as L_m but over the subpopulation $\mathscr{P} - (i)$

and $\mathscr{P} - (i, j)$ respectively. Sampford proposed three methods for achieving his sampling design.

(*a*) *Whole sample method:* Select a sample $s(i_i, ..., i_n)$ with probability

$$p(s) = k_n \sum_{u=1}^{n} p_{i_u} \sum_{v(\neq u)=1}^{n} \lambda_{i_v} \tag{2.2.21}$$

$$= nk_n \lambda_{i_1} \ldots \lambda_{i_n} \left(1 - \sum_{u=1}^{n} p_{i_u}\right) \tag{2.2.22}$$

where, $$K_n = \left(\sum_{t=1}^{n} \frac{t\,L_{n-t}}{n^t}\right)^{-1}$$

Thus for $n = 2$, $$p(s) = K_2 p_i p_j \left(\frac{1}{1-2p_i} + \frac{1}{1-2p_j}\right); \tag{2.2.23}$$

for $n = 3$, $$p(s) = K_3 p_i p_j p_k \left(\frac{1}{(1-3p_i)(1-3p_j)} + \frac{1}{(1-3p_j)(1-3p_k)} + \frac{1}{(1-3p_i)(1-3p_k)}\right)$$

(*b*) *Draw-by-draw method:* Conditional probabilities of selection for different draws are given so that $p(s)$ in (2.2.21) is realised. For $n = 4$,

$$p_1(i) = \text{prob. } i \text{ is selected at first draw}$$

$$= c_1 \lambda_i \sum_{k(\neq i)} \sum_{\substack{j<k \\ (\neq i)}} \lambda_j \lambda_k (1 - p_i - p_j - p_k)$$

$$p_2(j \mid i) = c_2 \lambda_i \sum_{k(\neq i,j)} \lambda_k (1 - p_i - p_j - p_k)$$

$$p_3(k \mid i, j) = c_3 \lambda_k (1 - p_i - p_j - p_k)$$

$$p_4(l \mid i, j, k) = c_4 p_l$$

where, c's are suitable constants to make the sum of the conditional probabilities unity.

It follows that $c_1 = \frac{1}{3}k_4$. For $n = 2$ this scheme coincides with Brewer's (1963) procedure.

(c) *Rejective sampling procedure:* Select the first unit with probability p_i and the subsequent $(n-1)$ units with probability proportional to λ_i and with replacement such that the selected units form a sample if all the n units are distinct; otherwise reject all the selected units and fresh selections are made until all the n units turn out distinct. This realises $p(s)$ given in (2.2.21) and (2.2.22). For this sampling design

$$\pi_i = np_i,$$

$$\pi_{ij} = K_n \lambda_i \lambda_j \sum_{t=2}^{n} \frac{\{t - n(p_i + p_j)\} L_{n-1}(\bar{i}\,\bar{j})}{n^{t-2}}$$

Also, $\pi_{ij} \le \pi_i \pi_j \; \forall \; i, j$ so that $v_{YG}(e_{HT}) \ge 0$ always.

Assuming N large and $p_i = 0\left(\frac{1}{N}\right) \; \forall \; i$, Asoke and Sukhatme (1976) have given the following approximate expressions,

$$\pi_{ij} = n(n-1)p_i p_j \left[1 + \left\{(p_i + p_j) - \sum p_t^2\right\} + \left\{2(p_i^2 + p_j^2) - 2\sum p_t^3 - (n-2)p_i p_j + (n-3)(p_i + p_j)\sum p_t^2 - (n-3)\left(\sum p_t^2\right)^2\right\}\right]$$

to $0(N^{-4})$, for $n \ge 3$.

$$\pi_{ij} = 2p_i p_j \left[1 + \left\{(p_i + p_j) - \sum p_t^2\right\} + \left\{2(p_i^2 + p_j^2) - 2\sum p_t^3 - (p_i + p_j)\sum p_t^2 + \left(\sum p_t^2\right)^2\right\}\right]$$

to $0(N^{-4})$ for $n \ge 2$.

They have shown both (Sampford's design, $e_{HT}) = H_1$ and (Goodman-Kish design, $e_{HT}) = H_2$ have the same variance to $0(N)$*, assuming that $V(\text{ppswr}, e_{HH}) = 0(N^2)$ and hence both the procedures have smaller variance than ppswr-strategy to $0(N)$. Again under their asymptotic analysis, to $0(N^0), V(H_1) \le V(H_2)$.

Gabler (1981) has given an exact comparison between H_1 and ppswr strategy. He has shown that H_1 has a smaller sampling variance than a ppswr strategy, both using the same n and t_i values.

For $n = 2$, the scheme coincides with Brewer (1963) and Durbin (1967) procedures.

---

* viz, $\frac{1}{n}\left[\Sigma P_i Z_i^2 - (n-1)\,\Sigma_i P_i^2 Z_i^2\right], Z_i = \frac{Y_i}{P_i} - Y$.

2.2.6 Midzuno (1950, 52) - Lahiri (1951) - Sen (1952) Sampling Design

The first unit is chosen with probability p_i, subsequent units by srswor from the remaining $(N - 1)$ units. Thus,

$$p(s) = \frac{1}{M_1} q_s$$

where, $M_1 = \binom{N-1}{n-1}$ and $q_s = \sum_s p_k$. Ordinary ratio estimator $e_R = \frac{\sum_s y_k}{\sum_s p_k}$ is unbiased for Y with

$$V(e_R) = \frac{1}{M_1}\left[\sum y_i^2 \sum_{s \ni i} \frac{1}{q_s} + \sum \sum y_i y_i' \sum_{s \ni i, i'} \frac{1}{q_s}\right] - Y^2$$

$$v(e_R) = e_R^2 - \frac{1}{q_s}\left\{\sum_s y_i^2 + \frac{N-1}{n-1} \sum \sum_{\substack{i \neq j \\ \in_s}} y_i y_j\right\}$$

Here,
$$\pi_i = \frac{N-n}{N-1} P_i + \frac{n-1}{N-1}$$

$$\pi_{ij} = \frac{(n-1)(N-n)}{(N-1)(N-2)} (P_i + P_j) + \frac{(n-1)(n-2)}{(N-1)(N-2)}$$

v_{YG} is always non-negative for this scheme (Sen, 1953).

Drawing the first unit with probability equal to revised size-measure $p_i^* = \frac{(N-1)np_i}{N-n} - \frac{n-1}{N-n}$ (in place of p_i), the scheme may be made πps, for which it is necessary

$$\frac{n-1}{n(N-1)} < P_i < \frac{1}{n} \ \forall\ i$$

The πps modification of the scheme has been studied by Rao (1963 *a*), Chaudhuri (1974), Mukhopadhyay (1974), Sankaranarayanan (1969) among others. Rao showed that for pps-modified scheme with $n = 2$ e_{HT} has a smaller variance than a ppswr-strategy. Chaudhuri (1974), Mukhopadhyay (1974) extended the result for any n. Recently, Mukhopadhyay and Bhattacharyya (1991), following Gabler (1981) considered the sufficient conditions for e_{HT} coupled with Midzuno scheme to be better than the ppswr-strategy (Exercise 8).

Prasad and Srivenkataramana (1980) suggested the following modification to HTE for Midzuno scheme. Since for this scheme y_i / π_i is not constant even when $y_i = \beta x_i$ $\forall\, i$, consider the transformation

$$z_i = y_i + \frac{(n-1)b}{N-1}$$

(so that z_i/p_i is constant when $y_i = \beta x_i$), where b is a suitably chosen constant.

$$\hat{Y}_1 = \sum_s \frac{z_i}{\pi_i} - \frac{N\,(n-1)b}{N-n}$$

is unbiased for Y with

$$V(\hat{Y}_1) = V(\hat{Y}_{HT}) + b^2\Delta_1 + 2b\Delta_2$$

where,
$$\Delta_1 = \left(\frac{n-1}{N-n}\right)^2 V(\hat{N})$$

$$\Delta_2 = \frac{n-1}{N-n}\sum\sum_{i<j}\left(\frac{y_i}{\pi_i} - \frac{y_j}{\pi_j}\right)\frac{(\pi_i\pi_j - \pi_{ij})\,(\pi_j - \pi_i)}{\pi_{ij}}$$

$\hat{N} = \Sigma_{is}\dfrac{1}{\pi_i}$. Optimum choice of b is

$$b_{\text{opt}} = -\Delta_2/\Delta_1 \tag{2.2.24}$$

and hence,
$$V_{\min}\,(\hat{Y}_1) = V(\hat{Y}_{HT}) - \frac{\Delta_2^2}{\Delta_1}.$$

Thus if $b \in [0, 2b_{\text{opt}}]$, $V(\hat{Y}_1) < V(\hat{Y}_{HT})$. Since y_i values are not known in practice, b_{opt} cannot be computed.

A reasonable choice of b is obtained from (2.2.24) by taking summation in both numerator and denominator over sampled units only and weighting each term in the summation by π_{ij}^{-1}.

2.2.7 Procedures (32)–(35)

These are all offshoots of Midzuno, Sen, Lahiri procedure. Sampling plan (32) is Midzuno's scheme, made πps and was foreshadowed by Horvitz and Thompson

(1952). In procedure (33), $p(s) \propto \Sigma_{\bar{s}} p_k, \bar{s} = \mathscr{P} - s$. Here, an unbiased estimator of $\bar{Y}$ is $\hat{\bar{Y}} = \bar{y}\bar{X}/\bar{x}_{\bar{s}}, \bar{x}_{\bar{s}} = \Sigma_{\bar{s}} x_k/(N-n)$. The scheme can be made πps by using the revised size-measure $p_i' = 1 - (N-1)p_i$ in place of p_i. Mukhopadhyay and Bagchi (1991) made some comparison between Deshpande (1978) strategies and ppswr-strategy under a superpopulation model.

In scheme (35), let V be a subset of $\mathscr{P}$ and $q = (q_1, ..., q_N)$ be a set of non-negative quantities, $\Sigma q_i = 1$. A unit i is drawn with probability q_i from $\mathscr{P}$. If $i \in V$ it is retained in the sample and a srswor of size $(n-1)$ is selected from the remaining units. If $i \notin V$, it is discarded and a srswor of size n from $\mathscr{P} - \{i\}$ is selected. These constitute the sample. For $V = \mathscr{P}$ and ϕ, this scheme reduces to Midzuno's scheme and Deshpande's (1982) scheme, respectively.

It is easy to see

$$P(s) = \frac{\sum_1 q_i}{\binom{N-1}{n-1}} + \frac{\sum_2 q_i}{\binom{N-1}{n}}$$

where Σ_1 is overall $i \in V \cap s$, Σ_2 overall $i \in V^c \cap s^c$, V^c, s^c denoting respectively complements of V and s. Thus,

$$\pi_i = \frac{n-q}{N-1} + \frac{N-n}{N-1} q_i, \; i \in V$$

$$= \frac{n-q}{N-1} - \frac{nq_i}{N-1}, \quad i \in V^c$$

where, $$q = \sum_{i \in V} q_i .$$

Also, $$\pi_{ij} = \frac{n-1}{N-2}(\pi_i + \pi_j) - \frac{n(n-1)}{(N-1)(N-2)}$$

so that if $\pi_i + \pi_j = \dfrac{n}{N-1}$, $V(e_{HT})$ is not estimable. If V contains $(N-k)$ units,

$$\sum_{i \in V^c} \pi_i = \frac{n(k-1)}{(N-1)} + \frac{n-k}{N-1} q$$

so that if $k = n$, $\Sigma_{i \in V^c} \pi_i$ is independent of q.

To make the scheme πps, assume without loss of generality $X_1 \le X_2 \le \ldots \le X_N$. Let also $X_{(r)} = \Sigma_{i=1}^{r} X_i, r = 1, \ldots, N$. Then $\pi_i = n\frac{X_i}{X}$ *iff* one of the following conditions hold:

(*a*) $X_{(n)} = \frac{n-1}{N-1}X$, when $V = (n + 1, n + 2, \ldots, N)$.

(*b*) There exists an integer $k \ne n$ such that

$$X_{k+1} \ge \left[X_{(k)} - \frac{n-1}{N-1}X\right]\frac{1}{k-n} \ge X_k$$

when $$V = (k + 1, k + 2, \ldots, N)$$

Recently Mukhopadhyay and Bagchi (1990) modified the above scheme by considering the subset V as a srswor of size m from U. For this modified scheme,

$$P(s) = \frac{m-n}{(N-n)\binom{N-1}{n-1}}\sum_{i\in s} q_i + \frac{N-m}{N\binom{N-1}{n}}$$

$$\pi_i = \frac{(m-n)q_i}{N-1} + \frac{Nn-m}{N(N-1)}$$

$$\pi_{ij} = \frac{(n-1)(m-n)}{(N-1)(N-2)}(q_i + q_j) + \frac{(n-1)(Nn-2m)}{N(N-1)(N-2)}$$

when $n = m$, the scheme reduces to srswor. For the scheme to be made πps, it is required

$$q_i = \frac{Nn-m}{N(n-m)} - \frac{n(N-1)p_i}{n-m} = q'_i \text{ (say) } (n \ne m)$$

$$\alpha < q'_i < \frac{Nn-m}{N(n-m)} \quad (n > m)$$

$$0 < q'_i < \alpha (n < m)$$

where, $$\alpha = \frac{N(N-1) - Nn + m}{N(m-n)}$$

This scheme, when made πps, has the same π_i, π_{ij} values as Midzuno scheme, Deshpande-Ajgaonkar generalisation of Midzuno scheme and hence shares their properties.

2.2.8 Scheme due to Gupta, Nigam and Kumar (1982)

This scheme uses a Balanced Incomplete Block (BIB) design. A BIB design with parameters v (no. of variates) = N, b (no. of blocks), r (no. of replications of each treatment), k (block size) ($\geq n$), λ (no. of times each pair of treatments occur together in a block) is chosen. From the totality of all blocks in the design, block s is selected with probability

$$P_b(s) = \frac{v\left(r\sum_{k\in s} p_k - \lambda\right)}{b(r-\lambda)}, \; s = 1, \ldots, b \tag{2.2.25}$$

where the survey units correspond to varieties and each variety is given an arbitrary labelling 1, ..., N. If $k = n$, this gives the desired sample. Otherwise n units are chosen at random out of k units. We have $\pi_i = np_i$ (using the relation $bk = vr$). The scheme is applicable only if

$$\sum_{k\in s} p_k > \frac{\lambda}{r} = \frac{k-1}{N-1} = \frac{n-1}{N-1} + \frac{k-n}{N-1}, \; \forall s \tag{2.2.26}$$

which is likely to be satisfied if $(k - n)$ is small.

Since for a given v and k there exist more than one BIB design, the above procedure gives rise to a family of sampling designs, of which only those are feasible for which (2.2.26) is satisfied. Again since the labelling of the units is done by the sampler, for a particular choice of labelling (2.2.26) may be satisfied for a chosen design while for another it may not.

For the unreduced BIB design where $b = \binom{N}{n}$, $r = \binom{N-1}{n-1}$, $\lambda = \binom{N-2}{n-2}$, *i.e.*, when all the possible samples of size n are considered, $P(s)$ reduces to

$$\frac{n!(N-n-1)!}{(N-2)!}\sum_{k\in s}\left(p_k - \frac{n-1}{n(N-1)}\right)$$

which is the same as in Sankaranarayanan's design.

This design has the advantage, therefore, that (unlike many other designs) it uses only a fraction of the totality of all possible samples, *i.e.*, a reduced sample space. However, no compact expression for π_{ij} is available in general.

In particular classes of doubly balanced designs where every triplet occurs a constant number, v times (a 3-design),

$$v(v-2) = \lambda(k-2)$$

and in such cases,

$$\sum_{s \ni i,j} \sum_{k \in s} p_k = (\lambda - \nu)(p_i + p_j) + \nu,$$

$$\sum_{s \ni ij} P_b(s) = k(k-1)\left\{p_i + p_j - \frac{1}{N-1}\right\}\frac{1}{N-2}$$

so that if $k > n$, when a srswor of size n from the selected block is chosen,

$$\pi_{ij} = n(n-1)\left\{p_i + p_j - \frac{1}{N-1}\right\}\frac{1}{N-2},$$

which coincides with the π_{ij} of Sankaranarayanan's scheme.

A block design is a t-design if any set of t units occur together in the same number λ_t of blocks where all the blocks may or may not be distinct (Raghava Rao, 1971). Bhattacharyya (1993) has shown if a t-design is used for selecting a πps design by this procedure,

$$\pi_{i_1, i_2, \ldots, i_{t-1}} = \frac{(N-t)!\,n!}{(N-2)!\,(n-t+1)!}\left[(p_{i1} + \ldots + p_{i_{t-1}}) - \frac{t-2}{N-1}\right] \quad (2.2.27)$$

2.2.9 Schemes due to Singh and Srivastava (1980)

Scheme A:

Step 1: Two units (i, j) are selected with probability proportional to $(x_i - x_j)^2$.

Step 2: $(n - 2)$ units are selected from the remaining $(N - 2)$ units in the population by srswor. It follows

$$P(s) = \frac{s_x^2}{M_0 S_x^2}$$

where $$M_0 = \binom{N}{n},\ s_x^2 = \frac{1}{(n-1)}\sum_s (x_k - \bar{x})^2,$$

$$S_x^2 = \frac{1}{(N-1)}\sum (x_k - \bar{X})^2.$$

Step 1 can also be achieved as follows:

Step 1 (a): Select i at first draw with probability proportional to

$$(x_i - \bar{X})^2 + \frac{(N-1)S_x^2}{N}$$

Step 1 (b): Select j at second draw with conditional probability

$$p_{j|i} \propto (x_j - x_i)^2$$

Under this scheme $E(b) = \beta$, $E(\hat{Y}_{lr}) = \bar{Y}$

where, $b = \frac{s_{xy}}{s_x^2}$, $\beta = \frac{S_{xy}}{S_x^2}$ $\hat{\bar{Y}}_{lr} = \bar{y} + b\,(\bar{X} - \bar{x})$, the classical linear regression estimator of $\bar{Y}$ and $S_{xy} = \frac{1}{N-1}\sum (x_k - \bar{X})(y_k - \bar{Y})$. Expressions for $V(\hat{\bar{Y}}_{lr})$ and $v(\hat{\bar{Y}}_{lr})$ have been given by the authors. For this scheme $\hat{S}_y^2(R) = \frac{s_y^2}{s_x^2} S_x^2$ is an unbiased estimator of S_y^2. It is easily seen that $v(\hat{\bar{Y}}_{lr})$ can often take negative values.

Lemma 2.1: (Bhattacharyya 1993). Under this scheme

$$\pi_{i_1 i_2 \ldots i_k} = 1 - \frac{N-n}{n}\sum_{q=1}^{k}\frac{S_x^2(i_q)}{S_x^2} + \frac{(N-n)_{(2)}}{N_{(2)}}\sum_{q_1<q_2=1}^{k}\sum \frac{S_x^2(i_{q_1}, i_{q_2})}{S_x^2} - \ldots + (-1)^k \frac{(N-n)_{(k)}}{N_{(k)}}\frac{S_x^2(i_1, \ldots, i_k)}{S_x^2} \tag{2.2.28}$$

where, $S_x^2(i_1, \ldots, i_q)$ is the variance of x on $\mathscr{P} - (i_1, \ldots, i_q)$.

Scheme B: Select unit i with probability $\alpha(x_i - \bar{X})^2$ at first draw. Select $(n - 1)$ units from the remaining $(N - 1)$ units by srswor. Here

$$p(s) = \sum_s (x_i - \bar{x}_s)^2 \Big/ \binom{N-1}{n-1} \sum (x_i - \bar{X})^2 \tag{2.2.29}$$

Under this scheme an unbiased estimator of $\bar{Y}$ is

$$\hat{\bar{Y}}^* = \frac{n(N-1)}{N(n-1)}\left(\bar{y} - \bar{x}'\frac{\Sigma_s x_i' y_i}{\Sigma_s x_i'^2}\right) \tag{2.2.30}$$

where $x_i' = x_i - \bar{X}$, $\bar{x}' = \frac{1}{n}\Sigma_s x_i'$.

Expressions for variance and unbiased estimator of variance have been given by the authors.

This is special case of Midzuno's sampling scheme and as such it shares, among others, the property $\pi_{ij} \le \pi_j \pi_i$ of Midzuno's scheme.

2.2.10 Sampling Scheme due to Chao

This is a sequential sampling scheme where the sample size is fixed at n but population size increases from n to N. Whenever the population increases by one unit we decide the probability by which the new unit should be included in the sample. One unit from the old sample is then replaced by the new unit with a certain probability, maintaining the sample size n.

Consider the population $\mathscr{P}_k = \{1, 2, ..., k\}, k \geq n$ at time k. Let X_i be the size of i, $\pi(k, i)$ inclusion probability of i at time k,

$$\sum_{i=1}^{k} \pi(k; i) = n \tag{2.2.31}$$

Further, let $$\pi(k; i) \, \alpha \, x_i \tag{2.2.32}$$

If $(nx_j / X) > 1, X = \sum_{t=1}^{k} x_t$, for some j, select unit j with the highest value of (nX_j / X) with probability one and recalculate (2.2.31) and (2.2.32) with $\mathscr{P}_k$ replaced by $\mathscr{P}_k - \{j\}$, k by $k - 1$. The process is repeated till $0 \leq \pi(k, i) \leq 1 \; \forall \; i$ when (2.2.32) is satisfied by those t for which $0 < p(k, t) < 1$.

Assume we have a sampling design of fixed size n which when applied to $\mathscr{P}_k$ realises $\pi(k, i)$ and gives a sample $s(k)$ (of size n). Whenever a new unit $k + 1$ occurs in the population, from $\mathscr{P}_{k+1}$ following sequences are computed.

$$A_k = \{i : \pi(k; i) = 1, \pi(k + 1; i) = 1, i \leq k\}$$

$$B_k = \{i : \pi(k; i) = 1, \pi(k + 1; i) < 1, i \leq k\}$$

$$C_k = \{i : \pi(k; i) < 1, \pi(k + 1; i) < 1, i \leq k\}$$

Let L_k = Number of units in $A_k U B_k$, $W_k = \pi(k + 1; k + 1)$,

$$T_{kj} = \{1 - \pi(k + 1; j)\}/W_k$$

for $j \in B_k$, $T_k = \Sigma_{j \in B_k} T_{kj}$ (= 0 if B_k is empty),

$$R_{kj} = \begin{cases} 0, & j \in A_k \\ T_{kj}, & j \in B_k \\ \dfrac{1 - T_k}{n - L_k}, & j \in C_k \end{cases}$$

Sample $s(k + 1)$ (of size n) from $\mathscr{P}_{k+1}$ is selected as follows: (*i*) Select $k + 1$ with probability W_k, (*ii*) if $k + 1$ is not selected retain the old sample, (*iii*) if $k + 1$ is selected, select one unit j from B_k with probability T_{kj}; if no unit from B_k is selected, select one unit at random from the remaining $n - L_k$ units in $s_k - (A_k U B_k)$. Replace this selected unit by $(k + 1)$ to obtain $s(k + 1)$.

For this procedure

$$\pi(k + 1; i) = \begin{cases} (1 - W_k R_{ki})\pi(k; i) & , \quad i \leq k \\ W_k & , \quad i = k + 1 \end{cases} \quad \{ \; \} \tag{2.2.33}$$

It follows that

$$\pi(k;\, i, j) \le \pi(k;\, i)\, \pi(k;\, j)\ \forall\ i < j = 1, \ldots, k$$

The values of second and higher order inclusion-probabilities depend on order of units in the population. Thus, for $N = 4$, $X_i = i(i = 1, \ldots, 4)$, $\pi(4; 1, 2) = 0$. If, however, $X_i = 5 - i$, $\pi(4; i, j) > 0\ \forall\ i < j$. A sufficient condition for $\pi(N; i, j) > 0\ \forall\ i < j$ is $0 < W_k < 1$, $k \ge n$ and B_n contains at least two elements. For $X_i = \frac{1}{k}\ \forall\ i$, the scheme reduces to Alan Waterman's algorithm (Kunth, 1973, p. 139) for drawing srswor of size n from $\mathscr{P}_k$, $k \ge n$.

Example 2.1: $\mathscr{P}_4$: {1, 2, 3, 4} with corresponding x values {1, 3, 4, 5}. $n = 3$. Since $3X_4 > X$, take $\pi(4; 4) = 1$ and recalculate $\pi(4; j), j = 1, 2, 3$ omitting the unit 4 and using sample size 2. This gives $\pi(4; 1) = 1/4$, $\pi(4; 2) = \frac{3}{4}$, $\pi(4; 3) = 1$. Hence, sample $s(4)$ is either (1, 3, 4) or (2, 3, 4). Suppose unit 5 with size 6 now occurs in the same population. Again $\pi(5; 1) = \frac{3}{19}$, $\pi(5; 2) = \frac{9}{19}$, $\pi(5; 3) = \frac{12}{19}$, $\pi(5; 4) = \frac{15}{19}$, $\pi(5; 5) = \frac{18}{19}$; $A_4 = \phi$, $B_4 = \{3, 4\}$, $C_4 = \{1, 2\}$, $L_4 = 2$, $R_{41} = R_{42} = R_{43} = \frac{7}{18}$, $R_{44} = \frac{4}{18}$, $T_4 = \frac{11}{18}$. $s(5)$ is selected as follows:

(*i*) Select 5 with probability $\frac{18}{19}$; (*ii*) If 5 is not selected take $s(5) = s(4)$; (*iii*) If 5 is selected, select one unit from {3, 4} with probabilities $\frac{7}{18}, \frac{4}{18}$ respectively; if none is selected then select the remaining unit in $s(4)$. Replace the unit thus selected by 5. It is easy to check that relations (2.2.33) hold.

Sengupta (1989) gave necessary and sufficient conditions for the π_{ij}'s to be strictly positive for some order of arrangement of the population units. He also proved the superiority of this scheme over the Hansen-Hurwitz strategy based on the same number of draws.

2.2.11 Sunter's (1977) List Sequential Sampling Plan I

Bobbington (1975) described the following method of drawing a srswor. Draw the first unit with probability $\frac{n}{N}$. Draw subsequent units $i = 1, 2, \ldots, N$ with probabilities $\frac{n - n_i}{N - i + 1}$ where n_i denotes the number of units selected up to the unit numbered $(i - 1)$. The process stops when $n_i = n$. Denote the event i is selected as a_i, not selected as $\bar{a}_i$. Thus, $P(a_{i_1}\, a_{i_2} \ldots a_{i_n})$, $i_1 < i_2 < \ldots < i_n$.

$$= P(\bar{a}_1 \ldots \bar{a}_{i_1-1}\, a_{i_1})\, P(\bar{a}_{i_1+1} \ldots \bar{a}_{i_2-1}\, a_{i_2} \mid n_{i_1+1} = \ldots = n_{i_2-1} = 1)$$

$$\ldots P(\bar{a}_{i_{n-2}+1} \ldots \bar{a}_{i_{n-1}-1}, \ldots a_{i_{n-1}} \mid n_{i_{n-2}+1} = \ldots = n_{i_{n-1}-1} = n-2)$$

$$P(\bar{a}_{i_{n-1}+1} \ldots \bar{a}_{i_n-1}, a_{i_n} \mid n_{i_{n-1}+1} = \ldots = n_{i_{n-1}} = n-1)$$

$$= \frac{1}{\binom{N}{n}} \text{ substituting } P(a_i) = \frac{n - n_i}{N - i + 1}.$$

Example 2.2:

$$N = 6,\ n = 3,\ P(s(2, 5, 6))$$

$$= P(s(\bar{1}, 2, \bar{3}, \bar{4}, 5, 6)) = \left[\left(1 - \frac{3}{6}\right) \cdot \frac{3}{6-2+1}\left(1 - \frac{3-1}{6-3+1}\right)\right.$$

$$\left.\left(1 - \frac{3-1}{6-4+1}\right)\frac{3-1}{6-5+1}\,\frac{3-2}{6-6+1} = \frac{1}{20} = \frac{1}{\binom{6}{3}}\right]$$

Sunter generalised the above procedure to unequal probability sampling. It is assumed units with $nx > X$ have been removed from the population and selected with probability 1.

(a) Order the remaining units as discussed below, (b) Choose the first unit with probability $\frac{nx_i}{X}$ (here n = sample size to be attained from the remaining population).

Choose the subsequent unit, i, say, with probability $\frac{(n - n_i)x_i}{X_i^*}$, $X_i^* = \sum_{j=i}^{N} x_j$ where n_i is the number of distinct units selected up to and including the trial for selection of unit $(i - 1)$.

(c) The process in step (b) terminates when

$n_i = n$ or $i = i^*$, whichever occurs first, when

$$i^* = \min\left\{\left(i : \frac{nx_i}{X_i^*} > 1\right); N - n + 1\right\}$$

We denote by $[i^*] = \{j; j \geq i^*\}$

(d) Choose a sample of size $n - n_i^*$ by SRS from the units in $[i^*]$.

The question remains as to the appropriate ordering of the units. The units i ($i < i^*$) are chosen by pps while those in $[i^*]$ by SRS with (equal) probability proportional to their average size. If the units are arranged by decreasing size there will be a relatively large number of small units in $[i^*]$. For a population arranged by increasing size the situation would be reverse.

Example 2.3: Sizes of 11 units arranged in decreasing size (8 8 8 8 1 1 1 1 1 1 1), $n = 2$. $P_1(1) = \frac{16}{39}$, $P_2\ (2|1) = \frac{8}{31}$, $P_2\ (2|\bar{1}) = \frac{16}{31}$, etc., $i^* = \min \left\{\left(i : \frac{nx_i}{X_i^*} > 1\right), N - n + 1\right\} = \min\ (4, 10) = 4.$

Therefore in SRS-set there will be 8 units, one with size 8, seven each with size 1.

If the units are arranged in increasing order of size, (1 1 1 1 1 1 1 8 8 8 8), $i^* = 10$ when SRS -set consists of 2 units each with size 8.

In most practical situations one will prefer ordering by decreasing size so that the smallest units are in the SRS-set. This is because often it is observed that the correlation between size-measures and the variable of interest becomes unstable for the smallest units in the population.

For this scheme,

$$\pi_i = \begin{cases} \dfrac{nx_i}{X}, & 1 \le i < i^*, \\ \dfrac{n\bar{x}^*}{X}, & i \in [i^*] \end{cases}$$

where $\quad \bar{x}^* = \dfrac{1}{N - i^* + 1} \sum_{i^*}^{N} x_i$

is the average size of the units in $[i^*]$ and

$$0 < \pi_{ij} < \pi_i \pi_j,\ i, j = 1, ..., N$$

$$\pi_{ij} = \frac{n(n-1)}{XC_{i+1}} x_i x_j \prod_{k=1}^{i-1} \left(1 - \frac{X_k}{C_{k+1}}\right),\ i < j < i^*$$

where, $C_i = \Sigma_i^N X_j$. For $j \ge i^*$ explicit formula for π_{ij} is not known.

There are many situations where getting a fixed sample size is not necessary, but it is sufficient to attain a sample size n' which is not very less than a known number n. In this case the scheme may be modified as follows.

Choose the sample according to rule (*b*), replacing X_i^* by $X_i^{**} = X_i^* + x'$ where x' is the smallest number such that $nx_i < X_i^{**}$ ($i = 1, ..., N$). The sampling is stopped when $n_i = n$ or the population is exhausted whichever occurs first. Here $n' \le n$,

$$E(n') = \frac{nX}{X^{**}},\ V(n') < n^2 \left(\frac{X}{X^{**}}\right)\left(1 - \frac{X}{X^{**}}\right)$$

$$\pi_i = \frac{nX_i}{X_i^{**}}$$

and $0 < \pi_{ij} < \pi_i \pi_j\ \forall\ i, j.$

Example 2.4: Size-measure of a population arranged in decreasing order (11, 6, 5, 5, 3, 1), $n = 3$. Here $x' = 6$.

Thus, $$X_3^{**} = 20.$$

If size-measures are arranged in increasing order (1, 3, 5, 5, 6, 11), $x' = 1$.

2.2.12 Gabler's (1987) Procedure

Gabler considered a measure of distance between two sampling designs. Let $(\pi_1^*, ..., \pi_N^*)$ be a set of inclusion probabilities which one wants to realise through a πps design $(\pi_i^* > 0 \ \forall \ i, \Sigma\pi_i^* = n)$. Suppose that from practical considerations, it would be ideal if one attains design $p_0(s)$. One should, therefore, follow a πps design say $\tilde{p}(s)$ for which $\pi_i(\tilde{p}) = \pi_i^* (i = 1, ..., N)$ and which is as close to $p_0(s)$ as possible in the sense that the distance

$$D(p_0, p) = \sum_{s \in \mathscr{S}_{p_0}} \frac{(p(s) - p_0(s))^2}{p_0(s)}$$

where, $\mathscr{S}_{p_0} = \{s : p_0(s) > 0\}$, the support of the design p_0, is minimum. Gabler termed the design $\tilde{p}$ as a nearest πps design. Obviously, in minimising the distance $D(p_0, p)$, one is confining to the design p with $\mathscr{S}_p \subseteq \mathscr{S}_{p_0}$. It has been shown that if $I I°$ is the matrix of first and second order inclusion probabilities of p_0, then the design $\tilde{p}$ is given by

$$\tilde{p}(s) = p_0(s) \sum_{i \in s} \lambda_i$$

provided $\tilde{p}(s) \geq 0 \forall s \in \mathscr{S}_{p_0}$, where λ_i's are given by the solution of the equation

$$\mathbf{II}°\lambda = \pi^*$$

$\lambda = (\lambda_1, ..., \lambda_N)'$, $\pi^* = (\pi_1^*, ..., \pi_N^*)'$. Midzuno's design modified to be made πps is found to be the πps design nearest to srswor. Sampford's (1967) design is shown to be nearest to a design with

$$p_0(s) \propto \prod_{i \in s} \pi_i^* / n(1 - \pi_i^*)$$

$\tilde{p}(s)$ can be realised using $p_0(s)$ as follows:

(*i*) Select i with probability $\lambda_i \pi_i°$, (*ii*) Select sample s with probability $p_0(s)$, and accept s as the desired sample if $s \ni i$; otherwise step (*ii*) is repeated. Gabler's procedure is very illuminating in the sense that it modifies an existing πps procedure to suit the need of a practically desirable sampling design. It establishes the intuitive result formally that Midzuno's scheme is the nearest to srswor design. Recently, Adhikary (1991) compared the performance of the generalised difference predictor based on Gabler's design with that of the same based on an arbitrary design from which Gabler's design is realised.

2.2.13 Procedures (52), (53)

Chaudhuri (1971) considered the problem of constructing a sampling design satisfying certain ordering of the rank values of π_{ij}'s when such information is available *apriori*, in order to control $V(e_{HT})$. Following Jessen (1969), Mukhopadhyay (1972) first considered the problem of constructing a sample design with a pre-assigned set of first and second order inclusion-probabilities $\pi^0(\lambda)$, $\pi^0(\lambda\lambda')$ (provided such a sampling design exists). His procedure consists in finding a sampling design D with selection probabilities,

$$P_D(s) = \sum_{t=1}^{m} p(t; s) \tag{2.2.34}$$

where $p(t; s)$ are some non-negative quantities, m is the number of stages (to be known during the process) and the inclusion-probabilities $\pi_D(\lambda)$, $\pi_D(\lambda\lambda')$ of the design D satisfy,

$$\begin{aligned} \pi_D(\lambda) &= \pi^0(\lambda), \quad \lambda = 1, ..., N \\ \pi_D(\lambda\lambda') &= \pi^0(\lambda\lambda'), \lambda \neq \lambda' = 1, ..., N \end{aligned} \tag{2.2.35}$$

a set of pre-assigned quantities.

A sample s here is an unordered set of n distinct units selected from a population of size N. His scheme is as follows:

$S(t)$: Collection of all feasible samples at stage t, $t = 1, ..., m$.

$p(t; s)$: A quantity (> 0) attached to a sample s at stage t, $s \in S(t)$; $p(t; s) = 0$ for $s \notin S(t)$.

$$D(t) = \sum_{s \in S(t)} p(t; s)$$

$$R(t) = R(t-1) - D(t), \; t = 1, ..., m; \; R(0) = 1$$

$$\pi(t; \lambda) = \sum_{s \ni \lambda \mid s \in S(t)} p(t; s), \lambda = 1, ..., N$$

$$\pi(t; \lambda\lambda') = \sum_{s \ni \lambda\lambda' \mid s \in S(t)} p(t; s), \lambda \neq \lambda' = 1, ..., N$$

$$\delta\pi^0(\lambda) = \pi^0(\lambda) - \pi(1; \lambda)$$

$$\delta^t\pi^0(\lambda) = \delta^{t-1}\pi^0(\lambda) - \pi(t; \lambda), \; t = 2, ..., m$$

$$\delta\pi^0(\lambda\lambda') = \pi^0(\lambda\lambda') - \pi(1; \lambda\lambda')$$

$$\delta^t\pi^0(\lambda\lambda') = \delta^{t-1}\pi^0(\lambda\lambda') - \pi(t; \lambda\lambda'), \; t = 2, ..., m.$$

Feasible samples at stage t and hence $S(t)$ is first enumerated. For this we must consider sample spaces $S(1)$ which are consistent wrt a given set of $\pi^0(\lambda)$, $\pi^0(\lambda\lambda')$ values and $S(t)$ which are consistent wrt $\delta^t\pi^0(\lambda)$, $\delta^t\pi^0(\lambda\lambda')$'s, $t = 2, \ldots, m$. A sample space S_γ is inconsistent wrt a set of $\pi(\lambda)$, $\pi(\lambda\lambda')$ values if there exists no set of values of $p(s) > 0$ for each $s \in S_\gamma$ s.t.

$$\sum_{s \in S_\gamma} p(s) = 1$$

$$\sum_{s \ni \lambda:\, s \in S_\gamma} p(s) = \pi(\lambda),\ \lambda = 1, \ldots, N$$

$$\sum_{s \ni \lambda\lambda':\, s \in S_\gamma} p(s) = \pi(\lambda\lambda'),\ \lambda \neq \lambda' = 1, \ldots, N$$

Similarly, a sample space $S(t)$ for the tth stage ($t = 2, \ldots, m$) is inconsistent wrt a set of $\delta^{t-1}\pi^0(\lambda)$, $\delta^{t-1}\pi^0(\lambda\lambda')$ values if there exists no set of values of $p(t; s)$, $s \in S(t)$ satisfying,

$$\sum_{s \ni \lambda:\, s \in S(t)} p(t; s) = \pi(t; s) \leq \delta^{t-1}\pi^0(\lambda),\ \lambda = 1, \ldots, N$$

$$\sum_{s \ni \lambda\lambda':\, s \in S(t)} p(t; s) = \pi(t; \lambda\lambda') \leq \delta^{t-1}\pi^0(\lambda\lambda'),\ \lambda \neq \lambda' = 1, \ldots, N$$

We notice that $\pi(t; \lambda)$, $\pi(t; \lambda\lambda')$, $R(t)$ will be subject to the following conditions:

$$\delta^t\pi^0(\pi) > \delta^t\pi^0(\lambda\lambda'), \tag{2.2.36}$$

$$\pi(1; \lambda\lambda') < \pi^0(\lambda\lambda'), \tag{2.2.37}$$

$$\pi(t; \lambda\lambda') < \delta^{t-1}\pi^0(\lambda\lambda'),\ t = 2, \ldots, m \tag{2.2.38}$$

$$R(t) > \delta^t\pi^0(\lambda),\ \lambda = 1, \ldots, N;\ t = 1, \ldots, m \tag{2.2.39}$$

The conditions (2.2.36) to (2.2.39) together with the condition,

$$R(m) = 0 \text{ (for some positive integer } m) \tag{2.2.40}$$

form a set of necessary and sufficient conditions for the attainment of a design D with the properties (2.2.35). It follows therefore any set of $p(t; s)$, $s \in S(t)$ which satisfy (2.2.36) – (2.2.40) will give the desired design through (2.2.34).

Rules have been given for enumerating $S(t)$.

After this we choose $p(1; s)$ such that

$$p(1; s) = p(1; s') \text{ for all } s \neq s' \in S(1) \tag{2.2.41}$$

In case $S(1) = S$ where S is a collection of all $\binom{N}{n}$ samples, conditions (2.2.36), (2.2.37), (2.2.39), respectively imply

$$D(1) < \frac{N(N-1)}{n(N-n)} \min_{(\lambda\lambda') \in (1,2,\ldots,N)} [\pi^0(\lambda) - \pi^0(\lambda\lambda')]$$

$$= D(1)^* \text{ (say)}, \tag{2.2.42}$$

$$D(1) < \frac{N(N-1)}{n(n-1)} \min_{(\lambda\lambda') \in (1,2,\ldots,N)} \pi^0(\lambda\lambda')$$

$$= D(1)^{**} \text{ (say)}, \tag{2.2.43}$$

$$D(1) < \frac{N}{N-n}\left[1 - \max_{\lambda \in (1,\ldots,N)} \pi^0(\lambda)\right]$$

$$= D(1)^{***} \text{ (say)} \tag{2.2.44}$$

and we take as a rule

$$D(1) = \min[D^*(1), D^{**}(1), D^{***}(1)] \tag{2.2.45}$$

and calculate $\qquad p(1; s) = D(1)/\binom{N}{n}$

In case S is not consistent wrt given set of $\pi^0(\lambda)$, $\pi^0(\lambda\lambda')$ values and some restricted sample S_γ is consistent we also take $p(1; s)$ such that (2.2.41) holds. Here conditions (2.2.36), (2.2.37), (2.2.39), respectively imply

$$D(1) < \min_{(\lambda\lambda) \in C(1)} \frac{M(1)}{n(1; \lambda) - n(1; \lambda\lambda')} \{\pi^0(\lambda) - \pi^0(\lambda\lambda')\}$$

$$= D'^*(1) \text{ (say)}, \tag{2.2.46}$$

$$D(1) < \min_{(\lambda\lambda') \in C'(1)} \frac{M(1)}{n(1; \lambda\lambda')} \pi^0(\lambda\lambda')$$

$$= D'^{**}(1) \text{ (say)}, \tag{2.2.47}$$

$$D(1) < \min_{(\lambda\lambda') \in C''(1)} \frac{M(1)}{M(1) - n(1; \lambda)} [1 - \pi^0(\lambda)]$$

$$= D'^{***}(1) \text{ (say)} \tag{2.2.48}$$

where, $C(1) = \{(\lambda\lambda')\} \mid \pi^0(\lambda) \neq \pi^0(\lambda\lambda')\}$,

$$C'(1) = \{(\lambda\lambda') \mid \pi^0(\lambda\lambda') > 0\},$$

$$C''(1) = \{(\lambda \mid \pi^0(\lambda) \neq 1\},$$

M_1 is the number of feasible samples, $n(1; \lambda)$ the number of samples containing λ, $n(1; \lambda\lambda')$ number of samples containing $\lambda\lambda'$ together in $S(1)$. We take as a rule,

$$D(1) = \min[D'^*(1), D'^{**}(1), D'^{***}(1)] \qquad (2.2.49)$$

and take $p(1; s) = D(1)/M(1)$. We then calculate $\delta\pi^0(\lambda)$, $\delta\pi^0(\lambda\lambda')$. $S(2)$ is then constructed and $D(2)$ is calculated by using $p(2; s) = p(2, s')$ for all $s, s' \in S(2)$ and by taking $D(2)$ as the greatest lower bound of all the three upper limits of $D(2)$ obtained by considering the inequalities (2.2.36), (2.2.38), (2.2.39). $R(t)$ will be reduced at each stage and after a finite number m of stages it will be zero. We shall then have $\delta^m\pi^0(\lambda) = 0\ \forall\ \lambda = 1, ..., N$ and $\delta^m\pi^0(\lambda\lambda') = 0\ \forall\ \lambda \neq \lambda' = 1, ..., N$ and the desired design with $\pi_D = \pi^0(\lambda)$, $\lambda = 1, ..., N$ and $\pi_D(\lambda\lambda') = \pi^0(\lambda\lambda')$, $\lambda = \lambda' \neq 1, ..., N$ is obtained by (2.2.34).

If the consistency of the sample space is broken at any stage, say, the tth, $D(t-1)$ and hence $\delta^{t-1}\pi^0(\lambda)$, $\delta^{t-1}\pi^0(\lambda\lambda')$ values may sometimes be suitably adjusted, *e.g.*, by taking $D(t-1)$ value less than its upper bound, by rounding off some of the $\delta^{t-1}\pi^0(\lambda)$, $\delta^{t-1}\pi^0(\lambda\lambda')$ values, to make $S(t)$ consistent wrt $\delta^t\pi^0(\lambda)$, $\delta^t\pi^0(\lambda\lambda')$ values.

The above scheme is a very general one, applicable for any n. Given a set of $\pi^0(\lambda)$, $\pi^0(\lambda\lambda')$ values for which a sampling design exists, we can always proceed with the above procedure, accounting for more and more of the probability mass until the remaining probability mass, $R(t)$ is zero, at least approximately, which is satisfactory for all practical purposes.

Sinha (1973) considered the same problem of constructing a sampling design as Mukhopadhyay (1972). His procedures are as follows:

(*a*) *Method of Extension:* From the given set of $\pi^0(\lambda)$, $\pi^0(\lambda\lambda')$ values, a fixed size (FS(N-2))-design is first constructed with values of first and second order inclusion probabilities $\pi^*(\lambda)$, $\pi^*(\lambda\lambda')$ given by

$$\pi^*(\lambda) = \frac{N-2}{n}\pi^0(\lambda)$$

$$\pi^*(\lambda\lambda') = \frac{(N-2)(N-3)}{n(n-1)}\pi^0(\lambda\lambda').$$

The extended design consists of $\binom{N}{n-2} = k$ (say) samples $s_1^*, ..., s_k^*$ with selection probabilities

$$p^*(s_j^*) = p_j^* \text{ (say)}$$
$$= 1 + \pi^*(\lambda\lambda') - \pi^*(\lambda) - \pi^*(\lambda')$$

whenever, $\lambda \notin s_j^*$, $\lambda' \notin s_j^*$. The selection probabilities of the final design is

$$p_D(s(i(1), \ldots i(n))) = \frac{1}{\binom{N-2}{n}} \sum_{(j \mid i(\lambda) \in s_j^*, 1 < \lambda \leq n)} p_j^* \tag{2.2.50}$$

$$1 < i(1) < \ldots < i(n) < N.$$

(*b*) *Method of Reduction:* From the given $\pi^0(\lambda)$, $\pi^0(\lambda\lambda')$ values a FS(2)-design is first constructed with the first order and second order inclusion-probabilities $\pi^{**}(\lambda)$, $\pi^{**}(\lambda\lambda')$ given by the solution of

$$\pi^0(\lambda) = \frac{n-2}{N-2} + \frac{N-n}{N-2}\pi^{**}(\lambda),$$

$$\pi^0(\lambda\lambda') = \frac{(n-2)(n-3)}{(N-2)(N-3)} + \frac{(N-n)(n-2)}{(N-2)(N-3)}(\pi^{**}(\lambda) - \pi^{**}(\lambda'))$$

$$+\left\{1 - \frac{2(n-2)}{N-2} + \frac{(n-2)(n-3)}{(N-2)(N-3)}\right\} \pi^{**}(\lambda\lambda'),\ 1 < \lambda \neq \lambda' < N.$$

The reduced design consists of k samples with selection probabilities,

$$p^{**}(s_j^*) = p_j^{**} \text{ (say)},$$
$$= \pi^{**}(\lambda\lambda'), \text{ if } \lambda, \lambda' \in s_j^{**}, \lambda < \lambda'$$

The selection probabilities of the final design are

$$p_D(s\ (i(1) \ldots i(n))) = \sum_{\substack{\lambda<\lambda' \\ (\lambda\lambda') \in (i(1) \ldots i(n))}} \pi^{**}(\lambda\lambda') \cdot \frac{1}{\binom{N-2}{n-2}} \tag{2.2.51}$$

$$1 \leq i(1) < \ldots < i(n) \leq N.$$

Both of these schemes are equivalent. It is easy to check that Sinha's schemes are not always applicable. In the example below, procedure (52) gives a solution whereas the present method of extension fails.

Example 2.5: $N = 7, n = 4$

$$\pi^0(1) = 0.37 \ \pi^0(2) = 0.76 \ \pi^0(3)$$

$$= 0.60 \ \pi^0(4) = 0.37$$

$$\pi^0(5) = 0.76 \ \pi^0(6) = 0.60 \ \pi^0(7) = 0.54$$

$$\pi^0(12) = 0.37 \ \pi(13) = 0.17 \ \pi^0(14)$$

$$= 0.05 \ \pi^0(15) = 0.20$$

$$\pi^0(16) = 0.17 \ \pi^0(17) = 0.15$$

$$\pi^0(23) = 0.41 \ \pi^0(24) = 0.20 \ \pi^0(25)$$

$$= 0.52 \ \pi^0(26) = 0.41 \ \pi^0(27) = 0.37$$

$$\pi^0(34) = 0.17 \ \pi^0(35) = 0.41 \ \pi^0(36)$$

$$= 0.35 \ \pi^0(37) = 0.29$$

$$\pi^0(45) = 0.37 \ \pi^0(46) = 0.17 \ \pi^0(47) = 0.15$$

$$\pi^0(56) = 0.41 \ \pi^0(57) = 0.37$$

$$\pi^0(67) = 0.29$$

Mukhopadhyay's scheme gives the following solution (Table 2.1):

We verify that $\pi_D(\lambda) = \pi^0(\lambda), = 1, ..., 7, \pi_D(\lambda\lambda') = \pi^0(\lambda\lambda'), \lambda \neq \lambda' = 1, ...,7.$

Results by Sinha's extension procedure are given in (Table 2.2). The probabilities for selection for some samples have turned out to be negative. Sinha's procedures fail to give any sampling design (with the stated properties) in this case.

We claim that procedure (52) is based on a type of programming techniques to find out values of $p(s)$'s satisfying certain inequalities observed by the π_i's and π_{ij}'s in general and some particular inequalities obeyed by the given set of π_i and π_{ij} values. This procedure is applicable to any consistent set of π_i and π_{ij} values and provides a desired sampling design.

Recently, Wright (1990) gave a πps scheme in situations where auxiliary variable x is not available but information on rank of the units in respect of the y-values are available before sampling, and size of a unit is taken proportional to its rank.

Table 2.1

Samples	*p(1; s)*	*p(2; s)*	*p(s)*
(1 2 3 5)	0.05		0.05
(1 2 3 6)	0.05	0.02	0.07
(1 2 3 7)	0.05		0.05
(1 2 4 5)	0.05		0.05
(1 2 5 6)	0.05		0.05
(1 2 5 7)	0.05		0.05
(1 2 6 7)	0.05		0.05
(2 3 4 5)	0.05		0.05
(2 3 5 6)	0.05	0.02	0.07
(2 3 5 7)	0.05		0.05
(2 3 6 7)	0.05	0.02	0.07
(2 4 5 6)	0.05		0.05
(2 4 5 7)	0.05		0.05
(2 5 6 7)	0.05		0.05
(3 4 5 6)	0.05	0.02	0.07
(3 4 5 7)	0.05		0.05
(3 5 6 7)	0.05	0.02	0.07
(4 5 6 7)	0.05		0.05
	0.90	0.10	1.00

2.3 SIMPLE RANDOM SAMPLING AND t-DESIGN

In a sample survey often all the possible samples are not equally preferable from the point of view of practical advantages. Thus, in an agricultural survey, the investigators may have a tendency to avoid grids which are located further away from the cell camps, located in inaccessible places, marshy land, etc. In such cases, the sampler would like to use only a fraction of the totality of all possible samples, allotting only a very small probability (including zero) to the non-preferred units. Such designs are called Controlled Sampling designs.

Chakraborty (1963) was the first to use a balanced incomplete block (BIB) design to obtain a controlled sampling design replacing a srswor design. For unequal probability sampling, BIB designs have been used by Das and Mahanty (1973) (procedure (39)), Gupta, *et al.* (1982) (procedure (37)), Nigam *et al.* (1984) (procedure (49)), Rao and Nigam (1989, 1990), Kumar (1991), Mukhopadhyay and Vijayan (1992), among others. We shall consider in this section use of a t-design to obtain a Controlled Sampling design in replacement of a srswor design.

An incomplete block design $D(v, b_0, k, t)$ with v, number of varieties, b_0, number of blocks, k, the block size is a t-design if any set of $u \le v$ varieties occur together a constant number λ_t of times. Thus a BIB design is a 2-design and hence a 1-design. The b_0 blocks may or may not be distinct. Let b_{0d} be the number of distinct blocks.

Table 2.2

Samples	*P(s)*
(1 2 3 4)	0.0067
(1 2 3 5)	0.0483
(1 2 3 6)	0.0017
(1 2 3 7)	0.0517
(1 2 4 5)	0.0400
(1 2 4 6)	0.0067
(1 2 4 7)	0.0100
(1 2 5 6)	0.0483
(1 2 5 7)	0.0450
(1 2 6 7)	0.0517
(1 3 4 5)	0.0067
(1 3 5 6)	0.0050
*(1 3 5 7)	– 0.0050
*(1 3 4 6)	– 0.0100
*(1 3 4 7)	– 0.0133
(1 3 6 7)	0.0183
(1 4 5 6)	0.0067
(1 4 5 7)	0.0100
*(1 4 6 7)	– 0.0133
*(1 5 6 7)	– 0.0050
(2 3 4 5)	0.0483
(2 3 4 6)	0.0050
*(2 3 4 7)	– 0.0050
(2 3 5 6)	0.0767
(2 3 5 7)	0.0600
(2 3 6 7)	0.0667
(2 4 5 6)	0.0483
(2 4 5 7)	0.0490
*(2 4 6 7)	– 0.0050
(2 5 6 7)	0.0600
(3 4 5 6)	0.0617
(3 4 5 7)	0.0517
(3 5 6 7)	0.0567
(3 4 6 7)	0.0183
(4 5 6 7)	0.0517

***Samples for which p(s) is negative.**

Here $$\lambda_t = \frac{\binom{k}{t}}{\binom{v}{t}} b_0$$

We let each variety correspond to a unit in the finite population, each block to a sample and select each sample with equal probability.

Thus, $v = N,\ k = n$ and $\pi_{i_1} \dots i_u = \dfrac{\binom{n}{u}}{\binom{N}{u}}$, $u = 1, \dots, t.$

Clearly $b_{0d} \le \binom{v}{k}$. In many t-designs, the b_0 blocks are all distinct but b_0 is much less than $\binom{v}{k}$. For example, for a D (11, 66, 5, 4) design $b_{0d} = 66 = \dfrac{1}{12}\binom{v}{k}$. The sample space of the sampling design corresponding to this 4-design is thus only $\dfrac{1}{12}$ as large as corresponding to a srswor (11, 5) design.

For any design, $E(\bar{y})$ depends only on π_i's, $V(\bar{y})$, $E(s^2)$ only on π_{ij}'s, $V(s^2)$ on $\pi ijkl$'s apart from $\mathbf{y} = (y_1, \dots, y_N)$ (Exercise 2.12). Hence a sampling design based on a t-design $D(N, b_0, n, t)$, $t \ge 4$ is equivalent to a srswor (N, n) design, providing the same values of π_{ijkl}'s and hence, same inference on $\bar{Y}$, S^2. Such a design uses only a fraction of the available sample space and thus many samples, specially the 'non-preferred' ones may be left out of the scope of survey. The 'non-preferred' samples may be those which are expensive, inconvenient, subject to inefficient handling etc. The use of a t-design in these circumstances would reduce the cost of survey and is expected to increase the overall efficiency (Sirvastava and Salah, 1985).

In Sections 2.2 and 2.3, we discussed some of the sampling designs. Our choice of the designs for discussion was mainly based on new ideas, techniques and motivations which some designs carried, generating new designs and modifications. Due to paucity of space we could not discuss many other interesting sampling designs,- systematic sampling, Hanurav's class of designs could not be considered in some details. For detailed discussion, the reader may refer to Brewer and Hanif (1982), Chaudhuri and Vos (1988), Mukhopadhyay (1991). In the next section, we consider estimation of variance of estimators of population total, mean.

2.4 VARIANCE ESTIMATION

2.4.1 Variance of Estimator Finite Population Total

This section deals with the problem of variance estimation of a finite population total. Since a variance is a non-negative quantity, it is desirable that an unbiased estimator of variance is always non-negative. Unfortunately this is not so. Sub-section 2.4.2 deals with the problem of non-negative unbiased estimation of mean square error of an estimator. The essential form of such an estimator is noted and studied for various sampling designs. Sub-section 2.4.3 discusses in details the non-negativity of v_{YG} and v_{HT}. In 2.4.4, we consider the problem of estimating the average value of V_{HT} under a superpopulation model in a bid to surmount the problem of calculating π_{ij} values. Sub-section 2.4.5 considers the stabilities of different variance estimators.

2.4.2 Non-negative Unbiased (NNU) Estimation of Variance

Since the mse (variance) of an estimator is always non-negative, it is desirable that its estimator should also always be non-negative. However, this is not the case in practice. For example, it is easy to see that $v_{YG}(e_{HT})$ can take negative values. We first consider the necessary form of a nnu-estimator of mse $(\hat{Y})$.

Let $\hat{Y} = \Sigma b_{si}\, y_i,\, b_{si} = 0$ for $s \not\ni i$, be a linear estimator of Y. Theorem 2.1 gives the form of a non-negative quadratic unbiased estimator (nnque) of mean square error of $\hat{Y}$. We first consider.

Lemma 2.2: (Vijayan, 1975; Mukhopadhyay and Vijayan, 1990). Let $Q = \mathbf{X}'\,\mathbf{AX}$ be a nnd-quadratic form in $\mathbf{X} = (x_1; \ldots, x_m)'$, where $\mathbf{A} = ((a_{ij}))$ and $Q = 0$ for $\mathbf{X} = \mathbf{V} = (v_1, \ldots, v_m)'$. Then $\mathbf{AV} = 0$ *i.e.*, $\Sigma_j a_{ij}\, v_i = 0,\ i = 1, \ldots, m.$

Proof: Since $\mathbf{A}$ is nnd there exists a matrix $\mathbf{H}$ such that $\mathbf{A} = \mathbf{HH}'$. Thus,

$$\mathbf{V'AV} = (\mathbf{H'V})'\,(\mathbf{H'V}) = 0$$

$$\Rightarrow \qquad \mathbf{H'V} = \mathbf{0} \quad \Rightarrow \quad \mathbf{HH'V} = \mathbf{0} \quad \Rightarrow \quad \mathbf{AV} = 0.$$

Theorem 2.1: (Vijayan, 1975; Rao, 1979). If mse $(\hat{Y}) = 0$ when $y_i \propto w_i$, w_i's being some known quantities ($i = 1, \ldots, N$), then

$$\text{mse}\,(\hat{Y}) = M(\hat{Y}) = -\frac{1}{2}\sum{}' w_i\, w_j\, (z_i - z_j)^2\, b_{ij} \tag{2.4.1}$$

where $\qquad z_i = y_i/w_i,\ b_{ij} = E(b_{si} - 1)(b_{sj} - 1).$

Further, a nnque of $M(\hat{Y})$ is necessarily of the form

$$m(\hat{Y}) = -\frac{1}{2}\sum_s{}' w_i\, w_j\, (z_i - z_j)^2\, b_{ij}(s) \tag{2.4.2}$$

$$= e_q(s, \mathbf{y})$$

where, $$E(b_{ij}(s)) = b_{ij},\ i < j \tag{2.4.3}$$

Proof:

$$M(\hat{Y}) = E\left\{\sum (b_{si} - 1)\, w_i z_i\right\}^2$$

$$= \sum_i \sum_j b_{ij} w_i w_j z_i z_j \tag{2.4.4}$$

When $z_i = c$ (a constant) $\forall\ i$, $M(\hat{Y}) = c^2\ \Sigma\Sigma b_{ij} w_i w_j = 0$ (by the condition of the theorem). By Lemma 2.2,

$$b_{ii} w_i^2 = -\sum_{j(\neq 1)} b_{ij}\ w_i\ w_j,\ \ i = 1, \ldots, N \tag{2.4.5}$$

(2.4.1) follows from (2.4.4) on substitution from (2.4.5). If $m(\hat{Y}) = \sum\sum_{ij \in s} b_{ij}\,(s)\, y_i y_j$ is nnque of $M(\hat{Y})$,

$$c^2 \sum\sum_{ij \in s} b_{ij}\,(s)\, w_i w_j = 0$$

By Lemma 2.2,

$$\sum_{j \varepsilon s} b_{ij}(s)\, w_i w_j = 0 \text{ for all } i \varepsilon s;$$

hence, $m(\hat{Y})$ reduces to $e_q(s, \mathbf{y})$ in (2.4.2). (2.4.3) is required for its unbiasedness.

Remark 2.1: For $n = 2$, a nnue of $M(\hat{Y})$ is necessarily of the form (2.4.2) coupled with (2.4.3), (Lanke, 1974).

Remark 2.2: For $n > 2$, the Theorem 2.1 can be extended to the class of polynomial estimators of $M(\hat{Y})$, following Vijayan (1975).

Corollary 2.1: If $\hat{Y}$ is unbiased, $b_{ij} = E(b_{si}b_{sj}) - 1 = h_{ij} - 1$ (say) (2.4.6)

If, further $b_{si} = b_i$ for all $s \ni i$, $b_{ij} = \dfrac{\pi_{ij}}{\pi_i \pi_j} - 1$, when (2.4.1) reduces to V_{YG}, provided the design is of fixed effective size n.

Corollary 2.2: If $b_{ij} = b$, $i < j$, $M(\hat{Y}) = -b \sum\sum_{i<j} w_i\, w_j\, (z_i - z_j)^2$

$$= -b\left(\sum \frac{y_i^2}{w_i^1} - Y^2\right), \quad w_i^1 = \frac{w_i}{w}, W = \sum w_i$$

Different Forms of NNQUE

For unbiased $\hat{Y} = \Sigma b_{si}\, y_i$, (2.4.1) gives

$$V(\hat{Y}) = \sum\sum_{i<j} g_{ij}\,(1 - h_{ij}) \tag{2.4.7}$$

where $g_{ij} = w_i w_j (z_i - z_j)^2$. Different forms of nnque of V are, therefore,

$$v_{kl} = \sum\sum_{\substack{i<j \\ \in s}} g_{ij}\, \hat{\alpha}^{(k)} - \sum\sum_{\substack{i<j \\ \in s}} g_{ij}\, h_{ij}^{(1)}, \; k, i = 0, 1, 2, 3 \tag{2.4.8}$$

where $\hat{\alpha}_{(0)} = \dfrac{b_{si}b_{sj}}{E(b_{si}b_{sj})}$, $\hat{\alpha}^{(1)} = \dfrac{1}{M_2 p(s)}$,

$$\hat{\alpha}^{(2)} = \frac{1}{\pi_{ij}}, \quad \hat{\alpha}^{(3)} = \frac{P(s \mid ij)}{p(s)} \qquad [2.4.9(a)]$$

and $$\hat{h}_{ij}^{(0)} = b_{si}\, b_{sj},\ \hat{h}_{ij}^{(1)} = \frac{h_{ij}}{M_2 p(s)},$$

$$\hat{h}_{ij}^{(2)} = \frac{h_{ij}}{\pi_{ij}},\quad \hat{h}_{ij}^{(3)} = \frac{h_{ij}\, P(s \mid ij)}{p(s)}$$

In practice many of the estimators v_{kl} would coincide. Theorem 2.1 gives only necessary conditions for nnque of $M(\hat{Y})$. In fact many of the estimators v_{kl} may not be non-negative for all values of **y**.

Remark 2.3: Writing $b_{ij} = (h_{ij} - a) + (a - 1) = m_{ij} + d$ (say), $(d = a - 1)$, a being some real constant, one may get till other forms of nnque $V(\hat{Y})$,

$$v_{kl}(a) = d \sum\sum_{i<j\, \varepsilon\, s} g_{ij}\, \hat{\alpha}^{(k)} - \sum\sum_{i<j\, \varepsilon\, s} g_{ij}\, \hat{m}_{ij}^{(1)},\ k, l = 0, 1, ..., 3$$

[2.4.9(*b*)]

Optimum choice of a for a given v_{kl} may depend on its sampling variance.

Remark 2.4: Other forms of $v(\hat{Y})$ are:

(*i*) $$v(\hat{Y}) = \hat{Y}^2 - est(Y^2)$$

(*ii*) If $$v(\hat{Y}) = \sum a_{ii} y_i^2 + \sum{}' a_{ij} y_i y_j$$

$$v_1(\hat{Y}) = \sum_s \frac{a_{ii} y_i^2}{\pi_i} + \sum_s{}' a_{ij} \frac{y_i y_j}{\pi_{ij}}$$

$$v_2(Y) = \sum \frac{a_i y_i^2}{M_1 p(s)} + \sum_s{}' \frac{a_{ij} y_i y_j}{M_2 p(s)}$$

However, these forms are derived without any consideration to non-negativity requirement of $v(\hat{Y})$.

Example 2.6: HH-strategy: $\hat{Y}_{pps} = \frac{1}{n} \Sigma(t_i y_i / p_i)$, t_i is the number of times i occur in $\mathscr{S}$; let $z_i = y_i / p_i$.

$$V_{pps} = \frac{1}{2n} \sum{}' (z_i - z_j)^2 p_i p_j$$

Some estimators of the form (2.4.8) are,

$$v_{00} = \frac{1}{n(n-1)} \sum_s t_i (z_i - \bar{z}_s)^2$$

$$2v_{01} = \sum_s{}' (z_i - z_j)^2 p_i p_j \left\{ \frac{t_i t_j}{n(n-1) p_i p_j} - \frac{n-1}{nM_2 p(s)} \right\}$$

$$2v_{02} = \sum_s{}' (z_i - z_j)^2 p_i p_j \left\{ \frac{t_i t_j}{n(n-1) p_i p_j} - \frac{n-1}{n\pi_{ij}} \right\}$$

$$2v_{12} = \sum_s{}' (z_i - z_j)^2 p_i p_j \left\{ \frac{1}{M_2 p(s)} - \frac{n-1}{n\pi_{ij}} \right\}$$

$$2v_{22} = \sum_s{}' (z_i - z_j)^2 \frac{p_i p_j}{\pi_{ij}}$$

$$2v_{32} = \sum_s{}' (z_i - z_j)^2 p_i p_j \left\{ \frac{P(s/ij)}{p(s)} - \frac{n-1}{n\pi_{ij}} \right\}$$

where $\pi_{ij} = 1 - (1 - p_i)^n - (1 - p_j)^n + (1 - p_i - p_j)^n$. v_{00} and v_{22} have been considered by Rao (1979).

Example 2.7: HT-estimator. Assume that $p \in \rho_n$.

$$e_{HT} = \sum_s \frac{Y_i}{\pi_i}, \text{ here } b_{si} = \frac{1}{\pi_i}(0),\ i \in (\notin)s,\ b_{ij} = \pi_{ij}(\pi_i \pi_j)^{-1} - 1.$$

Let $z_{ij} = (y_i / \pi_i - y_j / \pi_j)^2$. Some estimators of the form (2.4.8) are:

$$v_{00} = \sum_s{}' z_{ij} \left(\frac{\pi_i \pi_j}{\pi_{ij}} - 1 \right) = v_{02} = v_{20} = v_{22}$$

$$v_{01} = \sum_s{}' z_{ij} \left\{ \frac{1}{\pi_{ij}} = \frac{\pi_{ij}}{\pi_i \pi_j M_{2p}(s)} \right\} \pi_i \pi_j$$

$$v_{03} = \sum_s{}' z_{ij} \left\{ \frac{1}{\pi_{ij}} = \frac{\pi_{ij}}{\pi_i \pi_j} \frac{P(s/ij)}{p(s)} \right\} \pi_i \pi_j$$

$$2v_{10} = v_{12} = \sum_s{}' z_{ij}\, \pi_i \pi_j \left\{ \frac{1}{M_2 p(s)} - \frac{1}{\pi_i \pi_j} \right\}$$

$$2v_{30} = v_{32} = \sum_s{}' z_{ij}\,\pi_i\pi_j \left\{\frac{P(s/ij)}{p(s)} - \frac{1}{\pi_i\pi_j}\right\}$$

v_{00} is the well-known v_{SYG}, due to Sen, Yates and Grundy. Ajgaonkar (1967) considered v_{11}. Rao (1979) considered v_{10} and v_{30}. Chaudhuri (1981) suggested,

$$v_c = \sum_s{}' \left\{\pi_i\pi_j\left(\frac{d}{M_2 p(s)} + \frac{1-d}{\pi_{ij}}\right) - \pi_{ij}\left(\frac{e}{M_2 p(s)} + \frac{1-e}{\pi_{ij}}\right)\right\}$$

$$\left(\frac{y_i}{\pi_i} - \frac{y_j}{\pi_j}\right)^2$$

where (d, e) are fixed constants in [0, 1], as an estimator of V_{HT}. v_c is an estimator of the type (2.4.9).

Example 2.8: Midzuno strategy:

$$\hat{Y} = \hat{Y}_R = \frac{\sum_s y_i}{\sum_s p_i} = \frac{\sum_s y_i}{M_1 p(s)} = \sum_s \frac{y_i P(s/i)}{p(s)} \text{ since for this scheme}$$

$$\frac{P(s/i)}{p(s)} = \frac{1}{M_1 p(s)}. \text{ Also } P(s/ij) = \frac{N-1}{M_1(n-1)}. \text{ Let } \left(\frac{y_i}{P_i} - \frac{y_j}{P_j}\right)^2 = Z_{ij}$$

$$V(\hat{Y}_R) = \frac{1}{2}\sum{}' Z_{ij}\, P_iP_j \left\{1 - \frac{1}{M_1^2}\sum_{s \ni i,j}\frac{1}{q_s}\right\}$$

where $q_s = \sum_s p_i$.

Some of the estimators are:

$$v_{10} = v_{30} = \frac{1}{2}\sum_s{}' z_{ij}\, p_i p_j \left\{\frac{1}{M_2 p(s)} - \frac{P(s/i)P(s/j)}{p(s)^2}\right\}$$

$$= \frac{1}{2}\sum_s{}' z_{ij}\, p_i p_j \left[\frac{X}{x_s}\left(\frac{N-1}{n-1} - \frac{X}{x_s}\right)\right]$$

$$2v_{11} = \sum_s{}' z_{ij}\, p_i p_j \frac{(N-1)}{(n-1)} qs \left\{1 - \frac{1}{M_1}\sum_{s \ni i,j}\frac{1}{q_s}\right\}$$

$$2v_{20} = \sum_s{}' z_{ij}\, p_i p_j \left(\frac{1}{\pi_{ij}} - \frac{1}{q_s^2}\right)$$

$$2v_{22} = \sum_s{}' z_{ij} \frac{p_i p_j}{\pi_{ij}} \left\{ 1 - \frac{1}{M_1} \sum_{s \ni i,j} \frac{1}{q_s} \right\}$$

$$2v_{23} = \sum_s{}' z_{ij}\, p_i p_j \left\{ \frac{1}{\pi_{ij}} - \frac{1}{M_2 q_s} \sum_{s \ni i,j} \frac{1}{q_s} \right\}$$

$$2v_{13} = \sum_s{}' z_{ij}\, p_i p_j \left[\frac{1}{q_s} \left\{ \frac{N-1}{n-1} - \frac{1}{M_2} \sum_{s \ni i,j} \frac{1}{q_s} \right\} \right]$$

$$2v_{12} = 2v_{32} = \sum_s{}' z_{ij}\, p_i p_j \left[\frac{M_1}{M_2 q_s} - \frac{1}{\pi_{ij} M_1} \sum_{s \ni i,j} \frac{1}{q_s} \right]$$

Rao and Vijayan (1977) studied the stability and probability of taking a negative value by $v_{10}(= v_{30})$ and v_{22} empirically. Chaudhuri (1979) considered v_{22}, v_{11}, v_{23}, $v_{12}(= v_{32})$ [his estimators being v_2, v_3, v_4, v_5 respectively]. All the above estimators were considered by Padmawar (1982).

An Empirical Study: Vijayan, Mukhopadhyay and Bhattacharyya (1995) investigated the performance of v_{01}, v_{03} and v_{10} for HTE empirically. 21 populations (including the 14 populations considered by Rao and Vijayan (1977)) as shown in Table 2.3 and the cases $n = 3, 4, 5$ were considered. For the cases, $n = 4, 5$, to save computer time, samples were drawn from the modified populations, where the populations remain unchanged if $N \le 10$, but were restrieted to first ten units otherwise. Samples were drawn using Sampford's (1967) procedure. The performance of the biased non-negative variance estimators, v^* obtained by following Rao and Vijayan, where

$$v_s^* = v_s \text{ if } v_s \ge 0$$
$$= g_s X^2 \text{ (otherwise)},$$

g_s being the least square estimator of $\mathscr{V}(\hat{\beta}_s)$ under the superpopulation model:

$$Y_i = \beta x_i + e_i$$
$$\mathscr{E}(e_i \mid x_i) = 0,\ \mathscr{E}(e_i^2 \mid x_i) = \sigma^2 x_i^2 \qquad (2.4.10)$$
$$\mathscr{E}(e_i e_j \mid x_i, x_j) = 0,\ i \ne j$$

where Y_i is a random variable whose one particular realisation is y_i, $\hat{\beta}_s$ denotes the least square estimator β, and $\mathscr{V}$ denotes model - variance operator, were considered. This model is reasonable in the situation where the HT-estimator is appropriate. Their results for $n = 3$ have been reproduced in Table 2.4. For simplicity we have denoted v_{01}, v_{03} and v_{10} as v_1, v_2, v_3 respectively, with the corresponding modified estimators as v_i^*. In the table, p_i gives the probability of v_i taking negative values (given by the relative frequency of number of samples yielding negative variance estimates), E_0/E_i, the relative efficiency of Yates-Grundy variance estimator $v_0(= v_{00})$ over $v_i[E_i = 1/V(v_i)]$, b_i^*, the relative bias of v_i^*, $[b_i^* = | E(v_i^*) - V(e_{HT}) | / \{MSE(v_i^*)\}^{1/2}]$, and E_1^*/E_j^*, the relative efficiency of v_j^* wrt $v_1^*(j = 2, 3)[E_k^* = 1/MSE(v_k^*)]$. Their study showed that for $n = 3$, v_2 can be considered to be almost nnue of $V(e_{HT})$ for populations with coefficient of variation (of x), c_x not exceeding 0.5. It has got uniformly higher probability of being non-negative than both v_1 and v_3. Again, v_2 was, in general, found to be most efficient of all the three. This suggests that v_2 is the most preferable among the three estimators, both from the point of view of non-negativity and efficiency.

For the modified estimators, relative bias of v_2^* was almost always zero, while for 16 out of 21 populations b_3^* did not exceed $b_1^* \cdot v_2^*$ was found to be uniformly more efficient than v_1^*. The modified estimator v_2^* is most preferable of the three.

Their analysis suggested that

(*i*) for $n = 3, 4$, $v_2(v_2^*)$ is the most preferable among $\{v_i(v_i^*), i = 1, 2, 3\}$.

(*ii*) for $n = 5$, for populations with $c_x < 0.25$, $v_2(v_2^*)$ is recommendable, while for the other types of populations the choice should lie between v_2 and $v_3(v_2^*$ and $v_3^*)$.

2.4.3 On Non-negativity of v_{YG} and v_{HT}

Exercises 1, 5, 6, 10 of Section 2.5 underscore some sampling designs for which the sufficient conditions.

$$C : \pi_{ij} - \pi_i\pi_j \leq 0,\ i \neq j = 1, ..., N$$

for v_{YG} to be non-negative are always satisfied. Lanke (1974) called C as 'negative correlation' property as covariance

$$(I_{is}, I_{js}) = \pi_{ij} - \pi_i\pi_j.$$

Theorems 2.2 and 2.3 give the classes of sampling designs for which C holds. No such general conditions guaranteeing the non-negativity of v_{HT} is known. Consequently, v_{YG} is believed as taking non-negative values more often than v_{HT}.

Theorem 2.2: (Lanke, 1974; Mukhopadhyay, 1974). Let p' be a s.d. of size n'. After a sample s' is selected from $\mathscr{P}$ as per p', a sample of size n'' is selected from $\mathscr{P} - s'$ by srswor. Let $s = s'Us''$. If p' has the property $\pi'_{ij} \leq \pi'_i\pi'_j\ \forall\ i \neq j$, then the final design $p = p'U$ srswor $(n'' = n - n', N - n')$ has the property $\pi_{ij} \leq \pi_i\pi_j\ \forall\ i \neq j$.

Proof: Define exclusion-probability of i as $\pi_i^* = 1 - \pi_i$ and of (i, j) as $\pi_{ij}^* = 1 - \pi_i - \pi_j + \pi_{ij}$. Now $\pi_{ij}^* - \pi_i^*\pi_j^* = \pi_{ij} - \pi_i\pi_j$. We prove the result by induction taking $n'' = 1, 2, ...$

If $\pi_i'^*$, $\pi_{ij}'^*$ are exclusion-probabilities for p', for $p = p'U$ srswor $(1, N - n')$, exclusion-probabilities are

$$\pi_i^* = \pi_i'^* \left(1 - \frac{1}{N - n'}\right)$$

$$\pi_{ij}^* = \pi_{ij}'^* \left(1 - \frac{2}{N - n'}\right)$$

Thus, $$\pi_{ij}^* - \pi_i^* \pi_j^* = (\pi_{ij}'^* - \pi_i'^*, \pi_j'^*) \left(1 - \frac{1}{N - n'}\right)^2 - \pi_{ij}'^*(N - n')^{-2} < 0$$

if, $$\pi_{ij}'^* - \pi_i'^* \pi_j'^* \le 0 \text{ i.e., if } \pi'_{ij} - \pi'_i \pi'_j \le 0$$

The result for general n^2 follows by repeated application of the method. The above proof is due to Lanke (1974). Seth (1966) proved the result for $n¢ = 2$ by using an argument from the theory of probability. Mukhopadhyay (1974) proved the result by using an algebraic inequality.

Example 2.9: ppswr-sampling. Here $\pi_i^* = (1 - p_i)^n$, $\pi_{ij}^* = (1 - p_i - p_j)^n$, $\pi_i^* \pi_j^* - \pi_{ij}^* = (1 - p_i - p_j + p_i p_j)^n - (1 - p_i - p_j)^n > 0$. Thus $v_{YG} \ge 0 \ \forall \ \mathbf{y}$; however, this being a non-fixed size design, v_{YG} is not unbiased.

Theorem 2.3: (Lanke, 1974). In both rejective sampling procedures and generalised rejective sampling procedures (vide Section 1.3 for definitions), the negative correlation property C is satisfied. We now consider v_{HT}.

Theorem 2.4: (Vijayan, 1975). For any $p \in \rho_n$, with $\pi_{ij} > 0 (i \ne j)$, a nnue of $V(e_{HT})$ is necessarily of the form

$$-\frac{1}{2} \sum_s{}' (z_i - z_j)^2 \, a_{ij}(s) \tag{2.4.11}$$

where $z_i = y_i/\pi_i$, $a_{ij}(s)$ are independent of z_i-values, $a_{ij}(s) = 0$ if $(i, j) \notin s$ and $E(a_{ij}(s)) = \pi_{ij} - \pi_i \pi_j$.

Proof: This is a special case of Theorem 2.1.

Corollary 2.3: (Vijayan, 1975; Biyani, 1980; Lenz, 1989). For any $p \in \rho_n$, with $\pi_{ij} > 0 (\forall (i, j))$, a necessary condition for v_{HT} to be nn-definite is that,

$$v_{HT} = v_{YG}$$

We now consider the special case $n = 2$.

Theorem 2.5: (Vijayan, 1975; Lanke, 1974). Let $p \in \rho_2$ with $\pi_{ij} > 0$, $(i \neq j)$. Then there exists a nnue of V_{HT} *iff* $\pi_{ij} \le \pi_i\pi_j \; \forall \; i \neq j$. Furthermore, v_{YG} is the only possible nnue of V_{HT}.

Corollary 2.4: Let $p \in \rho_2$ with $\pi_{ij} > 0 \; \forall \; i \neq j$. Then v_{HT} is non-negative *iff* p is a srswor design.

Corollary 2.3 tells that for any fixed sample size design v_{HT} can be made to take negative values unless $v_{HT} = v_{YG} \ge 0$. Consequently, use of v_{HT} as a nnue of V_{HT} is very limited. Lenz (1989) considered the problem of nnu variance estimation in two stage sampling. He noted that 'generally' in two stage sampling $v_{YG} \ge 0$ for any design where sampling is by srs in the first stage and any other fixed size design (for a selected fsu) with the negative correlation property, $\pi_{ij} - \pi_i\pi_j \le 0$ at the second stage. He also considered asymptotic non-negativity of v_{YG} and v_{HT} in large scale stratified sampling.

Kharshikar (1980) considered the almost sure (a.s.) non-negativity of v_{HT} and v_{YG}. An estimator, $v(s, \mathbf{y})$ is a.s. nnue of $\mathbf{v}_{HT}$ if $v(s, y)$ is non-negative $\forall \; \mathbf{y} \in R_N^+$ and $E\{v(s, \mathbf{y})\} = V_{HT} \; \forall \; \mathbf{y} \in R_N^+$, R_N^+ denoting the non-negative part of the N-dimensional Eucledian space, R_N.

Now,
$$y'_{HT} = \sum Z_i I_{is},\; Z_i = \frac{y_i}{\pi_i}$$

$$V_{HT} = \mathbf{Z}' \wedge \mathbf{Z}$$

when
$$\mathbf{Z}' = (Z_1, \ldots, Z_N)',$$

where I_{is} is defined in Lemma (1.1).

$$\wedge = \pi - \pi\pi',\; \pi = ((\pi_{ij})),\; \pi = (\pi_1, \ldots, \pi_N)'.$$

$$v_{HT} = \sum{}' \lambda_{ij}^*(s)\, z_i z_j = \mathbf{Z}' \wedge^* (s)\mathbf{Z}$$

where,
$$\lambda_{ij}^*(s) = \begin{cases} (\pi_{ij} - \pi_i\pi_j)/\pi_{ij},\, (i, j) \in s, & i \neq j, \\ 0, & \text{otherwise} \end{cases}$$

and
$$\wedge^*(s) = (\lambda_{ij}^*(s)).$$

Similarly,

$$v_{YG} = \sum{}' \lambda_{ij}^{**}(s)\, z_i z_j = \mathbf{Z}' \wedge^{**} (s)\mathbf{Z}$$

where, $$\lambda_{ij}^{**}(s) = \begin{cases} (\pi_{ij} - \pi_i\pi_j)/\pi_{ij}, & (i, j) \varepsilon\, s, i \neq j \\ -\sum_{r(\neq i)\varepsilon\, s} \dfrac{\pi_{ir} - \pi_i\pi_r}{\pi_{ir}}, & i \varepsilon\, s, i = j \\ 0, & \text{otherwise} \end{cases}$$

and $$\wedge^{**}(s) = (\lambda_{ij}^{**}(s))$$

For v_{HT} to be a.s. non-negative, it is sufficient that $\wedge^*(s)$ be p.s.d. $\forall$ s. For v_{YG} to be a.s. non-negative, it is necessary and sufficient that $\wedge^{**}(s)$ be p.s.d. $\forall$ s. It is shown that for srswor both v_{HT} and v_{YG} are a.s. non-negative; for srswr, v_{HT} is a.s. non-negative while v_{YG} is not. For stratified s.d., $v_{HT}(v_{YG})$ is a.s. non-negative *iff* the s.d. in each stratum is such that the corresponding estimator in each possesses this property. He also examined the a.s. non-negativity of v_{YG} and v_{HT} in cluster sampling and multistage sampling.

Godambe and Joshi (1965) showed that for any fixed p with $\pi_{ij} > 0$ $(i \neq j)$ v_{HT} is admissible, in the class of all unbiased estimators (for a definition of admissibility of estimators see Section 3.5). Joshi (1970) proved admissibility of v_{YG} for $n = 2$ in the class of all unbiased estimators. Biyani (1980) gave a s.d. with $n \in \left[3, \frac{N}{2}\right]$ showing that v_{YG} is inadmissible within the class of all non-negative quadratic estimators. Sankaranarayanan (1980) proved that for any fixed $p \in \rho_n$, v_{YG} is admissible in the class of all nnque's of V_{HT}.

2.4.4 Estimating the Average Value of V_{HT}

The main limitations in estimating V_{HT} are in computation of π_i and π_{ij}. Though there are quite a large number of IPPS-schemes, computation of exact values of π_{ij}'s is quite involved and for many of these schemes only their asymptotic expansions are easily calculable.

When the finite population $\mathscr{P}$ may be considered as a random sample from a superpopulation ξ, one may contrive to estimate the average value of V_{HT} under ξ. A justification is, if $\mathscr{P}$ can be reasonably considered as a random sample from ξ, the value of V_{HT} for $\mathscr{P}$ can be taken on the whole to be represented by its model-expected value $\in V_{HT}$. Estimation of $\in V_{HT}$ will do away with the necessity of calculating the π_i, π_{ij} values. We shall assume $p \in \rho_n$ throughout this section.

Consider the superpopulation model as in (2.4.10) but with $\mathscr{V}(Y_i \mid x_i) = \sigma^2 x_i^g$. Call this model as ξ_g.

Then, $$\in [V_{YG}] = \frac{a}{n} \sum P_i^{g-1}(1 - nP_i) \tag{2.4.12}$$

For this model, Hanif and Brewer (1980) suggested the model-based variance estimator,

$$v_{HB} = \frac{n}{n-1}\left[1 - \frac{\sum \pi_i^g}{\sum \pi_i^{g-1}}\right] \sum_s \left(\frac{y_i}{\pi_i} - \frac{e_{HT}}{n}\right)^2 \qquad (2.4.13)$$

which for $g = 2$ reduces to Jessen's estimator, v_{JA} in (1.4.16).

Kumar, Gupta and Agarwal (1985) proposed another model-based estimator,

$$v_{KGA} = \frac{\sum P_i^{g-1}(1 - nP_i)}{2n(n-1)\sum P_i^{g-1}} \sum_s{}' \left(\frac{y_i}{\pi_i} - \frac{y_j}{\pi_j}\right)^2 \qquad (2.4.14)$$

which becomes identical with v_{HB} for $\pi_i = nP_i$. $v_{HB} = (v_{KGA})$ is *pm*-unbiased for any πps design under ξ_g, *i.e.*,

$$E \in (v_{HB}) = E \in (v_{KGA}) = \in (V_{YG}) \qquad (2.4.15)$$

(for definition of *pm*-unbiasedness see Chapter 4).

Hanif, Mukhopadhyay and Bhattacharyya (1993) suggested the model-unbiased estimator,

$$v_N(e_{HT}) = \frac{1}{n-1}\sum P_i^{g-1}(1 - nP_i)\sum_s{}' \left[\left(\frac{y_i}{\pi_i} - \frac{y_j}{\pi_j}\right)^2 \Big/ (P_i^{g-2} + P_j^{g-2})\right] \qquad (2.4.16)$$

for which $\in (v_N(e_{HT})) = \in (\mathbf{V}_{YG})\ \forall\ s : p(s) > 0$. v_N is however, design-biased and use of this estimator like v_{HB} $(= v_{KGA})$, presupposes the knowledge of g.

Hanif *et al.* studied the relative bias of v_{JA}, v_{HB}, v_N and stability of v_{YG}, v_{JA}, v_{HB}, v_N, for some natural populations depicted in Table 2.5 and n = 2, 3.4. The quantity

$$\Delta = |\ E\{v^*(e_{HT})/V_{YG}\} - 1\ | \cdot 100$$

was taken as a measure of the relative absolute bias of the estimator $v^* \cdot v_N$ was seen to involve less absolute relative bias than v_{KGA} for values of g = 2.0 onwords for n = 3, 4. For higher values of g, v_N seemed to have relative bias of the same magnitude for all values of n.

The parameter,

$$\delta = \frac{V\{v^*(e_{HT})\}}{[E\{v^*(e_{HT})\}]^2}$$

taken as a measure of the stability of the estimators was calculated for v_{YG}, v_{JA}, v_{HB} ($= v_{KGA}$), v_N. The smaller is the value of δ, greater is the stability of the estimator. v_N was found to have a smaller value of δ than the other estimators for $g = 0(0.5)1.5$ for all values of n considered, though its value increased in general as g increased. The estimators were seen to be insensitive to the values of g.

Hanif *et al*, therefore concluded that $v_N(e_{HT})$ (though design-biased), involved only a small amount of bias for all values of g and performed farely well both in respect of bias and stability, in comparison with v_{YG} (stability only), v_{JA}, $v_{KGA}(= v_{HB})$, specially, for higher values of g in the range [0, 3] and higher values of n. Its use was recommended in practice.

Kott (1988) considered the model (2.4.10) but with $\mathscr{V}(Y_i \mid x_i) = \sigma^2 v_i$, where v_i is a known function of x_i, model ξ'_v (say). They suggested the *pm*-unbiased estimators,

$$v_k = \left(\sum_s \frac{v_i}{p_i} - 2n\sum_s \frac{v_i}{p_i} + n^2 \sum v_i\right)\left(\sum_s \frac{v_i}{p_i^2}\right)^{-1} v_{pps} \tag{2.4.17}$$

$$v'_k = \left(\sum \frac{v_i}{p_i} - n\sum v_i\right)\left(\sum \frac{v_i}{p_i}\right)^{-1} v_{pps} \tag{2.4.18}$$

for which,

$$E_p \in (v_k) = E_p \in (v'_k) = \in (V_{HT})$$

for any $FS(n)$-πps design.

If v_i / x_i is a monotonically increasing function of x_i, *e.g.*, if $v_i = x_i^g (g \geq 1)$ or $v_i = c_1 x_i + c_2 x_i^2$, c_k (constant) $\in [0, 1]$, $k = 1, 2$, then

$$v'_k \leq \left(1 - \frac{n}{N}\right) v_{pps} = v''_k \tag{2.4.19}$$

when v''_k is a conservative estimator of $\in [V_{HT}]$.

When n is large.

$$v''_k = \left[\left\{\sum_s \frac{\hat{v}_i}{p_i^2}(1 - np_i)\right\} \Big/ \sum_s \frac{\hat{v}_i}{p_i^2}\right] v_{pps} \tag{2.4.20}$$

where $\hat{v}_i = r_i^2$, $r_i = y_i - n^{-1} x_i \Sigma_s (y_j / x_j)$, converges to v_k in large samples for many populations if the model is correctly specified and $\in [E(v''_k) - V_{HT}] \simeq 0$, asymptotically.

2.4.5 Stability of Variance Estimators

Rao and Bayless (1969) in their empirical study based on 7 artificial populations and 20 natural populations (with $N \in [4, 35]$, $cv(x) \in [0.14, 1.20]$, $\rho \in [0.59, 0.999]$)

considered the stability of v_{YG} in πps designs of Brewer (1963), Fellegi (1963) and Hanurav (1967), of v_D, v'_D in procedure of Des Raj (1967), v_M in Murthy (1957), v_{RHC} in Rao-Hartley-Cochran (1962), $v(e_R)$, $v'(e_R)[= v(e_R)[0]$ if $v(e_R) > 0$ [otherwise]] in Midzuno and v_{pps} in ppswr procedures for $n = 2$. The per cent gain in stability of a variance estimator over Brewer's variance estimator, $\left[\frac{cv^2 \text{ (Brewer's variance est.)}}{cv^2 \text{ (var. est.)}} - 1\right] 100$ was calculated as a measure of stability. The following conclusions were drawn:

1. v_{RHC} is more efficient than other variance estimators, though its gain over v_M is not large in general;

2. For the natural populations, stability of Brewer's, Fellegi's and Hanurav's variance estimators are almost equal.

3. Stability of v_D and v_M are essentially equal; v'_D is uniformly marginally more efficient than v_M.

4. $v(e_R)$, $v'(e_R)$ are consistently less efficient than the others.

5. Each of v_D, v'_D, v_M, v_{RHC} is more efficient than v_{pps}.

Considering the model ξ_g with the additional assumption of e_i's following normal distribution, model ξ'_g (say) they calculated,

$$\lambda = \frac{\{\in (Ev_i^2) - (\in V_i)^2\}}{(\in V_i)^2}$$

(where $E(v_i) = V_i$) as a measure of stability of estimator v_i for $g = 1.0, 1.5, 1.75, 2.00$ for the above-noted populations and the conclusions (1) and (3) were corroborated. It was further noted that,

6. v_M is considerably more efficient than the variance estimators of Hanurav, Fellegi and Brewer.

Fellegi's variance estimator is consistently and marginally more efficient than Brewer's; efficiencies of Hanurav's and Fellegi's variance estimators are essentially equal for $g \geq 1.75$; however, for $g = 2$, the latter is consistently more efficient than the former.

7. Stability of v_{pps} generally decreases as g; v'_D is more stable than v_{pps}; v_D, v_M are more efficient than v_{pps} for $g > 1$.

Bayless and Rao (1970) empirically studied the stability of v_{YG} in πps-schemes of Fellegi (1963), Sampford (1967), Carroll and Hartley (1964), v_D in Des Raj, v_M in Murthy, v_{RHC} in Rao-Hartley-Cochran, $v'(e_R)$ in Midzuno and v_{pps} in ppswr-procedures for $n = 3$ and $n = 4$. For $n = 3$ they considered 14 natural populations (with

$N \in [10,20]$, $cv(x) \in [0.14, 1.06]$, $\rho \in [0.50, 0.99)]$) and for $n = 4$, 10 out of these 14 populations with N reduced for some of them. The percent gain in stability of variance estimators over Sampford's variance estimators was calculated and the following conclusions were drawn:

(*i*) v_{RHC} is more efficient than v_M but the gain is small except when $cv(x)$ and n/N are large;

(*ii*) v_M is considerably more efficient than Sampford's, Carroll-Hartley's, Fellegi's variance estimators, the stability of the last three being essentially equal;

(*iii*) v_D is more efficient than v_{pps}, but is slightly less efficient than v_M;

(*iv*) $v(e_R)$ is considerably less efficient than other estimators.

Calculation of λ based on the model ξ'_g showed.

(*v*) v_{RHC} is more efficient than v_M for $g = 1.5, 1.75$, though the gain is small except when n/N and $cv(x)$ are large. Unlike in the case $n = 2$, v_{RHC} is consistently less efficient than v_M when $g = 2$, although losses are small;

(*vi*) v_M is consistently more efficient than Sampford's and Carroll-Hartley' estimators and gains are large when n/N and $cv(x)$ are larger; the last two estimators are essentially equally efficient and consistently more efficient than Fellegi's; v_M consistently and marginally more efficient than v_D.

From the above study we conclude that,

(*a*) v_{RHC} is generally recommendable as the most stable variance estimator followed by v_M;

(*b*) For $n = 2$, Brewer's, Fellegi's and Hanurav's variance estimators are equally efficient.

(*c*) For $n > 2$, Sampford's variance estimator is more stable than Fellegi's;

(*d*) There is little to choose between v_D and v_M from the point of view of stability, for $n = 2$, v'_D is more stable than either of the two.

(*e*) v_{pps}, $v(e_R)$, $v'(e_R)$ are of poor efficiency.

2.5 PROBLEMS AND EXERCISES

1. Show that for the sampling design where the first unit is selected with probability p_i, the second unit with probability normed values of the p_i of the remaining units and the rest $(n - 2)$ with srswor from the remaining units, $v_{YG}(e_{HT}) \geq 0$.

If the first m units are selected by any sampling design having inclusion-probabilities $\pi_i(m)$, $\pi_{ii}(m)$ and the rest $(n - m)$ units by srswor from the remaining $(N - m)$ units in the population, then $\pi_i(m)\pi_j(m) - \pi_{ij}(m) \geq 0 \; \forall i, j \Rightarrow \pi_i(n)\pi_j(n) - \pi_{ij}(n) \geq 0 \; \forall \; i, j (\pi_i(n), \pi_{ij}(n)$ are the inclusion-probabilities for the resulting design). [Rao, J.N.K. (1961), Seth, G.R. (1966), Mukhopadhyay, P. (1974)].

2. Consider Yates-Grundy sampling scheme with revised size-measure q_i and let $t_i = q_i/(1 - q_i)$, $T = \Sigma t_i$, q_i's chosen as to make $\pi_i = 2p_i \forall \; i$. Show that under this scheme $V(e_{HT}) = \sigma_y^2 - U$, where $\sigma_y^2 = V(e_{HH}) = \dfrac{1}{2}\sum P_i\left(\dfrac{Y_i}{P_i} - Y\right)^2$ and

$$2U = \sum t_i q_i \left(\frac{Y_i}{P_i} - Y\right)^2 - \sum t_i \left(\frac{Y_i}{P_i} - Y\right)^2 \sum q_j \left(\frac{Y_j}{P_j} - Y\right).$$ Also show $E(v_{pps}) = \sigma_y^2$ $+ U$ where $v_{pps} = \frac{1}{4}\left(\frac{y_1}{p_1} - \frac{y_2}{p_2}\right)^2$, unbiased estimator of $V(e_{HH})$ under PPSWR-scheme. [Brewer, K.R.W. and Undy G.C. (1962)].

3. Continuation. Show for Brewer's (1963) scheme, $E(v_{pps})$

$= \frac{1}{2}\sum P_i (1+U_i)\left(\frac{Y_i}{P_i} - Y\right)^2$ where U_i is given by $q_i / (1 - p_i)$ (vide subsection 2.2.2)

[Brewer, K.R.W. (1963)].

4. Consider these schemes: (*a*) Two units are selected wr, one with probability p_i and other with probability proportional to $\frac{p_i}{1-p_i}$. If the units are distinct, they constitute the sample. Otherwise they are rejected and the process is continued till a sample is selected.

(*b*) First unit is drawn with probability p_i. Then units are drawn with the same probabilities p_k's till a different unit occurs. Show that for both the schemes π_i and π_{ij}-values coincide with those of Yates-Grundy (1953) scheme. [Agarwal, S.K. and Goel, B.P.P.S. (1977)].

5. Show for the general class of rejective sampling designs where two units are selected wr, one with probability p'_i, other with probability p''_i such that if the units are different they constitute the sample, otherwise they are rejected and fresh selections are made till two different units are selected, $v_{YG}(e_{HT})$ is always non-negative. Hence, $v_{YG}(e_{HT}) \geq 0$, for the procedures 1, 15, 3, 17 and 6. (For the last three schemes take $p'_i \propto \frac{p_i}{1-2p_i}$, $p''_i = p_i$). [Agarwal, S.K. Kumar, P and Dey, A. (1982)].

6. Consider the following sampling scheme for $n = 2$, applicable if $p_i^2 < \frac{s}{2}$ $\forall\, i$, $s = \sum P_t^2$. Two draws are made by ppswr. If the units selected are distinct, they constitute the sample. Otherwise the first unit is retained and the second unit is selected with conditional probability

$$p_2(j \mid i) = \lambda p_j^2 \left(\frac{1}{s - 2p_i^2} + \frac{1}{s - 2p_j^2}\right),$$

$$\lambda^{-1} = 1 + \sum \frac{P_i^2}{s - 2P_i^2}.$$

Show that for this scheme

$$\pi_i = 2p_i, \ \pi_{ij} = 2p_i p_j \left\{1 + \lambda p_i \ p_j \left(\frac{1}{s - 2p_i^2} + \frac{1}{s - 2p_j^2}\right)\right\}$$

and $$\frac{1}{2} < \frac{\pi_{ij}}{\pi_i \pi_j} < 1. \text{ [Singh, P. (1978)].}$$

7. Show that for each sampling design there exists a rejective method of drawing a sample. Show also that for Hájek's method (1964), (where n units are drawn wr with probability p_i; if units are distinct they form sample; otherwise trials are repeated till a sample is selected) expected number of trials for drawing a sample is

$\left[\sum_{s\in\mathcal{S}} \{II_{i\in\mathcal{S}} p_i \ II_{j\notin\mathcal{S}} (1 - p_j)\}\right]^{-1}$. Show also that every rejective sampling scheme terminates with selection of the sample with probability one. [Deshpande, M.N. and Ajgaonkar, S.G.P. (1977)].

8. Show that a sufficient condition for a connected fixed size πps-strategy to have $V(e_{HT}) \le V(pps; e_{HH})$, both using the same P_i, $i = 1, ..., N$ and same n is

$$\sum_i \min_j \frac{\pi_{ij}}{\pi_j} \ge n - 1 \qquad \ldots(a)$$

Also show, by considering the example $N = 4$, $n = 2$, $\pi_{12} = \pi_{13} = \pi_{24} = \pi_{34} = 0.25$, $\pi_{14} = \pi_{23} = 0$, that result (a) above is not a necessary condition.

Show further that for Sampford's (1967) scheme, (a) is satisfied. Hence, show that a sufficient condition for Midzuno-strategy to be better than (ppswr, e_{HH}) is

$$p_i \ge \frac{1}{1 + \frac{(N-1)(Nn - n - 1)}{n^2 - 1}} \ \forall \ i$$

and $$\frac{n}{M_1} \sum_i p_i \min_j \sum_{s \ni i, j} \frac{1}{q_s} \ge n - 1$$

where $$M_1 = \binom{N-1}{n-1} \quad \text{and} \quad q_s = \sum_s pk \, .$$

[Gabler, S. (1984); Mukhopadhyay, P. and Bhattacharyya, S. (1991)].

9. Show, for $n = 2$, a necessary condition for a (πps, e_{HT}) strategy to have smaller variance than the PPSWR-strategy is $\pi_{ij} \le \pi_i \pi_j \ \forall \ i \ne j$. Give an example where $v_{YG}(e_{HT}) > 0$ and yet $V(e_{HT}) \nless V(e_{HH})$.

[Narain, R.D. (1951), Gabler, S. (1984)].

10. Show that for the following πps-schemes, $V(e_{HT}) \leq V(e_{HH})$, e_{HH} being based on a ppswr-scheme using the same p_i values.

(*a*) $n = 2$ (results are exact)

(*i*) Procedure (1)

(*ii*) Procedure (2) [Sen, A.R. (1953), Des Raj (1956), Brewer, K.R.W. and Undy, G.C. (1962)].

(*iii*) Midzuno's scheme modified to be πps [Rao, J.N.K. (1963), Chaudhury, A. (1974) (any n), Mukhopadhyay, P. (1974) (any n)].

(*iv*) Procedure (15) [Rao, J.N.K. (1963)].

(*v*) Procedures (3), (6), (21).

[Brewer, K.R.W. (1963), Gabler, S. (1981), Konijn, K.S. (1973), Sutradhar, B.C. (1984)].

(*vi*) Procedures (4), (43) [Fellegi, I.P. (1963), Samiuddin, M. and Asad, H, (1981)].

(*vii*) Procedure (40) [Singh, P. (1978)].

(*viii*) Procedure (14) [Fuller, W. (1971)].

(*b*) Results for the following schemes with $n > 2$ are asymptotic under the following assumptions: (1) $P_i \leq \frac{c}{N}$ $\forall\, i$, c a constant (2) $V(e_{HH}) = 0(N^2)$ (3) N large.

(*i*) Procedure (1) (modified to n draws using ppswor).

(*ii*) Procedure (15).

(*iii*) Procedure (9) [Rao, J.N.K. (1963); Hartley, H.O. and Rao, J.N.K. (1962)].

(*iv*) Procedure (21) [Asok, C. and Sukhatme, B.V. (1976)].

11. For any design p and any number t, show that

$$E(\bar{y}^t) = \frac{1}{n^t} \sum_{j_i=1}^{N} \cdots \sum_{j_t=1}^{N} y_{j_1} \cdots y_{j_t}\, \pi(j_1, \ldots, j_t)$$

where $\bar{y}$ is sample mean and $\pi(k_1, \ldots, k_t)$ is t^{th}-order inclusion probability of units $k_1, \ldots, k_t$.

Also prove for any positive real number q,

$$E\left\{\left(\sum_{j \in s} y_j^q\right) \bar{y}^t\right\} = \frac{1}{n^t} \sum_{j_0=1}^{N} \sum_{j_1=1}^{N} \cdots \sum_{j_t=1}^{N} y_{j_0}^q\, y_{j_1} \cdots y_{j_t}\, \pi(j_0, j_1, \ldots, j_t)$$

Hence or otherwise, prove that for any non-negative real number q,

if $$\phi_q = \sum_{j \in s} (y_j - \bar{y})^q$$

$$E(\phi_q) = \sum_{u=1}^{q-1} (-1)^u \binom{q}{u} \frac{1}{n^u} \left\{ \sum_{j_0=1}^{N} \sum_{j_1=1}^{N} \cdots \sum_{j_u=1}^{N} y_{j_0}^{q-u} \, y_{j_i} \cdots y_{j_u} \right.$$

$$\left. \pi(j_0, j_1, \ldots, j_u) \right\} + (-1)^q \frac{1}{n^{q-1}} \sum_{j_1=1}^{N} \cdots \sum_{j_q=1}^{N} y_{j_1} \cdots y_{j_q}$$

$$\pi(j_1, \ldots, j_q) + \sum_{j=1}^{N} y_j^q \pi(j)$$

[Srivastava, J. and Salah, F. (1985), *Utilitas Mathematica*].

Table 2.3

Popln. No.	*Source*	*Y*	*X*	*Popln. size (N)*
1.	Murthy (1967), p. 126 villages 1–8	No. of cultivators in 1961	Area in 1951 (in sq. Area in 1951 (in sq. miles)	8
2.	do	Workers at household industry in 1961		8
3.	do, village 9-16	No. of cultivators in 1961	No. of persons in 1961	8
4.	do	Workers at household industry in 1961	do	8
5.	Murthy (1967), p. 132 block no. 7	Timber volume	Strip length	13
6.	do, village 1–10, p. 178	Area under paddy	Geographical area	10
7.	do, p. 180	Catch of fish	No. of boats landing	12
8.	do, p. 228	Output	No. of workers	8
9.	Sukhatme and Sukhatme (1970), p. 51	Area under rice	Total cultivated area	10
10.	do, p. 166	No. of banana bunches	No. of banana pits	10
11.	Yates (1960), p. 159	No. of absentees	Total no. of persons	12
12.	Yates, p. 163	Volume of timber	Eye-estimate	9
13.	Yates, p. 232	Length coloured green	Length of line	8
14.	Cochran (1977), p. 182	No. of paralytic polio cases in 'placebo' group	No. of 'placebo' children	10
15.	Des Raj (1972), p. 70	No. of cattle	No. of farms	15
16.	Sukhatme and Sukhatme (1970), p. 183	Area under wheat (1973)	Area under wheat (1963)	20
17.	R.K. Som (1973), p. 255	Average no. of persons per h.h.	Average monthly expenditure on cereals per h.h.	16
18.	Jessen (1978), p. 151	Total no. of fish	No. of fish tagged	16
19.	do, p. 153	No. of h.h. in 1960	No of h.h. in 1950	13
20.	Konijn (1973), p. 389	Measurement obtained in reinterview	Measurement obtained in first interview	10
21.	do, p. 49	Food expenditure	Total expenditure	16

Table 2.4

Probabilities and Relative Efficiency of $v_i(v_i^*)$, $i = 1, 2, 3$ for 21 Real Populations

Popln.				Probabilities			Relative Efficiency			Relative Bias			Relative Efficiency	
No.	N	c_x	ρ	p_1	p_2	p_3	E_0/E_1	E_0/E_2	E_0/E_3	b_1^*	b_2^*	b_3^*	E_1^*/E_2^*	E_1^*/E_3^*
8	8	0.056	0.82	0.0	0.0	0.0	1.19	1.03	1.14	0.0	0.0	0.0	0.867	0.954
21	16	0.078	0.95	0.0	0.0	0.0	1.04	1.00	1.18	0.0	0.0	0.0	0.962	1.140
6	10	0.085	0.25	0.0	0.0	0.0	1.12	1.00	1.27	0.0	0.0	0.0	0.893	1.137
20	10	0.202	0.76	0.067	0.0	0.0	1.31	1.00	1.67	0.073	0.0	0.0	0.724	1.205
10	10	0.248	0.84	0.042	0.0	0.10	1.43	1.00	2.89	0.049	0.0	0.177	0.676	1.869
17	16	0.351	– 0.35	0.155	0.0	0.107	2.03	1.03	3.24	0.097	0.0	0.143	0.518	1.635
5	13	0.368	0.94	0.217	0.0	0.077	3.18	0.96	7.75	0.174	0.0	0.036	0.821	6.536
9	10	0.392	0.87	0.483	0.0	0.175	5.98	1.09	11.04	0.416	0.0	0.373	0.284	2.762
15	15	0.420	0.77	0.356	0.0	0.101	3.38	1.00	6.68	0.223	0.0	0.289	0.358	2.129
14	10	0.423	0.22	0.467	0.0	0.142	7.68	1.29	12.22	0.345	0.0	0.392	0.266	2.303
1	8	0.449	0.69	0.393	0.0	0.143	10.49	0.91	21.97	0.440	0.0	0.316	0.415	9.491
2	8	0.449	0.43	0.411	0.0	0.125	9.37	0.94	24.82	0.472	0.0	0.377	0.214	5.327
19	13	0.472	0.52	0.423	0.0	0.143	5.84	1.06	10.68	0.273	0.0	0.252	0.350	3.529
18	16	0.474	0.90	0.348	0.0	0.161	2.62	0.94	6.84	0.196	0.0	0.209	0.485	3.350
11	12	0.503	0.80	0.414	0.0	0.150	3.32	0.86	10.17	0.364	0.0	0.263	0.539	5.895
13	8	0.562	0.44	0.50	0.179	0.179	3.61	0.78	13.33	0.373	0.055	0.261	0.617	10.035
12	9	0.573	0.90	0.560	0.012	0.135	1.81	0.98	7.27	0.154	0.002	0.140	0.566	4.051
3	8	0.634	0.92	0.589	0.304	0.143	18.72	1.38	40.70	0.424	0.086	0.257	0.319	9.167
4	8	0.634	0.89	0.536	0.268	0.232	10.66	1.02	34.39	0.417	0.158	0.204	0.520	17.541
16	20	0.723	0.98	0.508	0.0	0.128	14.81	0.94	36.92	0.279	0.0	0.110	0.518	20.075
7	12	0.723	0.93	0.505	0.0	0.191	7.38	0.84	27.36	0.316	0.0	0.131	0.604	19.342

Table 2.5 Description of the Populations

Sl. No.	Source	Y	X	N	c.v(x)	c.v(y)	ρ
1.	Sampford (1967), p. 507	–	–	10	0.4050	0.2169	0.43
2.	Rao (1963), p. 207	Corn acreage in 1958	Corn acreage in 1960	14	0.3786	0.4176	0.93
3.	Murthy (1977), p.178	Area under wheat	Geographical area	20	0.1882	0.3675	– 0.07
4.	Murthy (1977), p.127	No. of cultivators 1961 census	Cultivated area 1951 census	17	0.5610	0.5784	0.88
5.	Murthy (1977), p.131	Volume of timber	Length of timber	13	0.3074	0.5569	0.92
6.	Murthy (1977), p. 228	Output	No. of workers	16	0.3422	0.2728	0.86
7.	Jessen (1978), p. 151	Total No. of fish	No. of fish tagged	16	0.4739	0.5708	0.90
8.	Jessen (1978), p.153	No. of h.h in 1960	No. of h.h. in 1950	13	0.4715	0.4279	0.52
9.	Konijn (1973), p. 389	Measurement obtained in reinterview	Measurement obtained in first interview	10	0.2020	0.1515	0.77
10.	Konijn (1973), p. 49	Food expenditure	Total expenditure	16	0.0781	0.115	0.95
11.	Murthy (1977), p. 398	No. of absentees	No. of workers	18	0.4775	0.5814	0.64
12.	Murthy (1977), p. 339	Area under wheat in 1964	Cultivated area in 1961	15	0.4938	0.6935	0.94
13.	Des Raj (1972), p. 70	No. of cattle	No. of farms	15	0.4197	0.4144	0.77
14.	Sukhatme (1970), p. 183	Area under wheat (1973)	Area under wheat (1936)	20	0.7231	0.7440	0.98
15.	Som (1973), p. 225	Average no. of persons per h.h.	Average monthly expenditure on cereals per h.h.	16	0.2427	0.4502	– 0.57

❑ ❑ ❑

CHAPTER 3

Some Problems of Inference Under a Fixed Population Set-up

3.1 INTRODUCTION

This chapter deals with some problems of inference about a finite population mean (total) through survey sampling under a fixed population set-up. Godambe (1955), his work being supplemented by Hanurav (1966), Hege (1965), Lanke (1974) among others, proved that there does not exist any uniformly minimum variance unbiased estimator (UMVUE) for population mean (total), in general. Attention was, therefore, drawn to the search for admissible estimators for a fixed sampling design in a class of estimators [Godambe and Joshi (1965), Joshi (1965 *a*, *b*, 1969), among others] and admissible strategies in a class of strategies [Joshi (1966), Godambe (1969), Scott (1975), Sekkappan and Thompson (1975) among others]. Godambe (1966) also observed that for a non-informative design, the likelihood function of the population parameter **y** is, in general, non-informative. The scale load approaches of Hartley and Rao (1968, 1969), however, provide some important exception. Basu (1958), Basu and Ghosh (1967) were first to consider the concept of sufficiency, and Rao-Blackwellisation in survey sampling. The criterion of minimaxity makes some well-known sampling strategies optimal.

In case the survey population is looked upon as a sample from a superpopulation, optimum sampling strategies are, however, available in certain classes and there exist lower bounds to the average variances of all unbiased strategies under some models [Godambe and Joshi (1965), Cassel *et al.* (1976), Tam (1984), among others]. If the finite population is considered as a member of an infinite sequence of monotonically increasing populations, the Horvitz-Thompson estimator has been noted to obtain some asymptotically optimum properties.

We take an overview of these concepts in this chapter. We, however, exclude from our coverage the concepts of uniform admissibility (or admissibility of strategies), hyperadmissibility, linear sufficiency, invariance etc., for a brief review of which interested readers may refer to Cassel *et al.* (1977).

3.2 THE p.d.f. OF DATA AND LIKELIHOOD FUNCTION OF y

Let $D'(D)$ be a random variable corresponding to data $d'(d)$. Let also $\Psi'(\Psi)$ be a random variable having values $S(s)$ on $\mathscr{P}$.

A data point $d'(d)$ is said to be consistent with a chosen parameter vector $\mathbf{y}$ if $d'(d)$ can be obtained from $\mathbf{y}$. For example, $d' = \{(2, 6), (1, 4), (2, 6), (4, 7)\}$ is consistent with $\mathbf{y} = \{4, 6, 3, 7, 11\}$ but not with $\mathbf{y}' = \{3, 6, 1, 8\}$.

Let for a given $d'(d)$, $\Omega_{d'}$, (Ω_d) be the set of $\mathbf{y}$ for which $d'(d)$ is consistent.

The p.d.f. of D' is

$$\begin{aligned} f_{D'}(d'; \mathbf{y}) &= P[\{(k, y_k), k \in S\} = d'; \mathbf{y}] \\ &= P[\Psi' = S]P[D' = d' \mid \Psi' = S; \mathbf{y}] \\ &= p(S)\ \delta(d'; \mathbf{y}) \end{aligned} \tag{3.2.1}$$

where $S(d'; \mathbf{y}) = 1(0)$ if d' is consistent (inconsistent) with $\mathbf{y}$. Thus

$$f_{D'}(d'; \mathbf{y}) = p(S)(0) \text{ for } y \in \Omega_{d'} \text{ (otherwise).}$$

Similarly, p.d.f. of D is

$$f_D(d; \mathbf{y}) = p(s)(0) \text{ for } \mathbf{y} \in \Omega_{d'} \text{ (otherwise).} \tag{3.2.2}$$

Likelihood Function of y: Given that data $D' = d'$, the likelihood function of the parameter vector $\mathbf{y}$ is

$$\begin{aligned} L(\mathbf{y} \mid d') &= f_{D'}(d'; \mathbf{y}) \\ &= p(S)(0) \text{ if } \mathbf{y} \in \Omega_d \text{ (otherwise).} \end{aligned} \tag{3.2.3}$$

Similarly, the likelihood of $\mathbf{y}$ given $D = d$. is

$$\begin{aligned} L(\mathbf{y} \mid d) &= f_D(d; \mathbf{y}) \\ &= p(s)(0) \text{ if } \mathbf{y} \in \Omega_d \text{ (otherwise).} \end{aligned} \tag{3.2.4}$$

The likelihood functions are, therefore, 'flat' taking values $p(s)$ $(p(S))$ for $\mathbf{y} \in \Omega_d(\Omega_{d'})$ and zero otherwise. There does not exist any unique maximum of the likelihood and hence no maximum likelihood solution of any parameter function $\theta(\mathbf{y})$. The likelihood functions only tell us that all $\mathbf{y} \in \Omega_d(\Omega_{d'})$ are equally probable and hence tell nothing about the unobserved components of the $\mathbf{y}$-vector. The likelihood functions (3.2.3) and (3.2.4), first considered by Godambe (1966), are therefore, non-informative. However, if a superpopulation model ξ is postulated for the population vector, the likelihood function may give more information (see Royall (1976), Brecking *et al.* (1990).

There are, however, some simple situations stated below, where even without assuming *a prior* ξ, likelihood function gives non-trivial information about $\mathbf{y}$.

Royall (1968), Hartley and Rao (1968, 1969) assumed that y is measured on a known scale with a finite set of points y_i^* $(i = 1, \ldots, T)$, $N_i(n_i)$ units in the population (sample) have each value y_i^*, the sample being drawn by srswor. Let $\mathbf{N} = (N_1, \ldots, N_T)$, $\mathbf{n} = (n_1, \ldots, n_T)$; $Y = \sum_t N_t y_t^*$. The likelihood function of $\mathbf{N}$ given $\mathbf{n}$ is multidimensional hypergeometric distribution,

$$L(\mathbf{N} \mid \mathbf{n}) = \begin{cases} \prod_{t=1}^{T} \binom{N_t}{n_t} \Big/ \binom{N}{n} \ \forall\, \mathbf{N} : N_t \geq n_t \ (t = 1, \ldots, T), \ \sum N_t = N \\ 0, \quad \text{otherwise} \end{cases} \tag{3.2.5}$$

The m.l.e. is $\hat{N}_t = \dfrac{Nn_t}{n}$, $\hat{Y} = \sum \hat{N}_t y_t^* = \dfrac{N}{n} \sum n_t y_t^* = N\bar{y}_s$ (3.2.6)

provided N/n is integral (otherwise m.l.e. of N_t is obtained by a rounding algorithm form $N\bar{y}$).

$N\bar{y}$ is *umvue* of Y among estimators depending only on scale loads in $\mathbf{n}$, since $\mathbf{n}$ is completely sufficient for the family in (3.2.5).

If a concomitant variable x with scale points $x_1^*, \ldots, x_H^*$ is also attached to each unit in the population and if only the population mean $\bar{X}$ is known, Hartley and Rao showed that m.l.e. of $Y = \sum_{t=1}^{T} \sum_{h=1}^{H} N_{th} y_t^*$ (where N_{th} is the number of elements in the population having values (y_t^*, x_h^*)) is

$$\hat{\hat{Y}} = N\{\bar{y} + b_1(\bar{X} - \bar{x})\}$$

where $b_1 = \sum_s y_i(x_i - \bar{X}) \Big/ \sum_s (x_i - \bar{X})^2$, $\bar{X} = \dfrac{X}{N}$.

In the case of stratified srs, the strata are regarded as separate populations, each described by its own set of parameters $(N_{h1}, \ldots, N_{hT}) = \mathbf{N}_h$. The customary estimator $\sum_{h=1}^{L} N_h \bar{y}_h = \sum_{h=1}^{L} N_h \sum_{t=1}^{T} n_{ht} y_t^* / n_h$, $N_h = \sum N_{ht}$, $n_h = \sum n_{ht}$ is the m.l.e. and *umvu* among the estimators depending only on the scale loads n_{ht}.

For ppswr-sampling, Hartley and Rao (1969) used the scale loads of $r_i = y_i/x_i$, assuming that r_i's are homogeneous and unrelated to x_i ($i = 1, \ldots, N$) and showed that e_{HH} is m.l.e. Sarndal (1976) showed that e_{HT} possesses the *umv* property in the scale load class for any F.S.-πps design. The scale-load approach was also considered by C.R. Rao (1971).

An objection due to Kempthorne (1969) against the use of $\hat{N}_t$ in (3.2.5) is that some of the scale points y_i^* may not be represented in the sample.

It appears, therefore in simple symmetric cases, a more informative likelihood than that obtained from a fixed population model can be obtained if the label part of the data is ignored.

3.3 SUFFICIENCY, RAO-BLACKWELLIZATION

The concept of sufficiency and Rao-Blackwellization in survey sampling in connection with resolving the problem—whether one is required to consider the whole body of data available through a sequence sample or the data obtained from a set sample is enough for making inference—was first considered by Basu (1958). As in the traditional statistical theory, if a sufficient statistic is available, any estimator can be improved upon by Rao-Blackwellization. In survey sampling D' forms the primary core of data. Any summarisation of data would have to be made necessarily over D'.

Definition 3.1: A statistic $u(D')$ is a sufficient statistic for $\mathbf{y}$ if the conditional distribution of D' given $u(D') = u_0$ is independent of $\mathbf{y}$, provided the conditional distribution is well defined.

Let $z(D')$ be a statistic defined over the range space of D' such that $z(D') = d$, *i.e.*, z reduces d' obtained through a sequence sample S to the data d for the corresponding set. As an example, if $d' = \{(2, 5), (4, 7), (2, 5)\}$, $z(d') = \{(2, 5), (4,7)\}$.

The following theorem, due to Basu and Ghosh (1967), was presented nicely by Cassel *et al.* (1977). We consider a later proof.

Theorem 3.1: For any ordered design p, the statistic $z(D')$ is sufficient.

Proof: Consider two points d' of D' and d of D. Assume the parameter vector $\mathbf{y} \in \Omega_{d'}$ and Ω_d. Otherwise the conditional probability will not be defined.

$$P_{\mathbf{y}}[D' = d' \mid z(D') = d] = \frac{P_{\mathbf{y}}[(D' = d') \cap (z(D') = d)]}{P_{\mathbf{y}}[z(D') = d]} \qquad \ldots(3.3.1)$$

provided $P_{\mathbf{y}}[z(D') = d] > 0$.

Case (*a*): $z(d') \neq d$. Here numerator in (3.3.1) is zero and hence, (3.3.1) has value zero.

Case (*b*): $z(d') = d$. The numerator in (3.3.1) is

$$P_y[D' = d'] = p(\mathbf{S}) \text{ by } (3.2.1)$$

The denominator is

$$\sum{}' p(S) = p(s)$$

where $\sum{}'$ is overall S corresponding to the set s (corresponding to d). Hence (3.3.1) is $p(S)/p(s)$.

In either case the conditional probability is independent of $\mathbf{y}$.

It follows that $z(D')$ is minimal sufficient for $\mathbf{y}$. For a proof the reader may refer to Cassel *et al.* (1977).

For any estimator $e(D')$ for θ define $e(d) = E\{e(D') \mid z(D') = d\}$. Since $z(D')$ is sufficient for $\mathbf{y}$, $e(d)$ is independent of any unknown parameter, and depends on D' only through $z(D')$ and as such can be taken as an estimator of θ.

Theorem 3.2: Let $e(D')$ be an estimator of θ. The estimator $e_1(d) = E\{e(D')/z(D') = d\}$ has the properties:

(*i*) $E(e) = E(e_1)$

(*ii*) $MSE(e_1) \leq MSE(e)$ with strict inequality $\forall\ \mathbf{y} \in R_N$ *iff* $P\{e \neq e_1; \mathbf{y}\} > 0$.

Proof:

$$E(e_1) = E[E\{e(D') \mid z(D') = d\}] = E\{e(D')\}$$

$$E(e - \theta)^2 = E(e_1 - \theta)^2 + E(e - e_1)^2 + 2E(e_1 - \theta)(e - e_1)$$

(The last term on the r.h.s.

$$= 2E[((e_1 - \theta)E\{(e - e_1) \mid z(D') = d\}] = 0$$

$$= MSE(e_1) + E(e - e_1)^2.$$

Rao-Blackwellization in survery sampling was first considered by Basu (1958). Some earlier works are due to Des Raj and Khamis (1958), Murthy (1957).

Example 3.1: For a S drawn by srswr with n draws, let $d = \{(i_1 > y_{i1}), \ldots, (i_v, y_{iv})\}$, $i_1 < \ldots < i_v$, corresponding to data d'. The customary sample mean $\bar{y}_S$ depends on the multiplicity of units in S.

$$E\{\bar{y}_S \mid d\} = E\left\{\frac{1}{n}\sum_{i=1}^{n} y_i \mid d\right\} = E(y_1 \mid d)$$

$$= \frac{1}{V}\sum_{j=1}^{V} y_{ij} = \bar{y}_v$$

Hence, $$E(\bar{y}_v) = E(\bar{y}_S) = \bar{Y}\,,\ V(\bar{y}_v) \le V(\bar{y}_S)\,.$$

Pathak (1961, 1962) showed $E\left(\dfrac{1}{v(s)}\right) = \dfrac{1}{N^n}\displaystyle\sum_{j=1}^{N} j^{n-1}$ and hence

$$V(\bar{y}_{v(s)}) = \frac{s^2}{N}\sum_{j=1}^{N-1}\left(\frac{j}{N}\right)^{n-1}$$

where $$s^2 = \frac{1}{N-1}\sum (Y_i - \bar{Y})^2\,.$$

Again, $$\left(\frac{1}{n} - \frac{1}{N}\right)s^2 \le E\left\{\frac{1}{v(s)} - \frac{1}{N}\right\}s^2 \le \frac{(N-1)s^2}{Nn} \tag{3.3.2}$$

Hence the strategy consisting of (srswr (n), $\bar{y}_{v(S)}$) is inferior to (srswor (n), $\bar{y}_s$), but is superior to (srswr (n), $\bar{y}_S$) in the smaller variance sense. Lanke (1975) discussed the inequality (3.3.2) thoroughly.

Example 3.2: In ppswr-sampling of n draws, let $d = \{((1), p_{(1)}, y_{(1)}), \ldots, ((v), p_{(v)}, y_{(v)})\}$ corresponding to data d'. The customary estimator $\hat{Y}_{pps}\left(= \dfrac{1}{n}\displaystyle\sum_{S} y_i / p_i\right)$ depends on S.

$$E\{\hat{Y}_{pps} \mid d\} = E\left\{\frac{1}{n}\sum_{i=1}^{n} \frac{y_i}{p_i}\middle| d\right\} = E\left(\frac{y_1}{p_1}\middle| d\right)$$

$$= \sum_{i=1}^{v} \frac{y_{(i)}}{p_{(i)}} P[y_i = y_{(i)} \mid d] = \sum_{i=1}^{v} \frac{y_{(i)}}{p_{(i)}} c_{(i)}$$

where,

$$c_{(i)} = \left[p_{(i)} \sum^{(2)} \frac{(n-1)!}{\alpha_{(1)}! \ldots \alpha_{(v)}!} p_{(1)}^{\alpha(1)} \cdots p_{(v)}^{\alpha(v)}\right]$$

$$\Big/ \left[\sum^{(1)} \frac{n!}{\alpha_{(1)}! \ldots \alpha_{(v)}!} p_{(1)}^{\alpha(1)} \cdots p_{(v)}^{\alpha(v)}\right]$$

where $\Sigma^{(1)}$ is overall integral α's such that $\alpha_{(i)} > 0$, $i = 1, \ldots, v$, $\Sigma\alpha_{(i)} = n$; $\Sigma^{(2)}$ is overall integral α's such that

$$\alpha_{(i)} \geq 0,\ \alpha_{(j)} > 0,\ j(\neq i) = 1, \ldots, v$$

and $$\sum \alpha_{(k)} = n - 1.$$

It may be noted, however, that the statistic D is not a complete sufficient statistic. Hence, there may exist more than one estimator $e(D)$ for Y. An example is that of an *HTE* which depends only on D and is unbiased for Y. For proof of the fact that D is not complete sufficient and for further examples in this area the reader may refer to Cassel *et al.* (1977).

3.4 UNIFORMLY MINIMUM VARIANCE UNBIASED ESTIMATION (UMVUE)

Godambe (1955) first observed that in survey sampling no UMV-estimator in the class of all linear unbiased estimators of population total exists for any p in general. The proof was subsequently improved upon by Hege (1965), Hanurav (1966), Ericson (1974), Lanke (1974). We shall consider here Lanke's proof. We consider first

Definition 3.2: A s.d. p is said to be a unicluster design if for any two samples s_1, s_2

$$\{p(s_1) > 0,\ p(s_2) > 0,\ s_1 \neq s_2\} \Rightarrow s_1 \cap s_2 = \phi,$$

i.e, either two samples are identical or are disjoint. A linear systematic sample is an example of a unicluster design.

Theorem 3.3: A s.d. p admits a UMVU-estimator in the class of all linear unbiased estimators *iff* p is a unicluster design with $\pi_i > 0\ \forall\ i$.

Proof: A s.d. p admits of an unbiased estimator of Y *iff* $\pi_i(p) > 0\ \forall\ i$.

Hence, one has to consider only the class of design with $\pi_i > 0\ \forall\ i$. Suppose that p is a non-unicluster design. We show that (*a*) if *e* is a UMV-linear unbiased estimator (UMVLUE), $e = e_{HT}$

(*b*) e_{HT} is not UMVLUE.

(*a*) Let $e = \sum_{i \in s} b_{si} y_i$, be uniformly better than e_{HT}. For $y = y^{(t)} = (0, \ldots, 0, 1, 0, \ldots 0)$,

$$V(e) - V(e_{HT}) = \sum_{s \ni i} \left(b_{si} - \frac{1}{\pi_i} \right)^2 p(s)$$

$$\left(\text{since} \sum_{s \ni i} b_{si} p(s) = 1 \, \forall \, i, \text{ by unbiasedness} \right)$$

$$\geq 0,$$

equality holds *iff* $\qquad b_{si} = \dfrac{1}{\pi_i} \forall s \ni i \quad \text{and} \quad \forall i.$

(*b*) Let s_1, s_2 be two non-identical samples with $i_1 \in s_1 \cap s_2$, $i_2 \in (s_1^c \cap s_2) \cup (s_1 \cap s_2^c)$.

Consider a modified estimator

$$e'(s, \mathbf{y}) = \begin{cases} e_{HT}(s, \mathbf{y}) + p(s_2)\, y_{i_1} & \text{for } s = s_1 \\ e_{HT}(s, \mathbf{y}) - p(s_2)\, y_{i_1} & \text{for } s = s_2 \\ e_{HT}(s, \mathbf{y}) & \text{otherwise} \end{cases}$$

$e'(s, \mathbf{y})$ is unbiased. At the point $\mathbf{y} = \mathbf{y}' = (0, \ldots, 0, y_{i_1}, 0, \ldots, 0 \;\; y_{i_2}, 0, \ldots, 0)$

$$V(e_{HT}) - V(e') = A_{i_1} y_{i_1}^2 + A_{i_1 i_2} y_{i_1} y_{i_2} \tag{3.4.1}$$

where $A_{i_1} = -p(s_1)p(s_2)\,(p(s_1) + p(s_2))$, $A_{i_1 i_2} = 2p(s_1)p(s_2)/\pi_{i_2}$. The sign of (3.4.1) can be made positive by proper choice of y_{i_2}. Hence e_{HT} is not UMVLUE.

If *p* is a unicluster design, an unbiased estimator $e(s, \mathbf{y}) = \sum_{i \in s} b_{si} y_i$ reduces to $\sum_{i \in s} y_i / p(s)$ (unbiasedness requires $b_{si} p(s) = 1 \, \forall \, i$) which is the HTE. Since there is only one LUE of *Y*, e_{HT} is UMVLUE of *Y*.

The following non-existence theorem was first proved by Godambe & Joshi (1965). The present proof is due to Basu (1971).

Theorem 3.4: For any non-census design *p* (with $\pi_i > 0 \, \forall \, i$), there does not exist any UMVUE of *Y* in the class of all unbiased estimators.

Proof: For a known parameter vector **a**, the generalized difference estimator $e_{GD}(\mathbf{a}) = \sum_s \dfrac{y_i - a_i}{\pi_i} + \sum a_i$ has zero variance at $\mathbf{y} = \mathbf{a}$. Since **a** is arbitrary, an estimator **e** is UMV *iff* is has zero variance at all $\mathbf{y} \in R_N$. This is possible *iff p* is a census design.

3.5 ADMISSIBILITY OF ESTIMATORS

Definition 3.3: For a fixed s.d. p, an estimator e is said to be an admissible estimator of Y within a class C of estimators *iff* there does not exist any estimator in C which is uniformly better than e.

Clearly, within the same class C there may exist more than one admissible estimator. Admissibility ensures than an estimator is uniquely best in C at least at some point y in the parameter space. In the absence of a UMVUE, one should choose an estimator within the class of all admissible estimators. However, a "slightly inadmissible" estimator may sometimes possess some practical advantages over an admissible estimator and may be used in preference to the latter.

It follows from the results of Theorem 3.1 that an estimator which depends on the order of selection and/or multiplicity of units in S is an inadmissible estimator.

The earliest results on admissibility are due to Godambe (1960) and Roy and Chakraborty (1960) who independently proved that with the squared error as loss function, the HTE is admissible in the class of all linear and unbiased estimators, whatever be the sampling design. This result was subsequently improved upon by Godambe & Joshi (1965) and Joshi (1965).

The following theorems (some stated without proof) state admissibility of some well-known estimators.

Admissibility of e_{GD}:

Theorem 3.5: For any s.d. p, with $\pi_i > 0$ $(\forall_i)$, the generalized difference estimator $e_{GD}(\mathbf{a}) = \sum_s \frac{y_i - a_i}{\pi_i} + A\left(A = \sum a_i\right)$ is admissible in the class of all unbiased estimators of Y.

(Godamde and Joshi (1965), Cassel *et al.* (1997)).

Proof: Let e_{GD} be not admissible. Then there exists an estimator

$$e(s, \mathbf{y}) = e_{GD} + \{e(s, \mathbf{y}) - e_{GD}\}$$
$$= e_{GD} + h(s, \mathbf{y}) \quad \text{(say)}$$

which is unbiased for Y and better than e_{GD}. Hence

$$\sum_{s \in \mathscr{S}} h(s, \mathbf{y}) p(s) = 0 \quad \forall\ \mathbf{y} \in R_N \tag{3.5.1}$$

and,

$$E(e_{GD} + h(s, \mathbf{y}))^2 \leq E(e_{GD}^2)$$

i.e.,

$$\sum_{s \in \mathscr{S}} h^2(s, \mathbf{y}) p(s) \leq -2 \sum_{s \in \mathscr{S}} e_{GD} h(s, \mathbf{y}) p(s)\ \forall\ \mathbf{y} \in R_N \tag{3.5.2}$$

Since $e(s, \mathbf{y})$ is an estimator, $h(s, \mathbf{y})$ is a function defined on $\mathscr{P} \times R_N$ such that for a given $\mathbf{y}$, its value depends on $\mathbf{y}$ only through those y_i for which $i \in s$. Let $\Omega_m = \{\mathbf{y}$: exactly m components of $\mathbf{y} - a$ are non-zero$\}$, where $\mathbf{a} = (a_1, \ldots, a_N)$.

We first prove.

Lemma 3.1: If $h(s, \mathbf{y})p(s) = 0 \ \forall \ (s, \mathbf{y})$ such that $p(S) > 0$ and $\mathbf{y} \in \Omega_k$ then $h(s, \mathbf{y})p(s) = 0 \ \forall \ (s, \mathbf{y})$ such that $p(s) > 0$ and $\mathbf{y} \in \Omega_{k+1}$, $k = 0, 1, 2, \ldots$.

Proof: Let $\mathbf{y}' \in \Omega_{k+1}$. Define $\mathscr{S}_i = \{s$: exactly i labels in s have $y_i - a_i \neq 0\}$.
Hence by (3.5.1),

$$\sum_{s \in \mathscr{S}} h(s, \mathbf{y}')p(s) = \sum_{i=1}^{k+1} \sum_{s \in \mathscr{S}_i} h(s, \mathbf{y}')p(s) = 0 \tag{3.5.3}$$

By (3.5.2),

$$\sum_{i=1}^{k+1} \sum_{s \in \mathscr{S}_i} h^2(s, \mathbf{y}')p(s) \leq -2 \sum_{i=1}^{k+1} \sum_{s \in \mathscr{S}_i} e_{GD} h(s, \mathbf{y}')p(s) \tag{3.5.4}$$

By assumption, $h(s, \mathbf{y})p(s) = 0 \ \forall \ (s, \mathbf{y})$ such that $p(s) > 0$ and $\mathbf{y} \in \Omega_k$; hence $h(s, \mathbf{y})p(s) = 0 \ \forall \ s \in \mathscr{S}_j (j = 1, \ldots, k)$ whenever, $\mathbf{y} \in \Omega_{k+1}$. Therefore, (3.5.3) and (3.5.4) imply respectively.

$$\sum_{s \in \mathscr{S}_{k+1}} h(s, \mathbf{y}')p(s) = 0 \tag{3.5.5}$$

$$\sum_{s \in \mathscr{S}_{k+1}} h^2(s, \mathbf{y}')p(s) \leq -2 \sum_{s \in \mathscr{S}_{k+1}} e_{GD}(s) h(s, \mathbf{y}')p(s) \tag{3.5.6}$$

Now, for $\mathbf{y} \in \Omega_{k+1}$ and $s \in \mathscr{S}_{k+1}$, e_{GD} is a constant e^0_{GD} (say), depending only on the units with non-zero y_i values. By (3.5.6)

$$\sum_{s \in \mathscr{S}_{k+1}} h^2(s, \mathbf{y}')p(s) \leq -2e^0_{GD} \sum_{s \in \mathscr{S}_{k+1}} h(s, \mathbf{y}')p(s)$$

$$= 0 \text{ by (3.5.5)}$$

Hence,

$$h(s, \mathbf{y}')p(s) = 0 \ \forall s \in \mathscr{S}_{k+1},$$

i.e.,

$$h(s, \mathbf{y}')p(s) = 0 \ \forall (s, \mathbf{y}) \text{ such that } p(s) > 0 \text{ and } \mathbf{y} \in \Omega_{k+1}.$$

Now for $\mathbf{y} \in \Omega_0$, $h(s, \mathbf{y}) = 0 \ \forall s$ with $p(s) > 0$. Hence, by repeated application of the lemma, $h(s, \mathbf{y}) = 0$ whenever $\mathbf{y} \in \Omega_1, \Omega_2, \ldots, \Omega_N$, $\forall s$ with $p(s) > 0$. Hence e_{GD} is admissible in the class of all unbiased estimators.

Corollary 3.1: For any p with $\pi_i > 0 (\forall i)$, e_{HT} is admissible in the class of all unbiased estimators of Y.

Proof: Put $\mathbf{a} = 0$ in $e_{GD}(\mathbf{a})$.

Corollary 3.2: For any p with $\pi_i > 0$ $(\forall i)$, the difference estimator $e_D = N[\bar{y}_s + d(\bar{x} - \bar{x}_s)]$ ('d' a constant) is admissible in the class of all admissible estimators of Y.

Proof: Put $\mathbf{a} = d_1$, where $\mathbf{1} = (1, ..., 1)$, in e_{GD}.

Remark 3.1: The proof of Theorem 3.5 depends solely on the fact that the value of e_{GD} does not depend on the units with $y_i = a_i$. Choosing in particular, $\mathbf{a} = \mathbf{0}$, we have

Theorem 3.6: Any function $g(s, \mathbf{y})$ defined on $\mathscr{S} \times R_N$ which depends on $(s, \mathbf{y})$ only through those (i, y_i) for which $y_i \neq 0$ is an admissible esimator of its expectation, $G(\mathbf{y}) = \sum_{s \in \mathscr{S}} g(s, \mathbf{y})\, p(s)$ for any given p.

Corollary 3.3: v_{HT}, the Horvitz-Thompson estimator of variance of e_{HT}, does not depend on the units $i \in s$ with $y_i = 0$. Hence v_{HT} is an admissible estimator of V_{HT} in the class of all unbiased estimators.

Remark 3.2: The proof of theorem 3.5 remains valid even if $\mathbf{y}$ belongs to some restricted parameter space $\bar{R}^N \subset R^N$, provided $\mathbf{a} \in \bar{R}^N$. In particular, e_{HT} remains admissible for Y if only the point $\mathbf{o} \in \bar{R}^N$. As a corollary, in srswor, sample proportion of units containing a particular character A is an admissible estimator of the population proportion in the class of all unbiased estimators, provided the region $\{-a_i \leq y_i \leq b_i, a_i \geq 0, b_i \geq 1\}$ is included in the parameter space.

Theorem 3.7: For any $p \in \rho_n$, any estimator $e_1(s, \mathbf{y}) = \sum_{k \in s} b_k y_k$ with

$$b_k \geq 1, \ \sum b_k^{-1} = n \tag{3.5.7}$$

is admissible in the class of all linear estimators of $\mathbf{y}$. (Godambe & Joshi (1965)).

Proof: If $e_1(s, \mathbf{y})$ be inadmissible, let $e(s, \mathbf{y}) = \sum_{i \in s} \beta_{si} y_i$ be such that

$$\sum_{s \in \mathscr{S}} (e(s, \mathbf{y}) - Y)^2 p(s) \leq \sum_{s \in \mathscr{S}} (e_1(s, \mathbf{y}) - Y)^2 p(s) \ \forall \ \mathbf{y} \tag{3.5.8}$$

with strict inequality holding for at least one $\mathbf{y}$.

Now,

$$e(s,\ \mathbf{y}) = e_1(s,\ \mathbf{y}) + h(s,\ \mathbf{y}) \tag{3.5.9}$$

where,
$$h(s,\ \mathbf{y}) = \sum_{k \in s} \alpha_{sk}\, y_k,\ \alpha_{sk} = \beta_{sk} - b_k\ .$$

From (3.5.8),

$$E(e_1 + h - Y)^2 \le E(e_1 - Y)^2$$

i.e.,
$$\sum_{s \in \mathscr{S}} h^2(s, \mathbf{y})\, p(s) + 2 \sum_{s \in \mathscr{S}} (e_1(s, \mathbf{y}) - Y) h(s, \mathbf{y})\, p(s) \le 0$$

i.e.,
$$\sum_{s \in \mathscr{S}} \left\{ \sum_{s} \alpha_{sk}\, y_k \right\}^2 p(s) + 2 \sum_{s \in \mathscr{S}} \left\{ \sum_{k \in s} (b_k - 1)\, y_k - \sum_{k \in \bar{s}} y_k \right\}$$

$$\left(\sum_{s} \alpha_{sk}\, y_k \right) p(s) \le 0, \tag{3.5.10}$$

$\bar{s} = \mathscr{P} - s$. (3.5.10) is a negative semi-definite quadratic form in y_k and coefficient of y_k^2 in (3.5.10) should be non-positive, *i.e.*,

$$\sum_{s \ni k} \alpha_{s_k}^2\, p(s) + 2 \sum_{s \ni k} (b_k - 1) \alpha_{sk}\, p(s) \le 0. \tag{3.5.11}$$

Let,
$$\sum_{s \ni k} \alpha_{sk}\, p(s) = \delta_k,$$

when
$$\sum_{s \ni k} p(s)(\alpha_{sk} - \delta_k)^2 - \sum_{s \ni k} \alpha_{sk}^2\, p(s) + \delta_k^2 (\pi_k \quad 2)\ .$$

Hence (3.5.11) reduces to

$$\sum_{s \ni k} (\alpha_{sk} - \delta_k)^2\, p(s) + 2\delta_k (b_k - 1) + \delta_k^2 (2 - \pi_k) \le 0. \tag{3.5.12}$$

Since $b_k \ge 1$, we must have $\delta_k \le 0\ \forall k$.

For the point $\mathbf{y} = \mathbf{y}' = (b_1^{-1}, \ldots, b_N^{-1})$, $e_1(s,\ \mathbf{y}') = n = Y$. Hence from (3.5.9),

$$h(s,\ \mathbf{y}') = 0\ \forall\ s : p(s) > 0.$$

Now,

$$\sum_{s \in \mathcal{S}} h(s, \mathbf{y}')p(s) = \sum \delta_k / b_k = 0;$$

since $b_k \geq 1$, $\delta_k = 0 \ \forall \ k$. Hence from (3.5.12)

$$\sum_{s \ni k} \alpha_{sk}^2 p(s) \leq 0 \quad \Rightarrow \quad \alpha_{sk} = \ 0 \ \forall \ k \text{ and all } s \ni k$$

i.e.,

$$e_1(s, \mathbf{y}) = e(s, \mathbf{y}) \ \forall \ \mathbf{y} \text{ and } \forall \ s : p(s) > 0.$$

Corollary 3.4: For any F.S.-design, e_{HT} is admissible in the class of all linear estimators of Y.

Theorem 3.8: For any varying sample size design p, $e_1(s, \mathbf{y})$ in Theorem 3.7 is inadmissible in the class of all estimators of Y. In particular, for any varying sample size design p, e_{HT} is inadmissible in the class of all estimators of Y. (Godambe & Joshi, 1965).

Theorem 3.9: For any wor-s.d. p, $N\,\overline{y}_s$ is admissible in the class of all estimators of Y. (Joshi, 1965*a*).

Theorem 3.10: For any s.d. p, ratio estimator $\hat{Y}_R = \dfrac{\sum_{i \in s} Y_i}{\sum_{i \in s} x_i} X$ is admissible in the class of all (measurable) estimators of Y. (Joshi, 1965*b*; 1966).

Theorem 3.11: If the loss function $V(t)$, where t is the numerical difference between the estimated and the true values, satisfies

(*a*) $V(t)$ is non-decreasing in $[0, \infty)$

(*b*) for every $K > 0$

$$\int_0^{\infty} V(t)\, e^{-kt^2/2}\, dt < \infty$$

then the sample mean and more generally the ratio estimator is always admissible for Y. (Joshi, 1969).

Theorem 3.12: For any s.d. p, the sample mean of distinct units is admissible in the class of all estimators (Joshi, 1965*a*).

The following theorem gives a class of admissible estimators in the class of all homogeneous linear estimators of Y.

Theorem 3.13: A linear estimator $e_2(s, \mathbf{y}) = \sum_{i \in s} b_{si} y_i$ with

$$b_{s_i} = 1 + \frac{k_i}{\sum_{i \in s} k_i q_i}\left(Q - \sum_{i \in s} q_i\right) \tag{3.5.13}$$

(where $(q_1, \ldots, q_N)$ is a set of non-zero constants, $Q = \Sigma q_i, (w_1, \ldots, w_N)$ is a set of positive numbers, $k_i = q_i / w_i$ is admissible in the class of all homogeneous linear estimators of Y. (Sengupta, 1980).

Proof: Let $e(s, \mathbf{y}) = \sum_{i \in s} b'_{si} y_i$ be an arbitrary linear estimator of Y.

$$MSE(e) = \sum y_i^2 \sum_{s \ni i} b'_{si} p(s) + \sum{}' y_i y_j \sum_{s \ni i, j} b'_{s_i} b'_{s_j} p(s)$$

$$- Y \sum y_i \sum_{s \ni i} b'_{s_i} p(s) + Y^2 \tag{3.5.14}$$

Let, $M_i = MSE(e)]\ \mathbf{y} = \mathbf{y}^{(i)}$,

where, $\mathbf{y}^{(i)} = (0, \ldots, 0, 1, 0, \ldots, 0)$ (3.5.15)

$$M_i = \sum_{s \ni i} b'^2_{s_i} p(s) - 2 \sum_{s \ni i} b'_{s_i} p(s) + 1$$

Minimization of $\sum M_i w_i$ w.r.t. b'_{s_i} subject to the condition $\text{MSE}(e)]_{\mathbf{q}} = 0$ where

$\mathbf{q} = (q_1, \ldots, q_N)$, *i.e.*

$$\sum_{i \in s} b'_{s_i} q_i = Q \quad \forall s : p(s) > 0 \tag{3.5.16}$$

gives $b'_{s_i} = b_{s_i}$, as given in (3.5.13).

Example 3.3: For $w_i = q_i = N^{-1}$, $e_2(s, \mathbf{y}) = N\overline{\mathbf{y}}_s$. Thus for any wor-s.d., $p, N\overline{y}_s$ is admissible in the class of all homogeneous linear estimators of Y (a more general result is in Theorem 3.9).

Example 3.4: For $w_i = q_i = x_i / X$ (size measure of unit i), $e_2(s, y) = \dfrac{\left(\sum_s y_i\right) X}{\left(\sum_s x_i\right)}$.

(This is a particular case of Theorem 3.10).

Example 3.5: Consider ppswor-design with $n = 2$. For $q_i = p_i$, $w_i = p_i^2/(1 - p_i)$, $s = (i, j)$, $e_2(s, \mathbf{y})$ reduces to Murthy's (1957) estimator e_M. Hence for ppswor-design (with $n = 2$), Murthy's estimator is admissible within the class of all linear estimators of Y.

Patel & Dharmadhikari (1978) proved the admissibility of e_M (for $n = 2$) within the class of all linear unbiased estimators. Sengupta (1980) proved the admissibility of e_M ($n = 2$) within the class of all unbiased estimators. Joshi (1970), Sengupta (1980) proved the admissibility of Murthy's (1963) estimator e'_M within the class of all unbiased estimators of Y.

Bellhouse & Joshi (1984) proved the admissibility of regression estimator

$$\hat{Y}_{lr}\left[= N\{\bar{y}_s + b(\bar{X} - \bar{x}_s)\},\ b = \sum_s (x_i - \bar{x}_s)(y_i - \bar{y}_s) \Big/ \sum_s (x_i - \bar{x}_s)^2\right]$$

for a single stage πps-sampling design and for double sampling.

3.6 MINIMAX STRATEGIES

Given a parameter space Ω, a strategy $H_m = (e_m, p_m)$ is said to be minimax in a class of strategies $\mathscr{H} = \{H : (p, e)\}$ if it minimises the maximum risk of strategies on Ω. Thus H_m should satisfy

$$\max_{\mathbf{y} \in \Omega} R(p_m, e_m; \mathbf{y}) = \min_{H \in \mathscr{H}} \max_{\mathbf{y} \in \Omega} R(p, e; \mathbf{y}) \quad \text{...(3.6.1)}$$

where $R(p, e; \mathbf{y}) = R(H; \mathbf{y})$ denotes the risk of the strategy H for the value of the parameter $\mathbf{y}$. In the case of non-existence of maximum on Ω, the maximum is to be replaced by the supremum. The corresponding risk is termed the minimax risk. If the sampling design (estimator) is kept fixed, (3.6.1) defines a minimax estimator (design). Minimax strategies depend very much on the parameter space Ω. In searching for a minimax strategy it is necessary that risk on Ω should be bounded. We recall a few definitions.

Let $\pi_1, \pi_2, \ldots, \pi_N$ be a permutations of integers 1, 2,..., N. We relable the units associating the label π_i to the unit originally labelled i. This change of label does not affect the variate value of a unit *i.e.*, with π_i is now associated the same variate value which was associated with the unit originally labelled i. If $\pi\mathbf{y}$ denotes the parameter vector obtained after relabelling, then

$$(\pi\mathbf{y})\pi_i = y_i \quad \Leftrightarrow \quad (\pi\mathbf{y})_i = \mathbf{y}_{\pi-1}(i) \quad \text{...(3.6.2)}$$

i.e., the π_i^{th} component of $\pi\mathbf{y}$ is identical with the ith component of $\mathbf{y}$. By relabelling, subsets s of labels is transformed into $\pi s = \{\pi_i : i \in s\}$. We denote by II the set of all $N!$ permutations.

Definition 3.4: A sampling design $\bar{p}$ is said to be symmetric if

$$\bar{p}(\pi s) = \bar{p}(s)\ \forall s : \bar{p}(s) > 0\ n\sqrt{N_j \bar{X}_j} \Big/ \sum \sqrt{N_k \bar{X}_k} \text{ and } \pi \in II.$$

This implies that $\bar{p}(s) = p(n(s)) \Big/ \binom{N}{n(s)}$ with $p(m) = \sum_{s \in \mathscr{S}_m} p(s)$, $\mathscr{S}_m$ being the collection of all samples of size m. If $p \in \rho_n$, it follows that $\bar{p}$ is a srswor design. p is the original design which is symmetrised to $\bar{p}$.

Definition 3.5: An estimator $\bar{e}(s, \mathbf{y})$ is termed symmetric if

$$\bar{e}(\pi s; \pi \mathbf{y}) = \bar{e}(s; \mathbf{y}) \ \forall \ s : p(s) > 0 \text{ and } \forall \ \pi \in \Pi.$$

From an arbitrary estimator e associated with a strategy (p, e) a symmetric estimator $\bar{e}_p$ can be constructed as follows (Royall, 1970*b*):

$$\bar{e}_p(s; \mathbf{y}) = \frac{\sum_{\pi} e(\pi s; \pi \mathbf{y}) p(\pi s)}{m!(N-m)!\, p(m)} \quad \text{if } n(s) = m \text{ and } p(m) > 0.$$

If $p(m) = 0$, $\bar{e}_p(s; \mathbf{y})$ may be assigned any arbitrary value.

Stenger (1979) considered the symmetrisation

$$\bar{e}_p(s; \mathbf{y}) = \sum_{\pi} e(\pi s; \pi \mathbf{y}) N!.$$

There are, however, other ways of constructing a symmetric estimator.

We speak of a random permutation model if all the permutations $\pi \mathbf{y}$ of the parameter $\mathbf{y}$ are equally probable. This is the case if the variate values on N units are considered as a random permutation of N values $(y_1, \ldots, y_N)$. The average risk of a strategy (p, e) overall the permutation $\pi \mathbf{y}$ is

$$\bar{R}(p, e; \mathbf{y}) = \sum_{s} R(p, e; \pi \mathbf{y}) / N! \tag{3.6.3}$$

Consider, now, the parameter space,

$$\Omega_x(\mathbf{y}) = \{\pi \mathbf{y} : \pi \in \Pi\} \tag{3.6.4}$$

Blackwell and Girshick (1954) showed that for any symmetric estimator $\bar{e}$,

$$\max_{\mathbf{y} \in \Omega_\pi(\mathbf{y})} R(p, \bar{e}; \mathbf{y}) \geq \max_{\mathbf{y} \in \Omega_\pi(\mathbf{y})} R(\bar{p}, \bar{e}; \mathbf{y}) \tag{3.6.5}$$

If $p \in \rho_n$, srswor is, therefore, a minimax design for any symmetric estimator provided $\mathbf{y} \in \Omega_\pi(\mathbf{y})$. Bellhouse (1984) also considered (3.6.5).

Using a symmetric sampling design $\bar{p}$, Royall (1970) proved that

$$\max_{\mathbf{y} \in \Omega_\pi(\mathbf{y})} R(\bar{p}, e; \mathbf{y}) \geq \max_{\mathbf{y} \in \Omega_\pi(\mathbf{y})} R(\bar{p}, \bar{e}; \mathbf{y}) \tag{3.6.6}$$

which justifies the use of symmetric estimator if the underlying sampling design is symmetric. Stenger (1979) proved that

$$\max_{\mathbf{y} \in \Omega_\pi(\mathbf{y})} R(p, e; \mathbf{y}) \geq \max_{\mathbf{y} \in \Omega_\pi(\mathbf{y})} R(\bar{p}, y_s; \mathbf{y}) \tag{3.6.7}$$

for all sampling designs p and for all linearly invariant estimators e (an estimator e is said to be linearly invariant if

$$e(s, \mathbf{y}) = \sum_{i} b_{si} y_i \text{ with } \sum_{i \in s} b_{si} = 1 \ \forall \ s : p(s) > 0)$$

(3.6.5)–(3.6.7) hold for any permutation-invariant parameter space and any convex loss function provided the risk is bounded.

Joshi (1979) showed that the strategy, H_0 = (srswor, $\bar{y}_s$) is the minimax strategy in the class of all unbiased strategies using $FS(n)$ designs if the parameter space $\Omega_1 = \{\mathbf{y} : y_i \in W, i = 1, ..., N\}$ where W is some set of real numbers, *i.e.*, Ω_1 is the N-fold product of the set W. Padmawar and Mukhopadhyay (1984) also used the same parametric space to obtain optimal strategies for symmetric functions of population parameters from which Stenger's result was obtained as a special case.

Works on the minimax solution of the population mean under srs also include Agarwal (1959), Chaudhuri (1969), Bickel and Lehmann (1981), Hodges and Lehmann (1982), among others.

Hodges and Lehmann (1982) showed that when $y_i \in [0, 1]$, $i = 1, ..., N$, a minimax estimator of $\bar{Y}$ under srswor is

$$\alpha\bar{y}_s + \beta$$

where,
$$\alpha = [1 + \sqrt{(N-n)/(N-1)n}]^{-1}, \ \beta = \frac{(1-\alpha)}{2}.$$

Bickel and Lehmann (1981) proved that under srswor, $\bar{y}_s$ is minimax for $\bar{Y}$ in the parametric space

$$\left\{\mathbf{y} : \sum_{i=1}^{N} (y_i - \bar{Y})^2 \le M, \text{for any arbitrary } M (> 0)\right\}$$

Scott and Smith (1975) were the first to consider the minimaxity of unequal probability sampling. We briefly discuss their results.

Suppose we want to estimate $\theta = \sum_{i=1}^{N} w_i y_i$ when $w_1, ..., w_N$ are arbitrary non-negative weights, $\sum_{j=1}^{N} w_j = 1$. A typical example is, y_i is the mean per element in the ith cluster and $w_i = M_i / M$, M_i being the number of elements in the ith cluster and $M = \sum M_i$, where θ is the population mean. For $n = 1$, if the unit i is sampled, a natural estimator of θ, in the absence of any prior knowledge to distinguish among the variate values, is $\hat{\theta} = y_i$. Let H_1 be the sampling strategy using the design which selects the unit i with probability $q_i = w_i$ and estimator $\hat{\theta}$. Let the parameter space be

$$\Omega_B = \{\mathbf{y} : 0 \le y_i \le B, \ i = 1, ..., N\} \tag{3.6.8}$$

where B is a pre-assigned known constant.

Lemma 3.2: Suppose there exists a subset $K_0 \subset \mathscr{P}$ such that $\sum_{i \in K_0} w_i = \frac{1}{2}$. Then for any sampling strategy using the estimator $\hat{\theta}$

$$\sup_{\mathbf{y}\in\Omega_B} R(H_1, \mathbf{y}) \le \sup_{\mathbf{y}\in\Omega_B} R(H:\mathbf{y})$$

i.e., $$\sup_{\mathbf{y}\in\Omega_B} \sum w_i\left(y_i - \sum w_j y_j\right)^2 \le \sup_{\mathbf{y}\in\Omega_B} \sum \left(y_i - \sum w_j y_j\right)^2 q_i$$

where $0 < q_i < 1$ $(i = 1, ..., N), \sum q_i = 1$.

Proof: Consider a point $\mathbf{y} = \mathbf{y}^* = (y_1^*, \ldots, y_N^*)$ such that $y_1^* = B(0)$ if $i \in (\notin) K_0$. Then

$$R(H; \mathbf{y}^*) = \sum q_i\left(y_i^* - \sum w_j y_j^*\right)^2$$

$$= B^2 \sum_{i\in K_0} q_i\left(\sum_{j\notin k_0} w_j\right)^2 + B^2 \sum_{i\notin K_0}\left(1 - \sum_{i\notin k_0} w_j\right)^2$$

$$= \frac{B^2}{4}.$$

Hence, $$\sup_{\mathbf{y}\in\Omega_B} R(H:\mathbf{y}) \ge \frac{B^2}{4}.$$

Consider now H_1. Since $R(H_1; \mathbf{y})$ is a non-negative definite quadratic form in $\mathbf{y}$ its maximum value in Ω_B is attained at some boundary point. For any subset $K \subset \mathscr{P}$, consider, therefore, a boundary point $\mathbf{y} = \mathbf{y}^0 = (y_1^0, \ldots, y_N^0)$ such that $y_i^0 = B(0)$ if $i \in (\notin)$ K. Then

$$R(H_1; \mathbf{y}^0) = B^2\left(\sum_{i\in K} w_i\right)\left(\sum_{j\notin K} w_j\right)$$

Hence, $$\max_{\mathbf{y}\in\Omega_B} R(H_1, \mathbf{y}) = \max_{K\subset\mathscr{S}} R(H_1, \mathbf{y}^0) = \frac{B^2}{4}$$

Hence the proof.

Replacing in θ, y_i by $y_i X/x_i$, w_i by x_i/X, one gets

Theorem 3.14: Suppose there exists a subset $K_0 \subset \mathscr{P}$ such that

$$\sum_{i\in K_0} x_i = \frac{X}{2} \qquad \ldots(3.6.9)$$

then for $n = 1$ the (Hansen-Hurwitz) HH-strategy is minimax in the class of all strategies using the estimator $y_i X/x_i$ in the sense that

$$\sup_{\mathbf{y}\in\Omega_B} \sum x_i (y_i - Y)^2 / x \le \sup_{y\in\Omega_B} \sum q_i (y_i - Y)^2$$

where $q_i \in [0, 1]$ $(i = 1, ..., N)$, $\Sigma q_i = 1$ and Ω_B is given by (3.6.8) with B replaced by Bx_i / X.

If the relation (3.6.9) is approximately satisfied, then HH-strategy is approximately minimax under the condition of the theorem. The result extends to any n.

For an excellent review of the literature on minimax strategies the reader is referred to Gabler (1990).

3.7 AVERAGE VARIANCE OF A STRATEGY UNDER A SUPERPOPULATION MODEL

3.7.1 Superpopulation Model

Assume that the value of y on i is a particular realization of a random variable $Y_i(i = 1, ..., N)$. Hence the value $\mathbf{y}$ of a survey population $\mathcal{P}$ may be looked upon as a particular realization of a random vector $\mathbf{Y} = (Y_1, ..., Y_N)$ having a superpopulation distribution ξ_θ, indexed by a parameter vector θ, $\theta \in \Theta$ (the parameter space). The class of priors, $(\xi_\theta, \theta \in \Theta)$ is called a superpopulation model. The model ξ for $\mathbf{Y}$ is obtained through one's prior belief about $\mathbf{Y}$. As an example, in agricultural survey of yield of crops in N plots, if acreages $x_1, ..., x_N$ under the crop on these plots are assumed fixed over years, one may assume that the yield $\mathbf{y}$ in a particular year is a random sample from a prior distribution ξ of $\mathbf{Y}$, which may depend, among others, on $x_1, ..., x_N$. In particular, one may assume that ξ, the joint p.d.f. of $\mathbf{Y}$ is such that $Y_1, ..., Y_N$ are independent with

$$\mathscr{E}(Y_i \mid x_i) = \beta x_i$$

$$\mathscr{V}(Y_i \mid x_i) = \sigma_i^2, \ i = 1, ..., N \tag{3.7.1}$$

where β, σ_i^2 (> 0) are constants. Here and subsequently $\mathscr{E}$, $\mathscr{V}$, $\mathscr{C}$ will denote expectation, variance, covariance wrt ξ.

The model ξ underlying in (3.7.1) is a regression model with zero intercept and one regressor variable. Superpopulation models of general use are transformation models, multiple regression models, polynomial models, exchangeable models, random permutation models. For a review of different superpopulation models the reader may refer to Cassel *et al.* (1977).

The use of an appropriate superpopulation model distribution in survey sampling is justified by the fact that in surveys of usual interest (agricultural surveys, industrial surveys, cost of living surveys, employment and unemployment surveys, traffic surveys, etc.), a $\mathbf{y}$ cannot take any value in R^N, but takes values only in a particular domain in R_N, some with higher probability. One may, therefore, postulate some reasonable superpopulation model ξ for $\mathbf{Y}$ and exploit ξ to produce suitable sampling strategies.

A good deal of inference in survey sampling emerges from the postulation of a suitable prior ξ for **Y** and methodologies have been developed to produce optimal sampling strategies based on ξ, *e.g.*, by using prediction-theoretic approach in survey sampling. The early uses of ξ are due to Cochran (1939, 1946), Yates (1949), Godambe (1955), Rao, Hartley & Cochran (1962), among others.

3.7.2 Average Variance Under ξ

Since in most cases the expressions for the variances of different strategies are complicated in nature and are not amenable to comparison, one may take the average values of variance under an assumed appropriate superpopulation model ξ and compare their average variances. In this section we consider ξ only for getting the average variance of a strategy.

The average variance (AV) of an unbiased strategy (p, e) under ξ is given by $\mathscr{E}\, V_p(e)$. A strategy H_1 will be better than an unbiased strategy $H_2 (H_1 \succ H_2)$ if $\text{AV}(H_1) < \text{AV}(H_2)$.

Lemma 3.3: For any strategy (p, e) for estimating, Y,

$$\mathscr{E}\{V(p, e)\} = E\{\mathscr{V}(e)\} + E\{\mathscr{B}(e)\}^2 + \mathscr{V}(Y) - 2\mathscr{E}\{(Y - \mu)E(e - \mu)\}$$

where $\mathscr{B}(e) = \mathscr{E}(e - Y)$ and $\mu = \mathscr{E}(Y)$. In particular, if (p, e) is unbiased for Y,

$$\mathscr{E}\{V(p, e)\} = E\{\mathscr{V}(e)\} + E\{\mathscr{B}(e)\}^2 - \mathscr{V}(Y).$$

The following theorem due to Godambe & Joshi (1965) shows that there exists a lower bound to the average variance of p-unbiased strategy under a very general superpopulation model ξ.

Theorem 3.15: Consider model $\xi : Y_1, \ldots, Y_N$ are independent with $\mathscr{E}(Y_i) = \mu_i$, $\mathscr{V}(Y_i) = \sigma_i^2$ $(i = 1, \ldots, N)$. For any unbiased sampling strategy (p, e), with value of first order inclusion probability π_i,

$$\mathscr{E}\, V(p, e) \geq \sum \sigma_i^2 \left(\frac{1}{\pi_i} - 1 \right) \tag{3.7.2}$$

Proof:

$$e(s, \mathbf{y}) = e_{HT}(s, \mathbf{y}) + h(s, \mathbf{y}) \tag{3.7.3}$$

where $h(s, \mathbf{y}) = e(s, \mathbf{y}) - e_{HT}(s, \mathbf{y})$; $h(s, \mathbf{y})$ depends on $\mathscr{S} \times R_N$ only through those y_i for which $i \in s$. By unbiasedness of $e(s, \mathbf{y})$,

$$\sum_{s \in \mathscr{S}} h(s, \mathbf{y}) p(s) = 0 \;\forall\; \mathbf{y} \tag{3.7.4}$$

Taking model-variance of both sides of (3.7.3), considering the random vector **Y** in place of **y**,

$$\mathscr{V}(e(s, \mathbf{Y})) = \mathscr{V}\{e_{HT}(s, \mathbf{y})\} + \mathscr{V}\{h(s, \mathbf{Y})\} + 2\mathscr{C}\{e_{HT}(s, \mathbf{Y}), h(s, \mathbf{Y})\}$$

Hence, $$E\mathcal{V}(e) = E\mathcal{V}(e_{HT}) + E\mathcal{V}(h(s, \mathbf{Y})) + 2E\mathcal{V}\{e_{HT}, h(s, \mathbf{Y})\} \tag{3.7.5}$$

Now, $$E\mathscr{C}\{e_{HT}, h(s, \mathbf{Y})\} = E\left\{\sum_{i\in s} \frac{Y_i - \varepsilon(Y_i)}{\pi_i} h(s, \mathbf{Y})\right\}$$

$$= \sum_{s\in\mathscr{S}} \mathscr{E}\left\{\frac{Y_i - \varepsilon(Y_i)}{\pi_i} \sum_{s\ni i} h(s, \mathbf{Y})p(s)\right\}$$

$$= -\sum_{s\in\mathscr{S}} \mathscr{E}\left\{\frac{Y_i - \varepsilon(Y_i)}{\pi_i} \sum_{s\not\ni i} h(s, \mathbf{Y})p(s)\right\} = 0$$

$\left(\text{since by (3.7.4.), } \sum_{s\ni i} h(s, \mathbf{y})p(s) + \sum_{s\not\ni i} h(s, \mathbf{y})p(s) = 0\right)$, $\frac{Y_i - \varepsilon(Y_i)}{\pi_i}$ and $\sum_{s\not\ni i} h(s, \mathbf{Y})p(s)$ being independently distributed. Hence, from (3.7.5),

$$E\mathcal{V}(e) \ge E\mathcal{V}(e_{HT}) = \sum \frac{\sigma_i^2}{\pi_i} \tag{3.7.6}$$

Hence, by Lemma 3.3

$$EV(e) \ge E\mathcal{V}(e) - \mathcal{V}(Y) \ge \sum \sigma_i^2\left(\frac{1}{\pi_i} - 1\right). \tag{3.7.7}$$

Corollary 3.5: The lower bound (3.7.7) is attained by e_{HT} applied to a $FS(n)$-design with $\pi_i \alpha\, \mathscr{E}(Y_i)$, $i = 1, \ldots, N$. In particular, if $\mu_i = \beta x_i$ (β a constant), any FS-πps design applied to e_{HT} attains the lower bound in (3.7.7).

Corollary 3.6: The lower bound in (3.7.7) is the minimum attainable value of average variance under ξ (in Theorem 3.15) in the class of all unbiased strategies using $p(\in \rho_n)$ with a fixed set of π_i-values. This bound is minimized by a $p(\in \rho_n)$ with $\pi_i \alpha \sigma_i$ when the minimum attainable value of $\mathscr{E}\,V(e)$ is

$$\frac{\left(\sum \sigma_i\right)^2}{n} - \sum \sigma_i^2. \tag{3.7.8}$$

Godambe (1955, 1982), Hájek (1959), Cassel *et al.* (1976), Tam (1984), Mukherjee & Sengupta (1989) considered optimal strategies (in the sense of minimum average variance) in the class of all linear unbiased estimators applied to $p \in \rho_n$.

Let L_u denote the class of linear unbiased estimators and H_n the class of strategies (p, e), $p \in \rho_n$, $e \in L_u$.

Theorem 3.16: Consider model ξ:

$$\mathscr{E}(Y_i) = \alpha_i + \beta x_i$$
$$\mathscr{V}(Y_i) = \sigma^2 f_i, \quad i = 1, \ldots, N \tag{3.7.9}$$
$$\mathscr{C}(Y_i, Y_j) = \rho\sigma^2 \sqrt{f_i f_j}\,, \ i \neq j = 1, \ldots, N$$

where β_i, α_i's are assumed known constants, f_i a known function of x_i, ρ, σ^2 are unknown $(-1/(N-1) \leq \rho \leq 1)$.

For any $p \in \rho_n$, $e \in Lu$,

$$\mathscr{E}E(e - Y)^2 \geq (1-\rho)\left\{\frac{1}{n}\left(\sum \sqrt{f_i}\right)^2 - \sum f_i\right\}\sigma^2 \tag{3.7.10}$$

$$= \mathscr{E}E_{p_0}(e_{GD}^0 - Y)^2$$

where $p_0 \in \rho_n$ is such that $\pi_i(p_0) = \dfrac{n\sqrt{f_i}}{\sum \sqrt{f_k}} = \pi_{i0}$.

$$e_{GD}^0 = \sum_{i \in s} \frac{Y_i - \alpha_i - \beta x_i}{\pi_i 0} + \sum (\alpha_i + \beta x_i) = e_0 \text{ (say)}.$$

Further, a pair (e_1, p_1) is optimal *iff* $(e_1, p_1) = (e_0, p_0)$. (Tam, 1984)

Proof: The inequality in (3.7.10) follows by standard calculations.

It follows that $\mathscr{E}E_{p_0}(e_{GD}^0 - Y)^2$ has the expression given in (3.7.10). Hence (p_0, e_{GD}^0) is optimal in H_n. To prove the necessity, suppose $\mathscr{E}E_{p_1}(e_1 - Y)^2 = \varepsilon E_{p_0}(e_0 - Y)^2$. It follows that $p_1 = p_0$. Denoting the expression in the middle of (3.7.10) as K, and applying the inequality in (3.7.10) to the estimator $(e_1 + e_0)/2$,

$$\mathscr{E}E_{p_0}\{(e_1 - Y)(e_0 - Y)\} \geq K.$$

Now,

$$E_{p_0}\mathscr{E}(e_1 - e_0)^2 = 2K - 2E_{p_0}\mathscr{E}(e_1 - Y)(e_0 - Y) \leq 0,$$

implying $\qquad e_1 = e_0.$

Godambe (1955, 1982) considered the model ξ in (3.7.10) for $\rho = 0$. Hájek (1959) considered for $\alpha_i = 0$, $\rho = 0$ and $\alpha_i = 0, f_i = x_i^2$. Cassel *et al.* (1976) considered $f_i = x_i^2$.

Mukherjee & Sengupta (1989) proved the optimality results under a more general model. We state their results in the following:

Theorem 3.17: Consider model ξ,

$$\mathscr{E}(Y_i) = \mu_i, \ i = 1, \ldots, N$$
$$\mathscr{C}(Y_i, Y_j) = \nu_{ij}, \ i \neq j = 1, \ldots, N \tag{3.7.11}$$

For a given $p \in \rho_n$, under model ξ,

$$\mathscr{E}E(e - Y)^2 \geq \mathbf{1}'\Phi^{-1}\mathbf{1} - \mathbf{1}'\mathbf{V}\mathbf{1} \quad (3.7.12)$$

for every $e = b_s + \sum_{i \in s} b_{s_i} y_i \in Lu$ with equality *iff* $e = e^*$. Here $\Phi = ((\Phi_{ij}))_{N\times N}$ when $\Phi_{ij} = \sum_{s \ni (i,j)} v_s^{ij} p(s)$, $V_s^{-1} = ((v_s^{ij}))$, V_s being $n \times n$ submatrix of $V = ((v_{ij}))$, obtained by considering the units in s; $\mathbf{1} = (1, \ldots, 1)'_{N\times 1}$, e^* is such that $b_s = \mu - \sum_{i \in s} b_{s_i}\mu_i \ \forall s : p(s) > 0$ and

$$\mathbf{b}_s = \mathbf{b}_s^* = V_s^{-1}\lambda_s$$

where $\mathbf{b}_s = (b_{s_i}; i \in s)$, a $n \times 1$ vector, λs is a $n \times 1$ subvector of λ given by the elements $i \in s$, $\lambda = \Phi^{-1}\,\mathbf{1}$ and $\mu = \sum \mu_i$.

Further, a strategy (p, e) is optimal in H_n provided $(p, e) = (p^*, e^*)$ where p^* is a s.d. which minimizes $\mathbf{1}'\Phi^{-1}\,\mathbf{1}$ w.r.t. $p \in \rho_n$.

In particular, when $v_{ij} = \rho\sqrt{v_{ii}v_{jj}}\,(1 \leq i \neq j \leq N)$, a strategy (p, e) is optimal (in the minimum average variance sense) *iff* $\pi_i(p) = \pi_{i_0}$ for every i and

$$e = \sum_s \frac{(Y_i - \mu_i)}{\pi_{i_0}} + \mu \quad (3.7.13)$$

where $\pi_{i_0} = \dfrac{n\sqrt{v_{ii}}}{\sum \sqrt{v_{ii}}}$.

Tam's results follow as particular cases of Theorem 3.17.

Remark 3.3: For $f_i = x_i^2, e_{GD}^0$ in (3.7.10) becomes $e_\alpha = X\left[\sum_s \frac{y_i - \alpha_i}{nx_i}\right] + \sum \alpha_i$ (which is free from β) and p_0 becomes a πps-design. If $f_i \neq x_i^2$ and β is unknown, use of $(e_\alpha, \pi ps)$ in place of (p_0, e_{GD}^0) results in loss of efficiency.

$$E_{\pi ps}\mathscr{E}(e_\alpha - Y)^2 - E_{p_0}\mathscr{E}(e_{GD}^0 - Y)^2$$

$$= (1-\rho)\times\left[\sum \frac{f_i}{nx_i}\right]\sigma^2 - \frac{\left(\sum\sqrt{f_i}\right)^2}{n}\{1 + (n-1)\rho\}\sigma^2 + \delta \quad (3.7.14)$$

where, $\delta = \rho\dfrac{X^2}{n^2}E_{p_0}\left[\sum \pi_i\sqrt{f_i}/x_i\right]\sigma^2 \leq \rho\dfrac{X^2}{n^2}\left\{\sum\sqrt{\dfrac{f_i}{x_i}}\right\}^2\sigma^2$.

One should decide, whether for the particular problem the requirement of p-unbiasedness warrants the loss of efficiency and if not consider constructing the asymptotically design unbiased (ADU) estimators (Brewer (1979), Isaki & Fuller (1982)).

3.8 COMPARISON AMONG SEVERAL STRATEGIES

As noted in Section 3.7, variances of strategies are not in general amenable for comparison and as such recourse is made to compare the average variances under an assumed appropriate superpopulation model.

Considering the superpopulation model

$$Y_i = \beta x_i + e_i,$$
$$\mathscr{E}(e_i) = 0,\ \mathscr{V}(e_i) = \sigma^2 x_i^g,\ \mathscr{C}(e_i, e'_i) = 0,\ i \neq i', \quad (3.8.1)$$

appropriate for the situation where y and x are highly correlated, with $g \in [0, 2]$ which occurs in a large class of socio-economic surveys [Scott *el al.* (1978)], Rao (1966*b*) compared the performance of the following strategies w.r.t. their average variances—Hansen-Hurwitz strategy: H_1, $(\pi ps, e_{HT})$: H_2, Rao-Hartley-Cochran strategy: H_3, Murthy's (1957) strategy for $n = 2$: H_4, Midzuno–Lahiri–Sen strategy: H_5, all the strategies involving the same $\{n, p_i\}$. It is concluded that for $g = 1$, H_4 is uniformly best among all these strategies when H_2, H_3, H_5 are equally efficient. $H_2 \succ (\prec)\ H_3$ and H_5 if $g > (<) 1$ (also Hanurav (1967), Rao (1967)); $H_2 \succ H_1\ \forall\ g$ and $H_5 \succ H_1$ for $g \leq 1$; for large N, assuming $\max_i p_i$ is $O(N^{-1}) H_4 \succ H_2 \succ H_3, H_2 \succ H_5$ for $1 < g < 2$ and for $g = 2$, H_2 and H_4 are equally efficient; for large n, $H_5 \succ (\prec)\ H_3$ if $g < (>) 1$. However, it follows from Theorem 3.14 that for $g = 2$, H_2 is better than all the other strategies considered. Chaudhuri and Arnab (1979) showed by exact calculations that $H_2 \succeq (\preceq)\ H_3 \succeq (\preceq)\ H_5$ for $g \geq (\leq) 1$.

In large scale sample surveys, where often one has to estimate the population totals (means) of several characteristics, a study variable may be unrelated (or poorly related) to the auxiliary variable x used in drawing the sample. In these situations, where x and y are poorly correlated, Rao (1966*a*) proposed alternative estimators, obtained by replacing y by $y_i p_i$, for use in connection with the above mentioned five sampling strategies and compared their performance with the corresponding usual strategies under the model:

$$Y_i = \mu + e_i,\ \mathscr{E}(e_i) = 0,$$
$$\mathscr{V}(e_i) = \sigma^2,\ \mathscr{C}(e_i, e_{i'}) = 0,\ i \neq i'. \quad (3.8.2)$$

appropriate for such situations. Denoting the alternative strategy corresponding to H_i as H_i^*, it was found $H_i^* \succ H_i$, $i = 1, 2, 3, 5$ under (3.8.2) (for H_4, the problem remains unresolved) (for H_2, approximate expressions for variances using Hartley-Rao scheme (1962) were used). Rao (1966*b*) showed that under (3.8.2) $H_2^* \succ H_4^* \succ H_3^* \succ (H_1^*, H_5^*)$ (no clear cut comparison between H_1^*, H_5^* being possible) and for large N, assuming $\max_i p_i$ is $O(N^{-1})$, H_2^* (equally efficient as H_4^*) $\succ H_3^* \succ (H_1^*, H_5^*)$ to $O(N^{-1})$. Further works in this area are due to Bansal and Singh (1985), Amahia *et al.* (1989).

Foreman and Brewer (1971) considered the model

$$Y_i = \alpha + \beta x_i + e_i,\ \mathscr{E}(e_i) = 0,\ \mathscr{V}(e_i) = \sigma^2 x_i^g,$$
$$\mathscr{C}(e_i, e_{i'}) = 0,\ i \neq i' \quad (3.8.3)$$

for comparing the strategies: (srswr, $\bar{y}_S$): H_6, (srswor $\bar{y}_s$): H_7, (srswr, $e_{R(S)}$): H_8, (srswor, e_R) : H_9 and H_1, H_2 w.r.t. their average variances where e_R denotes a ratio estimator. They showed that for large N, $H_8 \succ H_6$ and $H_9 \succ H_7$ provided $|\alpha| < |\beta\bar{X}|$; $H_1 \succ H_8$ and $H_2 \succ H_9$ if $g > 1$. If N is large and $\alpha = 0$, $H_2 \succ H_1$ to term $O(N^1)$ when $V_1 - V_2 = V_8 - V_9 = (n-1)\sigma^2 \, \Sigma x_i^g / n$, *i.e.*, the absolute reduction in expected variance in passing from swr to swor when sampling with pps is the same as when sampling with equal probability and ratio estimation.

Padmawar (1981) compared the strategies H_2, H_3 and (Midzuno-Sen design, e_{HT}): H_{10} under the model (7.1). He showed that a sufficient condition for $H_2 \succ H_{10}$ for a given value of $g(\geq 1)$ is

$$n \geq n_0 = \frac{N(2-g)p+(g-1)}{(2-g)p+(g-1)}$$

where $p = \max_i p_i$. Alternatively, for a given value of n such that $np \leq 1$, $H_2 \succ H_0$ if g satisfies

$$g \geq g_0 = 1 + p \Big/ \left(p + \frac{n-1}{N-n}\right).$$

Both these conditions are improvement over the conditions attained by Rao (1967). Further, sufficient conditions for $H_3 \succ H_{10}$, when $g > 1$ and N/n is an integer, is

$$n \geq n_1 = N - \frac{(N-1)(g-1-gp)}{(g-1)(1-p)^2}$$

3.9 SOME ASYMPTOTIC RESULTS

Consider the following framework of asymptotics, considered by Isaki and Fuller (1982), Robinson and Sarndal (1983) (vide Sections 7.6). Assume that there are M nested populations, $\mathcal{P}_1, \mathcal{P}_2, \ldots, \mathcal{P}_M$ with $\mathcal{P}_1 \subseteq \mathcal{P}_2 \subseteq \ldots \mathcal{P}_M$, $\mathcal{P}_t$ being of size N_t, $N_1 < N_2 < \ldots$ A sample s_t of size n_t is selected from $\mathcal{P}_t$ following the s.d. p_t with inclusion probabilities $\pi_{kt}, \pi_{kk't}$, etc; $n_1 < n_2 < \ldots$; however, samples are not nested, *i.e.*, s_t is not necessarily a subset of s_{t+1}. With population $\mathcal{P}_t$ there is associated a vector $y_t = (y_{1t}, \ldots, y_{N_t t})$ of y-values, $\bar{Y}_t = \sum_1^{N_t} y_{jt}/N_t$.

Assume that p_t is such that

$$0 < \lambda_2 < \pi_{it} < \lambda_1 \;\; \forall i \text{ and } t. \tag{3.9.1}$$

An estimator e is asymptotically design consistent (ADC) for $\bar{Y}_t$ if

$$\lim_{t\to\infty} P\{|\varepsilon_t - \bar{Y}_t| > \varepsilon\} = 0 \tag{3.9.2}$$

for every $\varepsilon(> 0)$, 'P' being in the sense of sampling design. The Theorems 3.18 through 3.22 are due to Isaki and Fuller.

Theorem 3.18: Under conditions,

(*i*) $(\pi_{it}\pi_{jt} - \pi_{ijt})/\pi_{it}\pi_{jt} \le \alpha_t$ (3.9.3.1)

with $\alpha_t \to 0$ as $t \to \infty$

$$(ii)\ \sum \left(\frac{y_j}{\pi_{jt}} - \frac{Y_t}{n_t}\right)^2 < A < \infty\ \forall\ t \tag{3.9.3.2}$$

e_{HT} is ADC for $\bar{Y}$.

Proof:

$$V(e_{HT(t)}) = V_t \text{ (say)} = \frac{1}{2N_t^2}\sum\sum_{i \ne j}\left(\frac{y_i}{\pi_{it}} - \frac{y_i}{\pi_{jt}}\right)^2 (\pi_{it}\pi_{jt} - \pi_{ijt})$$

$$\le \frac{\alpha_t}{2N_t^2}\sum_{i \ne j}\sum\left(\frac{y_i}{\pi_{it}} - \frac{y_i}{\pi_{jt}}\right)^2 \pi_{it}\pi_{jt} \tag{3.9.4}$$

$$= \frac{n_t\alpha_t}{2N_t^2}\sum\left(\frac{y_i}{\pi_{it}} - \frac{Y_i}{n_t}\right)^2$$

$\to 0$ as $t \to \infty$ under (3.9.3.1) and (3.9.3.2). Also, $e_{HT(t)} - \bar{Y}_t = O_p\left(\sqrt{\alpha_t}\right)$.

For srs, (3.9.3.1) holds.

Theorem 3.19: Let,

$$\left|\frac{y_i}{\pi_{jt}}\right| < A < \infty\ \forall\ j \text{ and } t. \tag{3.9.5.1}$$

also, $g_{ijt} = (\pi_{ii}\pi_{jt} - \pi_{ijt})[0]$ if $\pi_{it}\pi_{jt} - \pi_{ijt} \ge 0$ [otherwise] (3.9.5.2)

Assume also,

$$N_t^{-2}\sum\sum_{i \ne j} g_{ijt} = h_t \text{ (say)} = O(n_t^{-2\delta}),\ \delta \text{ (a fixed number)} > 0. \tag{3.9.5.3}$$

Then e_{HT} is consistent and $e_{HT} - \bar{Y}_t = O_p(n_t^{-\delta})$.

Proof:

$$V_t \le \frac{1}{2N_i^2}\sum\sum_{i \ne j} g_{ijt}\left(\frac{y_i}{\pi_{it}} - \frac{y_j}{\pi_{jt}}\right)^2$$

$$\leq \frac{2A^2}{N_i^2} \sum\sum_{i \neq j} g_{ijt} = 2A^2 O(n_t^{-\delta})$$

$\rightarrow 0$ and $t \rightarrow \infty$

Note: A design satisfying (3.9.5.3) is as follows. Consider a design with $\pi_{it} = n_t/N_t = 1/K$ (say) with K an integer. Divide $\mathcal{P}_t$ into n_t, strata of K adjacent units each and draw a sample of size one at random from each stratum. Thus $\pi_{ijt} = 0$ if i and j belong to the same stratum and is K^{-2} otherwise. Hence $h_t = (K-1)/N_t K^2 = O(n_t^{-1})$. For designs satisfying (3.9.5.3), condition (3.9.5.2) can be relaxed as in the following theorem.

Theorem 3.20: Consider sequence of stratified samples $\{s_t\}$ as at note above. Under conditions,

(i) $N_t^{-2} \sum\sum_{i \neq j} g_{ijt}^r = h_t^{(r)}$ (say) $= O(n_t^{-2\delta r})$

and

(ii) $N_t^{-1} \sum \left(\frac{y_i}{\pi_{it}} - \frac{Y_t}{n_t} \right)^{2k} < A < \infty,$

$\forall t$, when $r, k \geq 1$, $\frac{1}{r} + \frac{1}{k} = 1$, e_{HT} is ADC. Further, $e_{HT} - \bar{Y}_t = O_p(n_t^{-\delta})$.

Proof:

$$V_y \leq \frac{1}{2} N_t^{-2} \sum\sum_{i \neq j} g_{ijt} \left(\frac{y_i}{\pi_{it}} - \frac{y_i}{\pi_{jt}} \right)^2$$

$$\leq \left(\frac{1}{2} N_t^{-2} \sum\sum_{i \neq j} g_{ijt}^r \right)^{1/r} \left(\frac{1}{2} N_t^{-2} \sum\sum_{i \neq j} \left(\frac{y_i}{\pi_{it}} - \frac{y_j}{\pi_{jt}} \right)^{2k} \right)^{1/k}$$

(By Holder's inequality)

$$\leq \left(N_t^{-2} \sum\sum_{i \neq j} g_{ijt}^r \right)^{1/r} \left(\frac{4^k}{N_t} \sum \left(\frac{y_i}{\pi_{it}} - \frac{y_j}{\pi_{jt}} \right)^{2k} \right)^{1/k}$$

$$= O(n_t^{-2\delta})$$

For srs, $\quad h_t^{(r)} = O(n_t^{-r})$.

The following theorem demonstrates a sequence of design such that V_t decreases at the rate of n_t^{-1}.

Theorem 3.21: If $\lim_{t\to\infty} \bar{Y}_t = \bar{Y}$, $\lim_{t\to\infty} \sigma^2_{yt} = \sigma^2_y$, where $\sigma^2_{yt} = N_t^{-1}\sum (y_i - \bar{Y}_t)^2$ and π_{it}'s satisfy (3.9.1), there exists a sequence of designs such that $V_t = O(n_t^{-1})$.

Proof:

Isaki and Fuller (1981) considered the following s.d. Arrange the population units into natural order and form the cumulative sum of probabilities $T_{it} = \sum_1^i \pi_{jt}$. Divide the population into n_t strata by placing the boundary points at $T_{it} = 1, 2, ..., n_t$, when a unit may fall in two adjacent strata. One element is selected from each stratum using Fuller's (1970) sampling scheme with inclusion-probability π_{it} and such that the joint probability of including an element i that is completely within stratum q and an element j that is completely within a stratum at least $\lambda_1(1-\lambda_1)^{-1}$ strata distant from q is $\pi_{ijt} = \pi_{it}\,\pi_{jt}$.

Since $\pi_{ijt} > \lambda_2 \forall i \neq j$, for each element, there are at most $\lambda_2^{-1}\left[\frac{2\lambda_1}{1} - \lambda_1 + 1\right]$ elements for which $\pi_{it}\,\pi_{jt} - \pi_{ijt} \neq 0$. Again $|\,\pi_{it}\,\pi_{jt} - \pi_{ijt}\,| < \lambda_1$. Hence from (3.9.4),

$$V_t \leq 4\lambda_1\lambda_2^{-3}[2\lambda_1(1-\lambda_1)^{-1} + 1]N_t^{-1}(\sigma^2_{yt} + \bar{Y}_t^2) = O(n_t^{-1}).$$

A central limit theorem exists for such types of sampling designs.

Theorem 3.22: Let the assumptions of Theorem 3.20 hold and let

$$\lim_{t\to\infty} \frac{1}{N_t}\sum (y_i - \bar{Y}_t)^{2+\delta} = B \text{ for some } \delta > 0.$$

Let the sequences of s.d.'s be such that $V_t = O(n_t^{-1})$. Then $(e_{HT(t)} - \bar{Y}_t)/\sqrt{V_t^L}$ $\to N(0, 1)$.

Robinson (1982) proved that under conditions, both $N, n \to \infty$ (with $n < N$), (*i*) y_i's are bounded or (*ii*) $\lim_{N\to\infty} N^{-1}\sum y_i^2 < \infty$, e_{HT} is mean-square consistent for s.d.'s satisfying $\pi_{ij} \leq \pi_i\pi_j \forall i \neq j$.

❑ ❑ ❑

CHAPTER 4

Optimal Estimation of a Finite Population Total Under Regression Superpopulation Models

4.1 INTRODUCTION

In Section 3.7, we considered the concept of superpopulation models in finite population sampling and introduced its uses in finding the average variance of a strategy for comparison among several strategies. The approach there was necessarily based on a fixed population model, as we had to confine to strategies having some desirable properties, *e.g.*, unbiasedness, admissibility, sufficiency, -suggested by a fixed population set up. In all such cases superpopulation models were imposed at a later stage to give us an idea about the relative performance of strategies appropriate to the models.

Brewer (1963), Royall (1970*a*, 1976), Royall and Herson (1973) and their co-workers considered the survey population as a random sample from a superpopulation and attempted to draw inference about the population parameter from a prediction theorist's view point.

Assume that $\mathbf{y} = (y_1, \ldots, y_N)$ is a particular realisation of a random vector $\mathbf{Y} = (Y_1, \ldots, Y_N)$ having a joint density ξ_θ indexed by a parameter vector $\theta = (\theta_1, \ldots, \theta_k)$ $\theta \in \Theta$, parameter space. As noted before, ξ_θ may belong to a class of distributions $C = \{\xi_\theta\}$. C is called a superpopulation model. Given a sample s and the corresponding y-values, the problem here is that of drawing inference about the total, mean, etc. using the prior information as summarised in ξ_θ, estimating any unknown component of θ in the process. Looked in this way, $Y, \bar{Y}$ are random variables and not fixed quantities. One has, therefore, to predict $Y, \bar{Y}$ etc. on the basis of the data and ξ, *i.e.*, to estimate the model-expected values, $\mathscr{E}(Y)$, $\mathscr{E}(\bar{Y})$, etc. We shall use the same notation to denote the population total, mean etc. even when they are random variables.

4.2 PRINCIPLES OF INFERENCE BASED ON THEORY OF PREDICTION

We assume throughout that there is a superpopulation distribution ξ of $\mathbf{Y}$. Let $\hat{T}_s$ denote a predictor of Y or $\bar{Y}$ (the specific one being clear from the context) based on s.

Definition 4.1: $\hat{T}_s$ is model-unbiased or ξ-unbiased or m-unbiased predictor of $\bar{Y}$ if

$$\mathscr{E}(\hat{T}_s) = \mathscr{E}(\bar{Y}) = \bar{\mu} \text{ (say) } \forall\ \theta \in \Theta \text{ and } \forall s : p(s) > 0 \tag{4.2.1}$$

As before, $\mathscr{E}$, $\mathscr{V}$ and $\mathscr{C}$ will denote respectively the expectation, variance and covariance with respect to superpopulation distribution ξ.

Definition 4.2: $\hat{T}_s$ is design-model unbiased (or $p\xi$-unbiased or pm-unbiased) predictor of $\bar{Y}$ if

$$E\mathscr{E}\ (\hat{T}_s) = \bar{\mu}\ \forall\ \theta \in \Theta \tag{4.2.2}$$

Clearly, a m-unbiased predictor is necessarily pm-unbiased.

We have, therefore, three kinds of unbiasedness –p, m and pm-unbiasedness. Denoting the respective classes of estimators (predictors) as C_p, C_m, C_{pm}, note that both C_p, C_m are subsets of C_{pm}.

For a non-informative design where $p(s)$ does not depend on the y-values, order of summation E, $\mathscr{E}$ can always be interchanged, *i.e.*, $E\mathscr{E}(\cdot) = \mathscr{E}E(\cdot)$. We shall consider throughout non-informative designs only.

Consider the population total

$$Y = \sum_s Y_i + \sum_{\bar{s}} Y_i \tag{4.2.3}$$

where $\bar{s} = \mathscr{P} - s$. Give the data $d = \{(k, y_k), k \in s\}$, one can write

$$Y = \sum_s y_i + \sum_{\bar{s}} Y_i = \sum_s y_i + U_s \text{ (say)} \tag{4.2.4}$$

where the part $\sum_s y_i$ completely known and hence in predicting Y, one needs only to predict U_s.

$$\hat{T}_s = \sum_s y_i + \hat{U}_s$$

will be a m-unbiased predictor of Y if

$$\mathscr{E}\ (\hat{U}_s) = \mathscr{E}\left(\sum_{\bar{s}} Y_i\right) = \sum_{\bar{s}} \mu_i = \mu_{\bar{s}} \text{ (say) } \forall\ \theta \in \Theta, \forall s : p(s) > 0 \tag{4.2.5}$$

Two types of mean square errors (mse's) of a sampling strategy $(p, \hat{T}_s)$ for predicting Y has been proposed in the literature:

(a) $\mathscr{E}\, MSE(p, \hat{T}_s) = \mathscr{E}\ E(\hat{T} - Y)^2 = M(p, \hat{T})$ (say)

(b) $EMS\mathscr{E}\,(p, \hat{T}) = E\mathscr{E}\,(\hat{T} - \mu)^2$ where $\mu = \Sigma\mu_k = \mathrm{E}\,(Y)$

$= M_1(p, \hat{T})$ (say)

If $\hat{T}$ is p-unbiased for $Y(E(\hat{T}) = Y \forall \mathbf{y} \in R_N)$, M is model-expected p-variance of $\hat{T}$. If $\hat{T}$ is m-unbiased for Y, M_1 is p-expected model-variance $\hat{T}$.

It has been recommended that if one's main interest is in predicting the total of the current population from which the sample has been drawn, one should use M as the measure of uncertainty of $(p, \hat{T})$. If one's interest is in predicting the population total for some future population, which is of the same type as the present survey population (having the same μ), one is really concerned with μ, and here M_1 should be used (Sarndal, 1980). In finding an optimal predictor one minimises M or M_1 in the class of predictors of interest.

As noted in Lemma 3.3, the following relations among M or M_1 hold:

$$M(p, \hat{T}) = E\mathscr{V}(\hat{T}) + E\{\beta(\hat{T})\}^2 + \mathscr{V}(Y) - 2\mathscr{E}\{(Y - \mu)E(\hat{T} - \mu)\} \tag{4.2.6}$$

where $\beta(\hat{T}) = \mathscr{E}(\hat{T} - Y)$, model-bias in $\hat{T}$.

If $\hat{T}$ is p-unbiased.

$$M(p, \hat{T}) = E\mathscr{V}(\hat{T}) + E\{\beta(\hat{T})\}^2 - \mathscr{V}(Y) \tag{4.2.7}$$

If $\hat{T}$ is p-unbiased as well as m-unbiased,

$$M(p, \hat{T}) = M_1(p, \hat{T}) - \mathscr{V}(Y) \tag{4.2.8}$$

In finding an optimal $\hat{T}$ for a given p, one has to minimise $M(p, \hat{T})$ (or $M_1(p, \hat{T})$) in a certain class of predictors. Now,

$$M(p, \hat{T}) = E\mathscr{E}\left(\hat{U}_s - \sum_{\bar{s}} Y_k\right)^2$$

$$= E\mathscr{E}\left\{(\hat{U}_s - \mu_{\bar{s}}) - \left(\sum_{k \in \bar{s}} Y_k - \mu_{\bar{s}}\right)\right\}^2$$

$$= E\left[\mathscr{V}(\hat{U}_s) + \mathscr{V}\left(\sum_{\bar{s}} Y_k\right) - 2\mathscr{E}\left(\hat{U}_s, \sum_{\bar{s}} Y_k\right)\right] \tag{4.2.9}$$

If Y_i are independent (*i.e.*, if ξ is a product measure), $\mathscr{C}\left(\hat{U}_s, \sum_{\bar{s}} Y_k\right) = 0(\hat{U}_s$ being a function of Y_k, $k \in s$ only); hence, in this case, for a given p, $M\ (p, \hat{T})$ is minimum if $\mathscr{V}\ (\hat{U}_s)$ is minimum for each s, *i.e.*,

$$M(p, \hat{T}) \le M(p, \hat{T}')$$

$$\Leftrightarrow \quad \mathscr{V}(\hat{U}_s) \le \mathscr{V}(\hat{U}'_s) \ \forall x : p(s) > 0 \tag{4.2.10}$$

For a given s, the optimal m-unbiased predictor of Y (in the minimum $\mathscr{E}(\hat{T} - Y)^2$ – sense) is, therefore,

$$\hat{T}_s^+ = \sum_s y_k + \hat{U}_s^+ \tag{4.2.11}$$

where,
$$\mathscr{E}(\hat{U}_s^+) = \mu_{\bar{s}} \tag{4.2.12}$$

$$\mathscr{V}(\hat{U}_s^+) \leq \mathscr{V}(\hat{U}_s') \tag{4.2.12.1}$$

for any $\hat{U}_s'$ satisfying (4.2.12). It is clear that $\hat{T}_s^+$, when it exists, does not depend on the sampling design (unlike, say, e_{HT}). Its form is solely derived from the assumed model ξ.

An optimal design-predictor pair $(p, \hat{T}_s)$ in the class (p, τ) is, therefore, one for which

$$M(p^+, \hat{T}_s^+) \leq M(p, \hat{T}') \qquad \ldots(4.2.13)$$

for any $p \in p$, a class of sampling designs and $\hat{T}'$, any other m-unbiased predictor $\in \tau$.

After $\hat{T}_s$ has been derived via (4.2.11)–(4.2.12.1), an optimum sampling design is obtained through (4.2.13). The approach, here, is, therefore, completely model-dependent, the emphasis being on the correct postulation of a superpopulation model that will efficiently describe the physical situation at hand and generating $\hat{T}_s$. After $\hat{T}_s$ has been specified, one makes a pre-sampling judgement of efficiency of $\hat{T}_s$ with respect to different sampling designs and obtain p^+ (if it exists). The choice of a suitable sampling design is, therefore, relegated to a secondary importance in this prediction-theoretic approach.

Note 4.1: We have considered above finding the optimal strategies in the minimum $M(p, \hat{T}_S)$-sense. The analogy may, similarly, be extended to finding optimality results in the M_1 sense. This has been discussed in Section 4.5.

Note 4.2: One is finding, here, the best unbiased predictor (BUP), if it exists, in the class of all m-unbiased predictors C_m of Y, for a given p. Different results will be obtained, in general, if one attempts to find BUP (in the minimum M or M_1 sense) in the class C_p or C_{pm}. As an example, Scott and Smith (1974) obtained best linear p-unbiased, pm-unbiased predictors of Y under the following superpopulation models:

$$Y_i = \alpha + \beta x_i + e_i$$
$$\mathscr{E}(e_i | x_i) = 0,\ \mathscr{E}(e_i^2 | x_i) = \sigma_i^2$$
$$\mathscr{C}(e_i, e_i' | x_i, x_i') = 0,\ i \neq i' = 1, \ldots, N.$$

Our subsequent discussions in this chapter will focus on finding the optimal strategies under polynomial regression and multiple regression superpopulation models, these two models being most useful in socio-economic surveys. Some other superpopulation models used in survey sampling are transformation models (where some transformed variables $Z_k = (Y_k - b_k)/a_k$ are distributed with a common mean, variance and correlation coefficient, where a_k, b_k are known constants), random permutation models, product models, exchangeable models, etc. [the reader may refer to Cassel *et al.* (1977) for a review]. However, these will not be discussed separately in this treatise except for those cases when they occur as particular cases of polynomial and multiple regression models.

4.3 PREDICTION UNDER POLYNOMIAL REGRESSION MODELS

Suppose that with each $k \in \mathscr{P}$ there are two real numbers (y_k, x_k), values of a main variable 'y' and a closely related auxiliary variable 'x' respectively, x_k being known $\forall k$. We assume $\mathbf{y}$ to be a realisation of a random vector $\mathbf{Y}$ with a joint distribution ξ such that

$$\mathscr{E}(Y_k \mid x_k) = \sum_{j=0}^{J} \delta_j \beta_j x_k^j$$

$$\mathscr{V}(Y_k \mid x_k) = \sigma^2 v(x_k), \quad k = 1, \ldots, N \tag{4.3.1}$$

$$\mathscr{C}(Y_k, Y_k' \mid x_k, x_k') = 0, \quad k \neq k', 1, \ldots, N$$

where x_k's are assumed fixed (non-stochastic), $\beta_j (j = 0, 1, ..., J)$, σ^2 are unknown quantities, $v(x_k)$ is a known function of x_k, $\delta_j = 1(0)$ if the term x_k^j is present (absent) in μ_k. The polynomial regression model (4.3.1) will be denoted by $\xi(\delta_0, \delta_1, \ldots, \delta_j; v(x))$ following Royall and Herson (1973). As an example $\xi(1, 1; x^2)$ will denote model (4.3.1) with

$$\mu_k = \beta_0 + \beta_1 x_k, \; v(x_k) = x_k^2.$$

An optimal m-unbiased predictor of Y, for a given s, is

$$\hat{T}_s^+(\delta_0, \ldots, \delta_J; v(x)) = \sum_s y_k + \hat{U}_s^+ \tag{4.3.2}$$

where
$$\mathscr{E}(\hat{U}_s^+) = \mathscr{E}\left(\sum_{\bar{s}} Y_k\right)$$

$$= \sum_{j=0}^{J} \delta_j \beta_j \sum_{\bar{s}} x_k^j \tag{4.3.3}$$

and
$$\mathscr{V}(\hat{U}_s^+) \leq \mathscr{V}(\hat{U}_s') \tag{4.3.4}$$

for all $\hat{U}_s'$'s satisfying (4.3.3). Confining to the class of linear m-unbiased predictors (4.3.2), the best linear unbiased predictor (BLUP) of Y is

$$\hat{T}_s^*(\delta_0,\ldots,\delta_J;v(x)) = \sum_s y_k + \sum_{j=0}^{J} \delta_j \hat{\beta}_j^* \sum_{\bar{s}} x_k^j \tag{4.3.5}$$

where $\hat{\beta}_j^*$ is the BLUP of β_j, under $\xi(\beta_0, \ldots, \delta_J; v(x))$ as obtained from Guass-Markoff theorem.

Writing,

$$\mathbf{Y} = (Y_1, \ldots, Y_N)',$$

$$\mathbf{V}^0 = \text{Diag}\,(v(x_1), \ldots, v(x_N)),$$

$$\beta = (\beta_0, \ldots, \beta_J)'$$

$$\mathbf{X}^{(0)} = \begin{bmatrix} 1 & x_1 & x_1^2 & \ldots & x_1^J \\ - & - & - & \ldots & - \\ 1 & x_N & x_N^2 & \ldots & x_N^J \end{bmatrix}$$

$\mathbf{Y}_s$ as the column vector $(Y_k,\ k \in s)$, $\mathbf{X}_s^{(0)}$ as the $n \times (J+1)$ submatrix of $\mathbf{X}^{(0)}$ corresponding to $k \in s$, $\mathbf{V}_s^{(0)}$ the diag $(v(x_k),\ k \in s)$, we have

$$\hat{\beta}^* = (\mathbf{X}_s^{(0)'}\mathbf{V}_s^{(0)^{-1}}\mathbf{X}_s^{(0)})^{-1}(\mathbf{X}_s^{(0)'}\mathbf{V}_s^{(0)^{-1}}\mathbf{Y}_s), \tag{4.3.6}$$

$\hat{\beta}_j^*$ is given in the jth component of $\hat{\beta}^*$. Using (4.2.9)

$$M(p,\hat{T}^*) = E\left[\mathscr{V}\left(\sum_0^J \delta_j \hat{\beta}_j^* \sum_{\bar{s}} x_k^j\right) + \sigma^2 \sum_{\bar{s}} v(x_k)\right] \tag{4.3.7}$$

A particular case: One is often interested in the model $\xi(0, 1; v(x))$ where $\mathscr{E}(Y_k) = \beta x_k$, $\mathscr{V}(Y_k) = \sigma^2 v(x_k)$, $\mathscr{C}(Y_k, Y_k') = 0$, $k \neq k'(= 1, \ldots, N)$. Often $v(x_k) = x_k^g$, $g \in [0, 2]$ as found in many socio-economic surveys [*e.g.*, Jessen (1942), Mahalanobis (1946), Scott *et.al.* (1978)]. Here, using (4.3.5)–(4.3.7),

$$\hat{T}_s^*(0,1;v(x)) = \sum_s Y_k + \left\{\left(\sum_s x_k Y_k / v(x_k)\right)\left(\sum_s x_k^2 / v(x_k)\right)^{-1}\right\}\sum_{\bar{s}} x_k \tag{4.3.8}$$

$$\mathscr{E}(\hat{T}_s^* - Y)^2 = \sigma^2\left(\sum_{\bar{s}} x_k\right)^2 \Big/ \left\{\sum_s \frac{x_i^2}{v(x_i)}\right\} + \sigma^2 \sum_{\bar{s}} v(x_i) \tag{4.3.9}$$

It follows from (4.3.9), therefore, that if

(*a*) $v(x)$ is a monotonically non-decreasing function of x,

(*b*) $v(x)/x^2$ is a monotonically non-increasing function of x, (4.3.10)

the strategy $(p^*, \hat{T}^*)$ will have minimum average variance in the class of all strategies $(p, \hat{T})$, $p \in \rho_n, \hat{T} \in \mathscr{L}_m$, the class of all linear m-unbiased predictors under ξ, where the sampling design p^* is such that

$$p^*(s) = 1(0) \text{ for } s = s^* \text{ (otherwise)}, \tag{4.3.11}$$

s^* having the property,

$$\sum_{s^*} x_k = \max_{s \in \mathscr{L}_n} \sum_s x_k, \tag{4.3.11.1}$$

Hence, we have
$$\mathscr{L}_n = \{s : \nu(s) = n\} \tag{4.3.11.2}$$

Theorem 4.1: (Royall, 1970). Under $\xi(0,1; \nu(x))$ optimal m-unbiased predictor $\hat{T}^*$ is given by (4.3.8). Moreover,

$$M(p^*, \hat{T}^*) \le M(p, \hat{T})$$

$\forall p \in \rho_n, \hat{T} \in \mathscr{L}_m, p^*$ being given in (4.3.11)–(4.3.11.2), provided the conditions (4.3.10) hold.

Considering the particular case, $\nu(x) = x^g$ and writing $\hat{T}^*(0, 1; x^g)$ as $\hat{T}_g^*$,

$$\hat{T}_0^* = \sum_s Y_k + \left\{\left(\sum_s x_k Y_k\right)\left(\sum_{\bar{s}} x_k\right)\right\} \Big/ \sum_s x_k^2$$

$$\hat{T}_1^* = \sum_s Y_k + \left\{\left(\sum_{\bar{s}} Y_k\right)\left(\sum_{\bar{s}} x_k\right)\right\} \Big/ \sum_s x_k = \frac{\bar{Y}_s}{\bar{x}_x} X$$

$$\hat{T}_2^* = \sum_s Y_k + \left\{\sum_s (Y_k / x_k) \sum_{\bar{s}} x_k\right\} \Big/ \nu(s)$$

For each of these predictors p^* is the optimal design in $\rho_n \cdot \hat{T}_1^*$ is the ordinary ratio predictor $\hat{T}_R$ (say) and is optimal m-unbiased under $\xi(0, 1; x)$. This is in agreement with the design-based result in Cochran (1971), (pp.158-160), where the ratio estimator is shown to be the BLU-estimator if the population is such that the relation between y and x is a straight line through the origin and the variance of y about this line is proportional to the (fixed) value of x. Godambe and Joshi (1965) showed that under $\xi(0, 1; x^2)$ the strategy ($p_{\pi ps}$, e_{HT}) has the minimum average variance in the class of all p-unbiased strategies (Theorem 3.15). However, restriction is made here to the class of all m-unbiased strategies and hence the optimal strategy is a different one, viz. $(\hat{T}_2^*, p^*)$. However, in large samples, $|\hat{T}_2^*(s) - e_{HT}(s)|$ may not be very large with a high probability.

The most interesting result in this model-dependent approach is that this ushers in a purposive sampling design p^* as an optimal design as opposed to random sampling which are used under a fixed population model and are practised widely. It will, however, be seen and argued later that if an assumed superpopulation model turns out to be inaccurate (not describing the physical phenomena reasonably accurately), a purposive design may produce predictors which are highly biased and/or predictors with very low stability and in such situations a random sampling will protect the efficiency of an otherwise desirable sampling strategy.

4.4 PREDICTION UNDER MULTIPLE REGRESSION MODELS

Assume that apart from the main variable y we have $(r + 1)$ closely related auxiliary variables x_j $(j = 0, 1, \ldots, r)$ with known values x_{kj}, $\forall k = 1, \ldots, N$. Consider the following model: $(Y_1, \ldots, Y_N)$ have a joint distribution ξ such that

$$\begin{aligned} \mathscr{E}(Y_k \mid x_k) &= \beta_0 x_{k0} + \beta_1 x_{k1} + \ldots + \beta_r x_{kr} \\ \mathscr{V}(Y_k \mid x_k) &= \sigma^2 v(\mathbf{x}_k) = \sigma^2 v_k \\ \mathscr{C}(Y_k, Y'_k \mid x_k, x'_k) &= \sigma^2 v_{kk'} \end{aligned} \tag{4.4.1}$$

where $\mathbf{x_k} = (x_{k0}, x_{k1}, \ldots, x_{kr})$, $\beta_0, \ldots, \beta_r, \sigma^2 (> 0)$ are unknown parameters, $v_k (v_{kk}')$ is a known function of $\mathbf{x}_k$ (and $\mathbf{x}_k'$). If $x_{k0} = 1 \forall k$, the model has an intercept term β_0. Assuming without loss of generality that $s = (1, \ldots, n)$, we shall write

$$\mathbf{Y} = [\mathbf{Y}_s' \mathbf{Y}_{\bar{s}}']', \ \beta = (\beta_0, \ldots, \beta_r)'$$

$$\mathbf{X} = \begin{bmatrix} x_{10} & x_{11} & \cdots & x_{1r} \\ - & - & \cdots & - \\ x_{N0} & x_{N1} & \cdots & x_{Nr} \end{bmatrix} = \begin{bmatrix} \mathbf{X}_s \\ \mathbf{X}_{\bar{s}} \end{bmatrix}$$

$$\mathbf{V} = \begin{bmatrix} v_{11} & v_{12} & \cdots & v_{1N} \\ - & - & \cdots & - \\ v_{N1} & v_{N2} & \cdots & v_{NN} \end{bmatrix} = \begin{bmatrix} \mathbf{V}_s & \mathbf{V}_{sr} \\ \mathbf{V}_{rs} & \mathbf{V}_{\bar{s}} \end{bmatrix}$$

$\mathbf{X}_s$ being a $n(r + 1)$ submatrix of $\mathbf{X}$ corresponding to $k \in s$, $\mathbf{V}_{sr}$ being a $n(N - n)$ submatrix of $\mathbf{V}$ corresponding to $k \in s$ and $k' \in \bar{s}$ $(\mathbf{V}_{sr} = \mathbf{V}'_{rs})$, other submatrices being defined similarly. The multiple regression model (4.4.1) is, therefore,

$$\mathscr{E}(\mathbf{Y}) = \mathbf{X}\beta, \ \mathscr{D}(\mathbf{Y}) = \sigma^2 \mathbf{V} \tag{4.4.2}$$

$\mathscr{D}(\cdot)$denoting model-dispersion matrix of $(\cdot)$. We shall denote $X_j = \Sigma x_{kj}$, $x_{js} = \sum_{k \in s} x_{kj}$, $\bar{x}_{js} = x_{js}/n$, $\mathbf{x}_s = (x_{0s}, x_{rs})'$, $\bar{\mathbf{x}}_\mathbf{s} = (\bar{x}_{0s}, \ldots, \bar{x}_{rs})'$ and $x_{js}, \bar{\mathbf{x}}_s$ and $\bar{\mathbf{x}}_s$ similarly.

Consider the following three cases:

(*a*) Y_i's are uncorrelated so that $\mathbf{V}$ is diag $(v_k, k = 1, \ldots, N)$. Here $\mathbf{V}_{sr} = \mathbf{V}_{sr} = 0$. We shall denote this model as $\xi(\mathbf{X}, v)$. For a given s, the BLUP of Y is

$$\hat{T}_s^*(\mathbf{X}, v) = \sum_s Y_k + \mathbf{x}'_{\bar{s}}\hat{\beta}_s^* \tag{4.4.3}$$

where $\hat{\beta}_s^*$, generalised least square predictor is the BLUP of β.

$$\hat{\beta}_s^* = (\mathbf{X}_s'\mathbf{V}_s^{-1}\mathbf{X}_s)^{-1}(\mathbf{X}_s'\mathbf{V}_s^{-1}\mathbf{Y}_s) \tag{4.4.4}$$

Hence (Royall, 1971)

$$\hat{T}_s^*(\mathbf{X}, v) = [\mathbf{1}_\mathbf{n}' + (\mathbf{1}_{\mathbf{N}-n}'\mathbf{X}_{\bar{s}})(\mathbf{X}_\mathbf{s}'\mathbf{V}_s^{-1}\mathbf{X}_s)^{-1}\mathbf{X}_s'\mathbf{V}_s^{-1}]\mathbf{Y}_s \tag{4.4.5}$$

where $\mathbf{1}'_q = (1, \ldots, 1)_{q\times 1}$.

$$M(p, \hat{T}^*) = E\left[\mathscr{V}(x'_{\bar{s}}\hat{\beta}^*) + \mathscr{V}\left(\sum_{\bar{s}} Y_k\right)\right]$$

$$= \sigma^2 E\left[\left\{\mathbf{x}'_{\bar{s}}(\mathbf{X}_s'\mathbf{V}_s^{-1}\mathbf{X}_s)^{-1}\mathbf{x}_{\bar{s}}\right\} + \sum_{\bar{s}} v_k\right] \tag{4.4.6}$$

Example 4.1: For $\xi(\mathbf{X}, 1)$ with $\mathscr{E}(Y_k) = \beta_0 + \beta_1 x_k$, $\hat{\beta}_0^* = \bar{Y}_s - \hat{\beta}^*\bar{x}_s$, $\hat{\beta}^* = \left\{\sum_s (x_k - \bar{x}_s)Y_k\right\}\Big/\sum_s (x_k - \bar{x}_s)^2$, $\hat{T}^* = N[\bar{Y}_s + \hat{\beta}^*(\bar{X} - \bar{x}_s)]$, the usual linear regression predictor (estimator) in the classical set up with

$$M(p, \hat{T}^*) = NE\left[\frac{N-n}{n} + \frac{N(\bar{x} - \bar{x}_s)^2}{\sum_s (x_k - \bar{x}_s)^2}\right]\sigma^2 \tag{4.4.7}$$

An optimum sampling design is, therefore, one which selects with probability certainty the sample s consisting of n units, whose x-values minimise $\left\{(\bar{x} - \bar{x}_s)^2\Big/\sum_s (x_k - \bar{x}_s)^2\right\}$ among all samples of size n.

(*b*) $Y_1, \ldots, Y_N$ are not all pairwise uncorrelated and there is no constant correlation between Y_i and Y_j $(i = j = 1, \ldots, N)$.

Here V is an arbitrary positive semi-definite symmetric matrix and $\mathbf{V}_{sr} = (= \mathbf{V}'_{rs}) \neq \mathbf{0}$. We shall denote such a model as $\xi(\mathbf{X}, \mathbf{V})$. Let

$$\mathbf{K} = (\mathbf{V}_s, \mathbf{V}_{sr}),$$

Following Royall (1976), and Tam (1986), $\hat{T}^*(\mathbf{X}, \mathbf{V}) = h^{*\prime}\mathbf{Y}_s$ is the BLUP of $\mathbf{Y} = \mathbf{1}'\mathbf{Y}$ where $\mathbf{1} = \mathbf{1}_N$, if

(*i*) $\mathbf{h}^{*\prime}\mathbf{X}_s = \mathbf{1}'\mathbf{X}\ \forall s : p(s) > 0$ (4.4.8)

(*ii*) $\mathscr{E}(\mathbf{h}^{*\prime}\mathbf{Y}_s - \mathbf{1}'\mathbf{Y})^2 \le \mathscr{E}(\mathbf{h}'\mathbf{Y}_s - \mathbf{1}'\mathbf{Y})^2$ (4.4.8.1)

for all $\hat{T} = h'\mathbf{Y}_s$ satisfying (4.4.8). (4.4.8) is required for unbiasedness and (4.4.8.1) for ensuring minimum variance.

The following theorem stated without proof gives $\hat{T}^*(\mathbf{X}, \mathbf{V})$ and its average variance.

Theorem 4.2: (Royall, 1976; Tam, 1986). Under $\xi(\mathbf{X}, \mathbf{V})$, for any linear predictor $\mathbf{h}'\mathbf{Y}_s$ of $\mathbf{1}'\mathbf{Y}$

$$\mathscr{E}(\mathbf{h}'\mathbf{Y}_s - \mathbf{1}'\mathbf{Y})^2 = \mathscr{E}\{(\mathbf{h}' - \mathbf{1}'\mathbf{K}'\mathbf{V}_s^{-1})(\mathbf{Y}_s - \mathbf{X}_s\beta)\}^2$$

$$+\mathbf{1}'\mathbf{V}\mathbf{1} - \mathbf{1}'\mathbf{K}'\mathbf{V}_s^{-1}\mathbf{K}\mathbf{1} + \{(\mathbf{h}'\mathbf{X}_s - \mathbf{1}'\mathbf{X})\hat{\beta}\}^2 \quad (4.4.9)$$

Hence BLUP of $\mathbf{1}'\mathbf{Y}$ is $h^{*\prime}\mathbf{Y}_s = \hat{T}^*$ which satisfies (4.4.8.1) and

$$(h^{*\prime} - 1'K'V_s^{-1})(Y_s - X_s\hat{\beta}^*) = 0, \quad (4.4.10)$$

when,

$$\hat{T}^*(\mathbf{X}, \mathbf{Y}) = [\mathbf{1}'\mathbf{K}'\mathbf{V}_s^{-1} + (\mathbf{1}'\mathbf{X} - \mathbf{1}'\mathbf{K}'\mathbf{V}_s^{-1}\mathbf{X}_s)(\mathbf{X}'_s\mathbf{V}_s^{-1}\mathbf{X}_s)^{-1}\mathbf{X}'_s\mathbf{V}_s^{-1}]\mathbf{Y}_s \quad (4.4.11)$$

$$= [\mathbf{1}'_n\mathbf{Y}_s + \mathbf{1}'_{N-n}[\mathbf{X}_{\bar{s}}\hat{\beta}^* + \mathbf{V}_{r,s}\mathbf{V}_s^{-1}(\mathbf{Y}_s - \mathbf{X}_s\hat{\beta}^*)]$$

where
$$\hat{\beta}^* = (\mathbf{X}'_s\mathbf{V}_s^{-1}\mathbf{X}_s)^{-1}(\mathbf{X}'_s\mathbf{V}_s^{-1}\mathbf{Y}_s) \quad (4.4.12)$$

$$\mathscr{V}(\hat{T}_s^* - Y) = \sigma^2\mathbf{1}'_{N-n}[(\mathbf{V}_{\bar{s}} - \mathbf{V}_{rs}\mathbf{V}_s^{-1}\mathbf{V}_{sr}) + (\mathbf{X}_{\bar{s}} - \mathbf{V}_{rs}\mathbf{V}_s^{-1}\mathbf{X}_s)$$

$$(\mathbf{X}'_s\mathbf{V}_s^{-1}\mathbf{X}_s)^{-1}(\mathbf{X}_{\bar{s}} - \mathbf{V}_{rs}\mathbf{V}_s^{-1}\mathbf{X}_s)']\mathbf{1}_{N-n} \quad (4.4.13)$$

(*c*) Y_i, Y_j have constant non-zero correlation ρ, $i \neq j (= 1, ..., N)$: Here $v_{kk'} = \rho\sqrt{v_k v_{k'}}$, $-1/(N-1) \leq \rho \leq 1$, $k \neq k' = 1, ..., N$. Denote this model as $\xi(\mathbf{X}, \mathbf{V}, \rho)$. The following theorem follows from theorem 2 of Zyskind (1967).

Theorem 4.3: (Isaki and Fuller, 1982). Assume (*a*) $(\sqrt{v_1}, \ldots, \sqrt{v_N})' \in \mathscr{C}(\mathbf{X})$, column space of X, or (*b*) $\rho = 0$. The BLUP of Y, conditional on $\mathbf{X}_s$ is

$$\hat{T}^*(\mathbf{X}, \mathbf{V}, \rho) = \sum_s Y_k + \mathbf{x}'_{\bar{s}}\hat{\beta}^* \quad (4.4.14)$$

where,
$$\hat{\beta}^* = (\mathbf{X}'_s\mathbf{V}_s^{-1}\mathbf{X}_s)^{-1}(\mathbf{X}'_s\mathbf{V}_s^{-1}\mathbf{Y}_s) \quad (4.4.15)$$

the generalised least square predictor of β under $\xi(\mathbf{X}, v)$, $\mathbf{V}_s = \text{Diag}(v_k; k \in s)$.

Note that under conditions of Theorem 4.3, $\hat{T}_s^*(\mathbf{X}, V, \rho)$ becomes identical with $\hat{T}_s^*(\mathbf{X}, v)$.

Note 4.3: If $\mathbf{X}$ contains $\mathbf{1_N}$ (which is assumed for the present model) and v_k is a constant $\forall k$,

$$\hat{\beta}^* = (\mathbf{X}_s'\mathbf{X}_s)^{-1}\mathbf{X}_s'\mathbf{Y}_s \tag{4.4.16}$$

the ordinary least square predictor of β under $\xi(\mathbf{X}, 1)$.

If we assume further that $\mathbf{Y}$ follows a N-variate normal distribution N_N, then $\mathbf{Y}$ is said to follow an exchangeable general linear model EGLM (Arnold, 1979). In this case, $(\hat{\beta}^*, \delta)$ where $\hat{\beta}^*$ is given in (4.4.16) and

$$\delta = (n - r - 1)^{-1} \sum_s \left(Y_k - \sum_0^r \hat{\beta}_j^* x_{kj}\right)^2$$

is jointly complete sufficient for $(\beta, \sigma^2(1 - \rho))$. Hence,

Theorem 4.4: (Mukhopadhyay, 1991). Under EGLM with $\xi(\mathbf{X}, \mathbf{V}, \rho)$, the predictor $\hat{T}^*$ given in (4.4.14) is UMVUP in the class of all unbiased predictors of Y.

Example 4.2: $Y_k = \alpha + \beta' x_k + e_k\sqrt{v_k}$, $k = 1, ..., N$ with $\sqrt{v_k} = 1 + Ax_k$, $\mathbf{e} = (e_1, ..., e_N)$ following $N_N(\mathbf{0}, \Sigma)$ where $\Sigma = \sigma^2(1 - \rho)\mathbf{I}_N + \rho\sigma^2\mathbf{J}_N$, $I_q = \text{Diag}\,(1, ..., 1)_{q\times q}$, $\mathbf{J}_N$ is a $N \times N$ matrix with element 1 in each place, A is a known constant. Let $Z_k = Y_k/\sqrt{v_k}$, $Z_k = \alpha + \beta w_k + e_k$ with $\beta = \beta' - \alpha A$, $w_k = x_k/\sqrt{v_k}$. Here $(\hat{\alpha}, \hat{\beta}, \delta)$ is jointly completely sufficient where $\hat{\alpha} = \bar{Z}_{\bar{s}} - \hat{\beta}_{w_{\bar{s}}}$,

$$\hat{\beta} = \sum_s (Z_k - \bar{Z}_s)(w_k - \bar{w}_s) \Big/ \sum_s (w_k - \bar{w}_s)^2$$

$$\hat{\delta} = \sum_s (Z_k - \hat{\alpha} - \hat{\beta} w_k)^2 \Big/ (n-2)$$

$$\bar{w}_s = \sum_s w_k / n,$$

$$\bar{Z}_s = \sum_s Z_k / n$$

The UMVUP of $\bar{Y}$ is

$$\hat{T}^* = \left[\sum_s Y_k + \hat{\alpha}\sum_{\bar{s}} \sqrt{v_k} + \hat{\beta}\sum_{\bar{s}} x_k\right] \Big/ N$$

$$M(p, \hat{T}^*) = N^{-2} E\left[\frac{\left(\sum_{\bar{s}} \sqrt{v_k}\right)^2 \dfrac{(1-\rho)\sigma^2}{n} \left(\sum_{\bar{s}} w_k\right)^2}{\sum_s (w_k - \bar{w}_s)^2}\right.$$

$$\left. + \frac{(1-\rho)\sigma^2 \sum_{\bar{s}} x_k}{\sum_s (w_k - \bar{w}_s)^2} \left\{\sum_{\bar{s}} x_k - 2\bar{w}_s \left(\sum_{\bar{s}} \sqrt{v_k}\right)\right\} + \sigma^2(1-\rho)\sum_{\bar{s}} v_k\right]$$

If, in particular, $A = 0$ so that $v_k = 1 \forall k$.

$$M(p,\hat{T}^*) = \sigma^2 * (1-\rho)(1-f)^2 \left[\frac{1}{n(1-f)} + E\left\{ \frac{(\bar{x}_{\bar{s}} - \bar{x}_{\bar{s}})^2}{\sum_s (x_k - \bar{x}_s)^2} \right\} \right]$$

An optimal sampling design is, therefore, one which selects with certainty a sample of size n which minimises $\left\{ \frac{(\bar{x}_{\bar{s}} - \bar{x}_{\bar{s}})^2}{\sum_s (x_k - \bar{x}_s)^2} \right\}$. Thus Royall and Herson balanced sampling design (Section 5.3), if it exists, will be the optimal design.

Note 4.4: Recently Rodrigues *et al.* (1985) extended the concepts to develop a general theory of prediction in survey sampling which covers both linear and quadratic functions of population values. Skinner (1983) considered the multivariate prediction of population mean under the following model. With i are associated two vectors $\mathbf{y}_i = (y_{i1}, \ldots, y_{ip})'$, $\mathbf{x}_i = (x_{i1}, \ldots, x_{iq})'$, where $\mathbf{x}_i$ is known, $i = 1, \ldots, N$. It is assumed

$$\mathbf{Y}_i \sim N_p(\alpha + \beta \mathbf{x}_i, \Omega),\ i = 1, \ldots, N$$

where $\alpha = (\alpha_1, \ldots, \alpha_p)'$, $\beta = ((\beta_{ij}))_{p\times q}$, $\Omega_{p\times p}$ are matrices of parameters and $\mathbf{Y}_i$, $\mathbf{Y}_j$ $(i \neq j)$ are independent. The MVUP of $\mathbf{Y} = \sum_i \frac{\bar{\mathbf{Y}}_i}{N}$ is, for a given s,

$$\hat{\bar{\mathbf{Y}}}_s = \hat{\alpha}_s + \hat{\beta}\bar{\mathbf{X}},$$

where,

$$\bar{\mathbf{X}} = \frac{1}{N}\sum_{i=1}^{N} \mathbf{x}_i,\ \hat{\alpha}_s = \bar{\mathbf{y}}_s - \hat{\beta}\bar{\mathbf{x}}_s$$

$$\hat{\beta} = \mathbf{S}_{yx}\mathbf{S}_{xx}^{-1}$$

$$\bar{\mathbf{y}}_s = \frac{1}{n}\sum_s \mathbf{y}_i,$$

$$\hat{\mathbf{x}}_s = \frac{1}{n}\sum_s \mathbf{x}_i,\ \mathbf{S}_{xx} = n^{-1}\sum_s (\mathbf{x}_i - \bar{\mathbf{x}}_s)(\mathbf{x}_i - \bar{\mathbf{x}}_s)'$$

4.5 PREDICTING A SUPERPOPULATION MEAN

So far we have considered predicting $\sum_{\bar{s}} Y_k$ only, in predicting a population total (the part $\sum_{k\in s} y_k$ being completely known, given the data) and found optimal strategies that minimise $M(p,\hat{T})$. However, as noted in Section 4.2, in predicting the total of a finite population of the same type as the current survey population, one's primary interest is in estimating the superpopulation total $\mu = \Sigma\mu_k$. For a given s, a m-unbiased predictor of Y will, therefore, be

$$\hat{T} = \sum \hat{Y}_k \tag{4.5.1}$$

where, $$\mathscr{E}(\hat{Y}_k) = \mu_k,\ k = 1, ..., N \tag{4.5.2}$$

In $\hat{T}$ all the Y_k's (including those for which $k \in s$) are predicted and the data $\{(k, y_k); k \in s\}$ are used in this situation to estimate any unknown parameter θ involved in the superpopulation model through μ_k. For predicting the population total for any future point of time, the value of the auxiliary variable at that point of time (if available) may be used in estimating θ. The predictors (4.5.1) are, therefore, called 'projection predictors' (Cochran, 1977, p. 159; Sarndal, 1980).

If Y_k's ($k = 1, ..., N$) are independently distributed, an optimal predictor of Y is

$$\hat{T}^+ = \sum \hat{Y}_k^+$$

where, $$\mathscr{V}(\hat{Y}_k^+) \le \mathscr{V}(\hat{Y}_k')$$

for any other $\hat{\mathbf{Y}}_k'$ satisfying (4.5.2). Here one should use $M_1(p, \hat{T})$ as a measure of uncertainty and

$$M_1(p, \hat{T}^+) \le M_1(p, \hat{T}')$$

for any other m-unbiased $\hat{T}'$. Sarndal (1980), Isaki and Fuller (1982), Wright (1983), among others, considered optimal projection strategies for various superpopulation models.

Under $\xi(\delta_0, ..., \delta_J; v(x))$, BLU projection predictor of Y is

$$\hat{\hat{T}}^*(\delta_0, \ldots, \delta_J; v(x)) = \sum_{k=1}^{N} \sum_{0}^{J} \delta_j \hat{\beta}_j^* x_k^j \tag{4.5.3}$$

where $\hat{\beta}_j^*$ is given in (4.3.6). Under $\xi(\mathbf{X}, v)$,

$$\hat{\hat{T}}^*(\mathbf{X}, v) = \mathbf{x}_0'\hat{\beta}^* = \mathbf{1}'\mathbf{X}\hat{\beta}^* \tag{4.5.4}$$

where $\mathbf{x}'_0 = (x_{00}, x_{01}, ..., x_{0r})$, $x_{0j} = \sum x_{kj}$, $\hat{\beta}^*$ is given in (4.4.4).

Under $\xi(\mathbf{X}, \mathbf{V}, \rho)$,

$$\hat{\hat{T}}^*(\mathbf{X}, \mathbf{V}, \rho) = \mathbf{1}'\mathbf{X}\hat{\beta}^* \tag{4.5.5}$$

$\hat{\beta}^*$ being given in (4.4.16). For EGLM, $\mathbf{1}'\mathbf{X}\hat{\beta}^*$ is the best predictor in the entire class of unbiased predictors, $\hat{\beta}^*$ being given in (4.4.16).

Cassel *et al.* (1979) considered predicting a population total at a future point of time on the basis of a present survey, where the values of the explanatory variables x_{kj} are known only for $k \in s$.

4.6 RECONCILING $\hat{T}^*$ AND $\hat{\hat{T}}^*$

The conditions when $\hat{T}^*$ and $\hat{\hat{T}}^*$ coincide have been investigated by Sarndal (1980*a*, *b*). For this consider

Definition 4.3: A predictor T is consistent for Y in the classical sense *iff* $T_N = Y$, where T_N denotes the form taken by T in the 'census case', when $n = N$. When $T_N = Y$, we shall say "T_N converges" (Cochran, 1977, p. 21).

Theorem 4.5: Under $\xi(X, v)$, $P\{\hat{T}_s^* = \hat{\hat{T}}_s^* \ \forall s : p(s) > 0\} = 1$ *iff* $\hat{\hat{T}}^*$ converges.

Proof: If $\hat{\hat{T}}^*$ converges, from (4.5.1),

$$\sum (Y_k - \hat{Y}_k(N)) = 0 \tag{4.6.1}$$

where $\hat{Y}_k(N) = \mathbf{x}'_\mathbf{k}\hat{\beta}^*(N)$, $\hat{\beta}^*(N)$ denoting the value of $\hat{\beta}^*$ for $n = N$. (4.6.1) means all the N residuals $e_k(N) = Y_k - \hat{Y}_k(N)$ sum to zero. If (4.6.1) holds for $n = N$, it should also hold for any arbitrary sample s of size n. Thus $\sum_s (Y_k - \hat{Y}_k) = 0$, *i.e.*,

$$\hat{\hat{T}}^* = \sum_s Y_k + \sum_{\bar{s}} \hat{Y}_k = \hat{T}^*.$$

Corollary 4.1: For linear models with zero sum of residuals, $\sum e_k = 0, \hat{\hat{T}}^*$ converges and $\hat{\hat{T}}^* = \hat{T}^*$. In this case, for any s, $\mathscr{E}(\hat{\hat{T}}_s^* - Y)^2 \le \mathscr{E}(\hat{T}_s^* - Y)^2$ for any linear m-unbiased predictor $\hat{T}_s$ of Y.

Example 4.3: Under $\xi(\mathbf{X}, v)$,

$$\hat{T}_s^* - \hat{\hat{T}}_s^* = \mathbf{1}'_n(\mathbf{I}_n - \mathbf{J}_s)\mathbf{Y}_s \tag{4.6.2}$$

where

$$\mathbf{J}_s = \mathbf{X}_s(\mathbf{X}'_s\mathbf{V}_s^{-1}\mathbf{X}_s)^{-1}\mathbf{X}'_s\mathbf{V}_s^{-1} \tag{4.6.3}$$

Hence,

$$\hat{T}_s^* = \hat{\hat{T}}_s^* \forall s,$$

iff

$$\mathbf{1}'_n = \mathbf{1}'_n\mathbf{X}_s(\mathbf{X}'_s\mathbf{V}_s^{-1}\mathbf{X}_s)^{-1}\mathbf{X}'_s\mathbf{V}_s^{-1} = \mathbf{a}'_s\mathbf{X}'_s\mathbf{V}_s^{-1} \text{ (say)} \tag{4.6.4}$$

where,

$$\mathbf{a}_s = \mathbf{1}'_n\mathbf{X}_s(\mathbf{X}'_s\mathbf{V}_s^{-1}\mathbf{X}_s)^{-1} = (a_{0s}, a_{1s}, \ldots, a_{rs}) \tag{4.6.5}$$

(4.6.4) implies

$$v_k = \sum_{j=0}^{r} a_{js}\, x_{kj},\; k \in s$$

and since (4.6.5) should hold for all s, we must have $a_{js} = a_j\ \forall s$ and hence

$$v_k = \sum_{0}^{r} a_j x_{kj},\; k = 1, \ldots, N \tag{4.6.6}$$

i.e.,

$$\mathbf{V1}_N \in \mathscr{C}(\mathbf{X}) \tag{4.6.7}$$

The condition (4.6.7) was also noted by Isaki and Fuller (1982), Bolfarine and Rodrigues (1988). We shall call models satisfying (4.6.6) as models in subclass γ. Under $\xi(\mathbf{X}, v)$,

$$\mathscr{E}(\hat{\hat{T}}^* - Y)^2 = \mathscr{E}(\hat{T}^* - Y)^2 + K_1$$

$$= \sigma^2 \left\{ \sum_{\bar{s}} v_k + \sum_{\bar{s}} \sum_{\bar{s}} \mathbf{x}'_k (\mathbf{X}'_s \mathbf{V}_s^{-1} \mathbf{X}_s)^{-1} \mathbf{x}'_k \right\} + K_1 \tag{4.6.8}$$

where $K_1 \geq 0$.

$$\mathscr{E}(\hat{T}^* - \mu)^2 = \mathscr{E}(\hat{\hat{T}}^* - \mu)^2 + K_2$$

$$= \sigma^2 \left\{ \sum_{1}^{N} \sum_{1}^{N} \mathbf{x}'_k (\mathbf{X}'_s \mathbf{V}_s^{-1} \mathbf{X}_s)^{-1} \mathbf{x}'_k \right\} + K_2 \tag{4.6.9}$$

where $K_2 \geq 0$. It follows that

$$K_1 = K_2 = \sigma^2 \mathbf{1}'_n \{\mathbf{V}_s - \mathbf{X}_s (\mathbf{X}'_s \mathbf{V}_s^{-1} \mathbf{X}_s)^{-1} \mathbf{X}'_s\} \mathbf{1}_n$$
$$= 0 \textit{ iff } \xi \in \gamma$$

If $\xi \notin r$, $K_1 (= K_2)$ can still be expected to be small compared to the main term in (4.6.8) or (4.6.9) if f (= n/N) is small.

Some examples of models in r are: ξ such that (*i*) $\mathscr{E}(Y_k) = \mu$, $\mathscr{V}(Y_k) = \sigma^2$, (*ii*) $\mathscr{E}(Y_k) = \beta x_k$, $\mathscr{V}(Y_k) = \sigma^2 x_k$, (*iii*) $\mathscr{E}(Y_k) = \beta_0 + \beta x_k$, $\mathscr{V}(Y_k) = \sigma^2$; Y_k and $Y_{k'}$, are uncorrelated in all such cases. In (*i*) predictor of $\bar{Y}$ is $\bar{Y}_s$; in (*ii*) and (*iii*) they are

$$\frac{(\bar{X}\bar{Y}_s)}{\bar{x}_s} \text{ and } \bar{Y}_s + \hat{\beta}^*(\bar{X} - \bar{x}_s)\quad \hat{\beta}^* = \left\{ \sum_{s} (x_k - \bar{x}_s) Y_k \right\} \Big/ \sum_{s} (x_k - \bar{x}_s)^2 \text{ respectively.}$$

In many situations, specially, if N is large and the models fit well, $|\hat{T}^* - \hat{\hat{T}}^*|$ is likely to be numerically small and practically unimportant.

Example 4.4: Model $\xi(\delta_0, ..., \delta_J; v_k)$; hence, $\hat{T}^* = \hat{\hat{T}}^*$ if $v_k = \sum_0^J c_j \delta_j x_k^j$ for some constants $c_0, c_1, ..., c_J$; $\delta_j = 1(0)$ according as x_k^j is present (absent) in v_k.

Example 4.5: Stratified population $\mathscr{P} = \bigcup_1^L U_h, U_h \cap U'_{h,} , = \phi,\ h \neq h' = 1, ..., L$. A sample s_h of size n_h is drawn from U_h by srs, $x = \bigcup_1^L s_h; \bar{Y}_{s_h} = \sum_{s_h} \frac{Y_k}{n_h}$.

Model: $\mathscr{E}(Y_k) = \beta_h, k \in U_h, \mathscr{V}(Y_k) = \sigma^2 v_k, k = 1, ..., N$. $\hat{\hat{T}}^* = \hat{T}^*$ *iff* $\hat{\hat{T}}^*$ converges *i.e.*,

$$\sum \hat{Y}_k = \sum Y_k \text{ where } \hat{Y}_k = \hat{\beta}_h(N), k \in U_h.$$

$$\hat{\beta}_h(N) = \left(\sum_{U_h} \frac{Y_k}{v_k}\right)\left(\sum_{U_h} \frac{1}{v_k}\right)^{-1}.$$

Thus if $v_k = v_h$ for all $k \in U_h$, $\hat{\hat{T}}^* = \hat{T}^* = \frac{1}{N}\sum_h N_h \bar{Y}_{s_h}$.

4.7 DISCUSSIONS

The approach in this chapter is completely model-dependent. If an assumed superpopulation model is known to be reasonably correct for a specified survey population, optimal sampling strategies, using a purposive sample most often, are available. However, if the assumed model is miserably wrong or ill-specified, the optimal strategies are often seriously biased and inefficient. Statisticians faithful in the philosophy of this approach tend to avoid inferences based on design-randomisation even if the sample is randomly selected. Even if the assumed model is correct, there is a major difference in estimating the mean square error of the predictors in this approach from the one based on the classical randomisation theory. One is estimating here the model-based variance, which is different from the variance overall possible samples. Model-based variance estimation has been considered in Chapter 11.

❑ ❑ ❑

CHAPTER 5

Robustness of Model Dependent Optimal Strategies Under Alternative Polynomial Regression Models

5.1 INTRODUCTION

In the previous chapter we have considered optimal sampling strategies based on some superpopulation models. Here we consider the robustness of these strategies under some alternative models. Royall and Herson (1973*a*, *b*), Mukhopadhyay (1977*a*, *b*, 1985), Scott, Brewer and Ho (1978), among others, addressed the question of robustness of BLU-predictors $\hat{T}^*$ under various polynomial regression models $\xi(\delta_0, \ldots, \delta_j; v(x))$. We consider these results in this chapter.

5.2 BIAS AND MSE UNDER ALTERNATIVE MODELS

If it is exactly known beforehand that the true superpopulation model is ξ, one would use the optimal predictor $\hat{T}^*(\xi)$ and the appropriate sampling plan $p^*(\xi)$ (if it exists). In actual situations, however, the form of the underlying superpopulation model is assumed from practical considerations and one's assumption about the model may not be completely correct. One may, therefore, face the following consequences:

(*a*) Suppose from practical considerations we assume that the model is $\xi(\delta_0, \ldots, \delta_j; v(x))$ and use the predictor $\hat{T}^*(\xi)$ which is BLU under ξ. This predictor will not, in general, be unbiased if the true model is something different, ξ' (say) and the ξ'-bias involved for a particular sample s is

$$\mathscr{E}\,\xi'\{\hat{T}_s^*(\xi) - Y\} = B\{\hat{T}_s^*(\xi) - \xi'\}$$

$$= B_{\delta_0', \ldots, \delta_J'; v'(x)}(\hat{T}_s^*(\delta_0, \ldots, \delta_J; v(x))) \qquad \ldots(5.2.1)$$

ξ'-bias of $\hat{T}^*(\xi)$ for a particular sampling design p is

$$\sum_{s \in \mathscr{S}} \mathscr{E}\,\xi'\{\hat{T}_s^* - Y\}p(s) \qquad \ldots(5.2.2)$$

To preserve the property of unbiasedness of $\hat{T}^*(\xi)$ even when the true model is ξ', we may choose the sampling design in such a way that $\hat{T}^*(\xi)$ remains also unbiased under ξ'. With this end in view, as explained subsequently, Royall and Herson (1973*a*) introduced the concept of balanced sampling design.

Another way to deal with the situation may be as follows. Of all the predictors belonging to a subclass $\bar{\hat{T}}(\xi)$ say, we may choose one $\hat{T}^0(\xi)$, which is least subject to bias even when the true model is ξ'. This for a given s, we should use $\hat{T}^0(\xi)$ such that

$$|B(\hat{T}_s^0(\xi), \xi')| < |B(\hat{T}_s(\xi), \xi')|$$

$\forall \hat{T}(\xi) \in \bar{\hat{\tau}}(\xi)$, the choice of the subclass $\bar{\hat{\tau}}(\xi)$ being made from other considerations, *e.g.*, from the point of view of mse, etc.

(*b*) The expression for the mse of $\hat{T}^*(\xi')$ under ξ' would, in general, be different from the one which one obtains under ξ. One may compare the mse of $\hat{T}^*(\xi')$ under ξ' with the variance of BLUP $\hat{T}^*(\xi')$ under ξ' and consider situations under which $\hat{T}^*(\xi')$ would fare better than $\hat{T}^*(\xi')$ when the true model is ξ', in the mse sense.

5.3 BIAS OF $\hat{T}_g^*$

The following analysis gives a post-sample selection of bias-robust predictor in a class of predictors with respect to alternative models $\xi'(\delta'_0, \ldots, \delta'_j; v(x))$. We assume throughout

$$\left.\begin{array}{l} (a)\ \beta_j \geq 0, j = 0, \ldots, J; \\ (b)\ x_k \geq 0, k = 1, \ldots, N \\ (c)\ x_k, k \in s \text{ are not all equal.} \end{array}\right\} \quad (5.3.1)$$

Now,

$$B\{(\hat{T}_g^*, \xi'(\delta'_0, \ldots, \delta'_j; v(x))\} = \sum_{j=0}^{J} \delta'_j H_g(j, s)\beta_j$$

where,

$$H_g(j, s) = \left[\sum_s x_k^{j+1-g} \sum_{\bar{s}} x_k - \sum_s x_k^{2-g} \sum_{\bar{s}} x_k^g\right] \Big/ \sum_s x_k^{2-g} \quad (5.3.2)$$

$$= I_g(j, s) \Big/ \sum_s x_k^{2-g} \quad (5.3.3)$$

which is independent of the form of the variance function in ξ'. Note that $H_g(1, s) = 0$. We first consider $B(\hat{T}_g^*)$, $g = 0, 1, 2$ under the following class of alternatives:

(A) $\xi(1, 1; v(x)) = \xi_1$ (say)

(B) $\xi(0, 0, \delta_2, \ldots, \delta_J; v(x))$ with $\delta_2, \ldots, \delta_J$ not all zero; $= \xi_2$ (say).

We shall denote $B_{\xi_1}(\hat{T}_g^*)$ as $B_1(\hat{T}_g^*)$ and B_{ξ_2} as $B_2(\hat{T}_g^*)$. If for a given s, a predictor $\hat{T}_1$ involves less absolute bias than $\hat{T}_2$ under model ξ', we will write $\hat{T}_{1s} \overset{b:\xi'}{\succ} \hat{T}_{2s}$.

Lemma 5.1: Under (5.3.1) (b, c), $H_g(0, s)$ is a monotonically increasing function of g, a real quantity; also $H_g(j, s)$ is a monotonically decreasing function of g, for $j \geq 2$.

Proof: Consider Callebaut's inequality: Let $(A_1, \ldots, A_M)$, $(B_1, \ldots, B_M)$ be two sets of positive quantities which are not proportional. Then

$$\phi = \left(\sum_1^M A_i^{w+z} + B_i^{w-z}\right)\left(\sum_1^M A_i^{w-z}\, B_i^{w+z}\right) \tag{5.3.4}$$

increases with increasing value of $|z|$ for any real w.

Now, for $\epsilon > 0$,

$$H_{g+\epsilon}(0, s) - H_g(0, s)$$

$$= \frac{\left(\sum_s x_i\right)}{\left(\sum_s x_i^{2-g-\epsilon}\right)\left(\sum_s x_i^{2-g}\right)}$$

$$\left[\sum_s x_i^{2-g} \sum_s x_i^{1-g-\epsilon} - \sum_s x_i^{2-g-\epsilon} \sum_s x_i^{1-g}\right]$$

Let in C-inequality, $M = n$, $B_k = 1$, $A_k = x_k \forall k \in s$, $w = (3 - 2g - \epsilon)/2$, $s = (1 - \epsilon)/2$ and $(1 + \epsilon)/2$ respectively. It is found that $H_g(0, s)$ is a monotonically increasing function of g. The second part can be proved similarly.

Theorem 5.1: Under (5.3.1), for s for which

$$(a)\ I_{g'}(0, s) > 0,\ \hat{T}_g^* \overset{b:\xi_1}{\succ} \hat{T}_g^* \text{ with } g > g' \tag{5.3.5}$$

$$(b)\ I_{g'}(0, s) < 0,\ (5.3.5) \text{ holds with } g < g' \tag{5.3.6}$$

$$(c)\ I_{g'}(0, s) = 0,\ B_1(\hat{T}_{g''}^* < 0),\ B_1(\hat{T}_{g'}^*) = 0,\ B_1(\hat{T}_{g'''}^*) > 0 \tag{5.3.7}$$

with $g'' < g' < g'''$.

In particular for s with

$$(a)\ I_0(0, s) \geq 0, B_1(\hat{T}_2^*) > B_1(\hat{T}_1^*) > B_1(\hat{T}_0^*) \geq 0 \tag{5.3.8}$$

$$(b)\ I_2(0, s) \leq 0, |B_1(\hat{T}_2^*)| < |B_1(\hat{T}_1^*)| < |B_1(\hat{T}_0^*)| \tag{5.3.9}$$

$$(c)\ I_1(0, s) = 0, B_1(\hat{T}_0^*) < 0, B_1(\hat{T}_1^*) = 0, B_1(\hat{T}_2^*) > 0 \tag{5.3.10}$$

Theorem 5.2: Under (5.3.1), for an optimal sampling plan p^*, (5.3.9) holds.

Proof: It is sufficient to prove $I_2(0, s^*) \le 0$, where s^* is given in (4.3.12)–(4.3.13). Assume without loss of generality, $x_1 >, \ldots, > x_N$ when $s^* = (1, \ldots, n)$. Clearly,

$$\sum_{s^*} x_k^{-1} \sum_{\bar{s}^*} x_k - \frac{n}{x_n} \sum_{1}^{N-n} x_{n+i} < n(N-n)\,.$$

Definition 5.1: (Royall and Herson, 1973). A sampling design $\bar{p}(L)$ is a balanced s.d. of order L (if it exists) if $\bar{p}(s) = 1(0)$ for $s = s_b(L)$ (otherwise), where $s_b(L)$, called a balanced sample of order L is such that

$$\bar{x}^{(j)}_{s_b(L)} = \bar{x}^{(j)}_{\bar{s}_b(L)},\ j = 1, \ldots, L$$

where,
$$\bar{x}^{(h)}_{s} = \sum_{s} \frac{x_k^h}{n},$$

$$\bar{x}^{(h)}_{\bar{s}} = \sum_{s} \frac{x_k^h}{(N-n)} \tag{5.3.11}$$

If there are K such samples (5.3.11), $\bar{p}$ chooses each such sample with probability K^{-1}.

Theorem 5.3: Under (5.3.1), for $\bar{p}\,(1)$, (5.3.10) holds and

$$n - n^{-1} \sum_{s_b(1)} x_k \sum_{s_b(1)} x_k^{-1} (<0) < n\,\frac{|[|B_1(\hat{T}_0^*)| - B_1(\hat{T}_2^*)]|}{(N-n)\beta_0}$$

$$< n - \left(\sum_{s_b(1)} x_k\right)^2 \left(\sum_{s_b(1)} x_k^2\right)^{-1} (>0) \tag{5.3.12}$$

The following theorem readily follows from (5.3.2) and Lemma 5.1.

Theorem 5.4: Under (5.3.1), for s for which

(*a*) $I_{g'}(j, s) \le 0,\ \forall j = 2, \ldots, J$ with at least one $I_{g'}(j, s)$ strictly negative,

$$\hat{T}_g'^{*} \overset{\xi_2}{\succ} \hat{T}_g^{*} \tag{5.3.13}$$

with $g > g'$.

(*b*) $I_{g'}(j, s) \ge 0\ \forall j = 2, \ldots, J$ with at least one $I_{g'}(j, s)$ strictly positive, (5.3.13) holds for $g < g'$.

(*c*) $I_{g'}(j, s) = 0\ \forall j = 2, \ldots, J$

$$B_2(\hat{T}_{g''}^*) > 0,\ B_2(\hat{T}_{g'}^*) = 0,\ B_2(\hat{T}_{g'''}^*) < 0 \tag{5.3.14}$$

with $g'' < g' < g'''$.

In particular, for s with,

$$(d)\ I_0(j, s) \le 0\ \forall\, j = 2, \ldots, J,\ |B_2(\hat{T}_0^*)| < |B_2(\hat{T}_1^*)| < |B_2(\hat{T}_2^*)| \quad (5.3.15)$$

$$(e)\ I_2(j, s) \ge 0\ \forall\, j = 2, \ldots, J,\ B_2(\hat{T}_0^*) > B_2(\hat{T}_1^*) > B_2(\hat{T}_2^*) \ge 0 \quad (5.3.16)$$

$$(f)\ I_1(j, s) = 0\ \forall\, J = 2, \ldots, J,\ B_2(\hat{T}_0^*) > 0, B_2(\hat{T}_1^*) = 0, B_2(\hat{T}_2^*) < 0 \quad (5.3.17)$$

Theorem 5.5: Under (5.3.1), for p^*, (5.3.16) holds.

Proof: It is sufficient to prove that $I_2(j, s) \ge 0$, $\forall j \ge 2$. Denoting $\dfrac{\left(\sum_s x_k^{j-1}\right)}{\left(\sum_{\bar{s}} x_k^j\right)} = \mu(j, s)$,

it is sufficient to prove $\mu(j, s)$ is an increasing function of $j = 2, \ldots, J$.

$$[\mu(j+1, s) - \mu(j, s)]\left(\sum_{\bar{s}} x_k^{j+1} \sum_{\bar{s}} x_k^j\right)$$

$$= \sum_s x_k^j \sum_{\bar{s}} x_k^j - \sum_s x_k^{j-1} \sum_{\bar{s}} x_k^{j+1}$$

$$> \sum_x x_k^{j-1}\left(n^{-1}\sum_s x_k \sum_{\bar{s}} x_k^j - \sum_{\bar{s}} x_k^{j+1}\right) > 0$$

$$(\because \ \text{under } p^*, \bar{x}_s \ge x_k \ \forall\, k \in \bar{s})$$

Theorem 5.6: Under (5.3.1), for $\overline{p}(J)$, $B_2(\hat{T}_1^*) = 0$, and

$$2(N-n)n^{-2}\sum_2^J \delta_j\beta_j\left[\sum_s x_k^{j-1}\sum_s x_k - n\sum_s x_k^j\right]$$

$$(<0) < B_2(\hat{T}_0^*) - |B_2(\hat{T}_2^*)|$$

$$< (N-n)\left(\sum_s x_k^2\right)^{-1}\sum_2^J \delta_j\beta_j\left[\sum_s x_k^{j+1} - n^{-1}\sum_s x_k^j \sum_s x_k\right] \ (>0)$$

Example 5.1: Let $N = 5$, $\mathbf{x} = (2, 3, 4, 5, 8)$, $n = 3$. Denoting samples as $s_1 = (1, 2, 3)$, $s_2 = (1, 2, 4)$, $s_3 = (1, 2, 5)$, $s_4 = (1, 3, 4)$, $s_5 = (1, 3, 5)$, $s_6 = (1, 4, 5)$, $s_7 = (2, 3, 4)$, $s_8 = (2, 3, 5)$, $s_9 = (2, 4, 5)$, $s_{10} = (3, 4, 5)$, we have $I_0(0, s) > 0$, for $s = s_1, s_2, s_4, s_7$; $I_2(0, s) < 0$ for $s = s_6, s_8, s_9, s_{10}$; for s_3, s_5, $I_0(0, s) < 0$ and $I_2(0, s) > 0$. For s_1, s_2, s_4, s_7, $\hat{T}_1^*$ is the most robust, for s_6, s_8, s_9, s_{10}, $\hat{T}_2^*$ is the most robust among $\{\hat{T}_g^*, g = 0, 1, 2\}$ wrt $\xi(1, 1; v(x))$ while for s_5, s_6 nothing can be definitely concluded.

Again, for s_1, s_2, s_4 and s_7, $I_0(j, s) < 0$, for s_5, s_6, s_8, s_9, s_{10}, $I_2(j, s) > 0$ while for s_3; $I_0(j, s) > 0$, $I_2(j, s) < 0$ for $j = 2, 3, \ldots$. Thus for s_1, s_2, s_4, s_7, $\hat{T}_1^*$ is the most robust, for s_5, s_6, s_8, s_9, s_{10}, $\hat{T}_2^*$ is the most robust wrt ξ_2, while for s_3 nothing can be definitely concluded.

5.4 MEAN SQUARE ERRORS OF $\hat{T}_g^*$

An appeal to the Lemma 5.1 gives the following:

Theorem 5.7: Under (5.3.1), for samples with $I_2(0, s) \leq 0$,

$$\hat{T}_2^* \succ \hat{T}_1^* \succ \hat{T}_0^* \tag{5.4.1}$$

when the true model is $\xi(1, 1; x^2)$, where $\hat{T}_1 \succ \hat{T}_2'$ denotes $\hat{T}$ is a better predictor than $\hat{T}_2$ both in the sense of having less bias and less mse. Clearly the result holds for p^*.

Theorem 5.8: If the true model is $\xi(1, 1; x)$, under (5.3.1), for samples with

$$(i)\ I_2(0, s) \leq 0,\ \hat{T}_1^* \succ \hat{T}_0^* \tag{5.4.2.1}$$

$$(ii)\ I_0(0, s) \geq 0,\ \hat{T}_1^* \succ \hat{T}_2^* \tag{5.4.2.2}$$

Clearly, under p^*, (5.4.2.1) holds. Under $\overline{p}(1)$, both (5.4.2.1) and (5.4.2.2) hold.

Theorem 5.9: If the true model is $\xi(1, 1; 1)$ and (5.3.1) holds, for samples with $I_0(0, s) \geq 0$,

$$\hat{T}_0^* \succ \hat{T}_1^* \succ \hat{T}_2^*.$$

If $\overline{p}(1)$ is used

$$\hat{T}_1^* \succ \hat{T}_2^*$$

Remark 5.1: When from preliminary observations one assumes that the underlying superpopulation model is $\xi(0, 1; v(x)), v(x)$ being not completely known, but reasonably assumed to be of the form, x^g, $g = 0, 1, 2$, one has at one's disposal three predictors, $\hat{T}_0^*, \hat{T}_1^*, \hat{T}_2^*$, each of which is m-unbiased for Y. If, however, the assessment of the true model is wrong and it is $\xi(1, 1; v(x))$, each of these predictors will be biased, unless the design is $\overline{p}(1)$ in which case $\hat{T}_1^*$ is unbiased for Y. However, in such situations, to control the mse, one uses the s.d. p^* and under this scheme,

$$\hat{T}_2^* \overset{b:\xi_1}{\succ} \hat{T}_1^* \overset{b:\xi_1}{\succ} \hat{T}_0^* \tag{5.4.3}$$

Thus, for taking protection against the class of alternatives $\xi(1, 1; v(x))$, $v(x) = x^g$, $g = 0, 1, 2$ if one is interested in controlling the bias of the predictors only, without paying any heed to mse, one should use $\hat{T}_1^*$ and a balanced sampling design; if, however, one wants to control the mse also, $\hat{T}_2^*$ would be the most useful of the three.

Remark 5.2: The same conclusion holds also for the alternatives in ξ_2. If one is interested only in keeping his predictor bias-free, one should use $\hat{T}_1^*$ along with balanced s.d. $\overline{p}(J)$. If however, wanting to control the mse, he uses p^*, he should use $\hat{T}_2^*(0, 1; x^2)$, which is most robust wrt bias among $\{\hat{T}_g^*, g = 0, 1, 2\}$.

Remark 5.3: No such specific conclusion can be drawn regarding the behaviour of bias $\hat{T}_g^*$ wrt general class of alternatives $\xi(\delta_0, \ldots, \delta_J; v(x))$, where $\delta_0 = 1$ and at least one $\delta_2, \ldots, \delta_J$ is non-zero.

5.5 BALANCED SAMPLES AND $\hat{T}^*(0, 1; v(x))$

Consider the bias of $\hat{T}^*(0, 1; v(x))$ under $\xi(\delta_0, \ldots, \delta_J; V(x))$ which is

$$\sum_0^J \delta_j \beta_j \left[\left(\sum_{\bar{s}} x_k\right)\left(\sum_s x_k^{j+1} \Big/ v(x)_k \Big/ \sum_s x_k^2 \Big/ v(x_k)\right) - \sum_{\bar{s}} x_k^j\right] \tag{5.5.1}$$

Hence, if a sample $s^*(J)$ satisfies

$$\frac{\left(\sum_s x_k^j\right)}{\left(\sum_{\bar{s}} x_k\right)} = \frac{\left(\sum_s x_k^{j+1} \Big/ v(x_k)\right)}{\left(\sum_s x_k^2 \Big/ v(x_k)\right)},$$

$$j = 0, 1, \ldots, J \tag{5.5.2}$$

the predictors $\hat{T}^*(0, 1; v(x))$ remains unbiased under $\xi(\delta_0, \ldots, \delta_J; V(x))$ on $s(J)$ for any $V(x)$.

Samples $s^*(J)$ satisfying (5.5.2) may be called "generalised balanced samples of order J" (Scott, Brewer and Ho, 1978).

For $v(x) = x$, $s^*(J)$ reduces to $s_b(J)$, Royall and Herson-balanced sample of order J, which preserves the unbiasedness of ratio predictor $\hat{T}_1^*$ under $\xi(\delta_0, \ldots, \delta_J; V(x))$.

For $v(x) = x^2$, (5.5.2) reduces to

$$n^{-1} \sum_s x_k^{j-1} = \frac{\left(\sum_{\bar{s}} x_k^j\right)}{\left(\sum_{\bar{s}} x_k\right)}, \; j = 0, 1, \ldots, J \tag{5.5.3}$$

samples satisfying which have been called "over balanced samples" $s_0(J)$ by Scott, *et al.* (1978).

The following theorem shows $\hat{T}^*(0,1;v(x))$ becomes BLUP under $\xi(\delta_0, \ldots, \delta_J; V(x))$ for $s = s^*(J)$ when $V(x)$ is of a particular form.

Theorem 5.10: (Scott *et al.*, 1978). For $s = s^*(J)$, $\hat{T}^*(0,1;v(x))$ is BLUP under $\xi(\delta_0, \ldots, \delta_J; V(x))$ provided

$$V(x) = v(x)\sum_0^J \delta_j a_j x^{j-1} \tag{5.5.4}$$

where a_j's are arbitrary non-negative constants.

Proof: Under $\xi(0, \ldots, 0, \delta_j = 1, 0, \ldots, 0; v(x)x^{j-1}) = \xi_j$ (say), on $s = s^*(J)$, BLUP is $\hat{T}^*(0,1;v(x))$.

Consider a more general model $\xi(\delta_0, \ldots, \delta_{j-1}, \delta_j = 1, \delta_{j+1}, \ldots, \delta_J; v(x)x^{j-1}) = \xi_{(j)}$ (say). For any linear $\xi_{(j)}$-unbiased predictor $\hat{T}$ of Y, $\mathscr{E}_{\xi(j)}(\hat{T}-Y)^2$ depends only on the variance function $v(x)x^{j-1}$ and not on the coefficients $\beta_0, \beta_1, \ldots, \beta_J$. Thus on $s^*(J)$, $\hat{T}^*(0,1;v(x))$ is BLUP under $\xi_{(j)}$.

When $\hat{T}$ is a unbiased linear predictor under $\xi(\delta_0, \ldots, \delta_J; V(x)) = v(x)\sum_0^J a_j\delta_j x^{j-1})$ $= \xi$, say

$$\mathscr{E}_\xi(\hat{T}-Y)^2 = \sum_0^J \delta_j a_j \mathscr{E}_{\xi(j)}(\hat{T}-Y)^2$$

Since under $\xi_{(j)}$, for $s^*(J)$, $\hat{T}^*(0,1;v(x))$ is BLUP, $j = 0, 1, \ldots, J$, the theorem follows. (For a different proof see Example 5.3).

Corollary 5.1: For $v(x) = x$, $s = s_b(J)$, $\hat{T}_1^*$ reduces to mean per unit predictor $N\bar{Y}_s$ which is BLUP under $\xi\left(\delta_0, \ldots, \delta_J; \sum_0^J a_j\delta_j x^j\right)$, a_j's being some non-negative constants (Royall and Herson, 1973*a*; Tallis, 1978) (see also Example 5.2).

Note 5.1: Royall and Cumberland (1981) defined a π-balanced sample of order J, $s_\pi(J)$ as one for which $\Delta_j(s) = 0$, $j = 0, \ldots, J$ where

$$\Delta_j(s) = \bar{X}\bar{x}_s^{j-1} - \bar{X}^{(j)}$$

$\bar{X}^{(j)} = \Sigma x_k^j / N$. For a πps sampling plan, π-balanced samples are met in expectation *i.e.*, $E(\Delta_j(s)) = 0$. We denote a π-balanced sampling design of order J as $p_\pi(J)$ (where $p(s_{\pi(J)}) = K^{-1}$ if there are K such samples). Under $\xi(\delta_0, \delta_1, \ldots, \delta_J; v(x))$ $\hat{Y}_{HT}$ at π-balance, $\hat{T}_1^*$ at $s_b(J)$, $\hat{T}_2^*$ at $s_0(J)$, are all unbiased for Y. When $v(x) \propto x^2$,

$$\mathcal{V}(\hat{Y}_{HT} - Y) \leq \mathcal{V}(\hat{T}_2^* - Y) \leq \mathcal{V}(\hat{T}_1^* - Y)$$

and when $v(x) \propto x$,

$$\mathcal{V}(\hat{T}_2^* - Y) \leq \mathcal{V}(\hat{Y}_{HT} - Y) = \mathcal{V}(\hat{T}_1^* - Y)$$

at the respective balanced samples. Kott (1986 *b*) showed that under model $\xi(0, \delta_1, \delta_2; 1)$ the mean of the ratio predictor $\frac{1}{n}\sum_s \frac{Y_i}{x_i}$ coupled with $p_{\pi(2)}$ design provides the BLUP strategy.

Note 5.2: Kott (1986) defined an asymptotically balanced sample as follows. Consider a sequence of nested populations $\{\mathcal{P}_j\}$ of size N_j, $j = 1, 2, \ldots$, the sequence $\{N_j\}$ being unbounded. A sample s_j of size n_j is drawn from $\mathcal{P}_j$ such that $n_j \to \infty$ with $N_j/n_j = k(>1)$, a fixed integer. Assume $X_1 \leq X_2 \leq \ldots$ for each population. The population is stratified into n strata with the first k units forming the first stratum, the next k units the second stratum and so on. s_j is selected from $\mathcal{P}_j$ by choosing one unit from each stratum, either adhering to the same position in each stratum, giving $(t, t+k, \ldots, t+\overline{n+1}k)$, $t \in [1, k]$ being chosen at random, as s_j or by choosing one unit randomly from each stratum. The sequence of samples $\{s_j\}$ so obtained is not necessarily nested. The sample thus obtained is called asymptotically balanced on x, provided the x's are bounded, *i.e.*, $|x_i| < B_x$, $i = 1, 2, \ldots$

Consider the model: $\mathcal{E}(Y_i) = \beta x_i$, $\mathcal{V}(Y_i) = \sigma^2 v_i$, $\mathcal{C}(Y_i, Y_{i'}) = 0$, $i \neq i'$ and assume

$$0 < B_1 < v_i < B_2 < \alpha \ \forall i \tag{5.5.5}$$

Now,

$$\mathcal{E}(\bar{Y}_s - \bar{Y}_N)^2 = \beta^2(\bar{x}_s - \bar{X}_N)^2 + \sigma^2\left(\frac{1}{n^2} - \frac{2}{nN}\right)\sum_s v_i + \frac{\sigma^2}{N^2}\sum v_i \tag{5.5.6}$$

$\bar{Y}_N, \bar{X}_N$ being population means based on size N. The first term in (5.5.6) is the contribution of the bias of $\bar{Y}_s$ to its model mse and the remaining terms are the contribution of model s.d. of $\bar{Y}_s$ to its model mse. If the sample is drawn by srswor and if $\sum (x_i - \bar{X})^2$ converges to a finite positive value as $N \to \alpha$, then under (5.5.5) and by virtue of the fact that x_i's are bounded, contribution of both these terms to the average mse $E\mathcal{E}(\bar{Y}_s - \bar{Y}_N)^2$ is of order n^{-1}.

Under asymptotic balanced sample as above $|\bar{x}_s - \bar{x}_N| < 2B_x/n$ when contribution of the model bias of $\bar{Y}_s$ to its model mse in (5.5.6) is bounded by $4\beta^2 B_x^2/n^2$. As $N \to \alpha$, this contribution $\to 0$, faster than the model mse itself which is of order n^{-1}.

5.6 EFFICIENCY OF A BALANCED SAMPLE

The efficiency of a balanced sample may be assessed in terms of the ratio of the minimum attainable variance of the appropriate predictor to the variance of this predictor under the balanced sample.

Now for a model with

$$\mathscr{E}(Y_i) = \mu_i,\ \mathscr{V}(Y_i) = \sigma^2 v_i,$$
$$\mathscr{C}(Y_i, Y_{i'}) = 0,\ i \neq i',$$

$$\mathscr{E}(\hat{T}_1^* - \bar{Y})^2 = \left[\left(\sum_{\bar{s}} x_i\right)\left(\sum_{s} \mu_i\right) \Big/ \sum_{s} x_i - \sum_{\bar{s}} \mu_i\right]^2$$

$$+ \left[\left(\sum_{\bar{s}} x_i \Big/ \sum_{s} x_i\right)^2 \sum_{s} v_i + \sum_{\bar{s}} v_i\right]\sigma^2 \qquad (5.6.1)$$

Therefore, for $\xi = (1, 1; x)$, on $s_b(1)$,

$$\mathscr{E}(\hat{T}_1^* - Y)^2 = \frac{N}{f}(1-f)\sigma^2 \bar{X},\ f = \frac{n}{N}; \qquad (5.6.2)$$

again, on $s^*(x)$,

$$\mathscr{E}(\hat{T}_1^* - Y)^2 = \frac{N}{f}(1-f)\sigma^2 \bar{X}\, \bar{x}_{\bar{s}*} / \bar{x}_{s*} \qquad (5.6.3)$$

Hence, efficiency of $s_b(1)$ is

$$\frac{\bar{x}_{\bar{s}*}}{\bar{x}_{s*}} \times 100\% \qquad (5.6.4)$$

Thus if the $s_b(1)$ and the ratio predictor is used and the true model is $\xi(1, 1; v(x))$, no bias is involved; however, there is some loss of efficiency compared to the use of purposive sample s^* if the true model is $\xi(0, 1; x_j)$, the efficiency being given in (5.6.4). The loss in efficiency is at the cost of gaining protection against $\xi(1, 1; v(x))$.

For $s_b(J)$, the first term in the right hand side of (5.6.1) vanishes when $\mu = \sum_0^J \delta_j \beta_j x^j$. Hence the efficiency of the strategy $(\hat{T}_1^*, s_b(J))$ under $\xi(\delta_0, \ldots, \delta_J;$ $v(x))$ compared to $(\hat{T}_1^*, s^*)$ under $\xi(0, 1; x)$ remains the same as in (5.6.4), *i.e.*, there is no further loss in efficiency in using $s_b(J)$ over $s_b(1)$.

Note 5.3: Royall and Herson (1973, *b*) proposed stratification on an auxiliary variable to improve the efficiency of balanced sampling design. Herson (1976), in a numerical study, empirically assessed the efficiency of ratio prediction of the finite population total using conventional unrestricted random sampling, extreme and balanced sampling plans, for sample sizes ranging from small to large. The balanced sampling plan was seen to be as much as 30% more efficient than the corresponding unrestricted random sampling plan regardless of sample size, but the extreme sample was found to be inferior to both of these plans.

Note 5.4: By Theorem 5.10, under $\xi(\delta_0, \ldots, \delta_J; V(x))$, where $V(x) = \sigma_1^2 x + \sigma_2^2 x^2$, both $(\hat{T}_1^*, s_b(J)) = H_1$ (say) and $(\hat{T}_2^*, s_0(J)) = H_2$ (say) are BLU strategies. However,

$$\mathscr{E}\{(\hat{T}_1^* - Y)^2 \mid s_b(J)\} = N(N-n)(\sigma_1^2 \bar{X} + \sigma_2^2 \bar{x}^{(2)})/n = M_1(\text{say}),$$

$$\mathscr{E}\{(\hat{T}_2^* - Y)^2 \mid s_0(J)\} = N(N-n)\bar{x}_{\bar{s}}(\sigma_1^2 + \sigma^2 \bar{X})/n = M_2(\text{say}),$$

$$< \frac{N(N-n)\bar{X}}{n}(\sigma_1^2 + \sigma_2^2 \bar{X}),$$

where $\bar{x}^{(2)} = \sum \frac{x_k^2}{N}$, using (5.5.3) for $j = 1$. Thus

$$M_1 - M_2 > \frac{N(N-n)}{n}\sigma_2^2(\bar{x}^{(2)} - \bar{X}^{(2)}) > 0$$

showing that H_1 is less efficient than H_2, the loss in efficiency will be small in general if σ_1^2 dominates σ_2^2 but can be substantial if σ_2^2 is relatively large.

Note 5.5: Consider the same model as in Note 5.4. The population is stratified in L strata with the N_1 units with the smallest x-values forming stratum 1, the N_2 units, units with the next higher x-values forming stratum 2, etc. and over-balanced samples $s_h = s_{0(J)}^{(h)}$ satisfying

$$n_n^{-1} \sum_{s_h} x_{kh}^{j-1} = \left(\sum_{\bar{s}_h} x_{kh}^j\right)^{-1} \left(\sum_{\bar{s}_h} x_{kh}\right), \; j = 0, 1, \ldots, J$$

are drawn from stratum h (= 1, ..., L), n_h being the sample size, x_{kh} the value of x for the kth unit in stratum h. Hence for $s = \bigcup_1^L s_{0(J)}^{(h)}$,

$$\hat{T}_{st} = \sum_h \hat{T}_2^*$$

is unbiased with

$$\mathscr{E}(\hat{T}_{st} - Y)^2 = \sum_1^L N_h(N_h - n_h)\bar{x}_{\bar{s}_h}(\sigma_1^2 + \sigma_2^2 \bar{X}_h) n_h^{-1} = M_3 \text{ (say)}$$

M_2 and M_3 cannot be compared since the sample will not be the same in general. However, if f is small and no group of x's dominate the others,

$$\bar{x}_s \simeq \bar{X} \quad \text{and} \quad M_2 \simeq N(N-n)\bar{X}(\sigma_1^2 + \sigma_2^2 \bar{X})n^{-1}.$$

Since $\bar{x}_{s_h} \leq \bar{X}_h$ for over-balanced sampling,

$$M_3 \leq \sum_1^L N_h n_h^{-1}(N_h - n_h)\bar{X}_h(\sigma_1^2 + \sigma_2^2\bar{X}_h) \tag{5.6.5}$$

The right hand side of M_3 is minimised for a fixed sample size n when

$$n_h \simeq N_{\hat{h}}(\sigma_1^2\bar{X}_h + \sigma_2^2\bar{X}_h^2)^{1/2}.$$

This allocation requires knowledge of σ_2^2/σ_1^2. However, if σ_2^2 dominates σ_1^2, we may take $n_h \simeq N_h\bar{X}_h$ when

$$M_2 - M_3 \geq \sigma_2^2\sum_1^L N_h(\bar{x}_h - \bar{X})^2 \geq 0$$

so that with this allocation $\hat{Y}_{st}$ and stratified over-balanced sampling is more efficient than $\hat{T}_2^*$ in unstratified over-balanced sampling.

Recently Mukhopadhyay and Bhattacharyya (1994) considered the performance of strategies H_1, H_2 and $(s_{\pi(J)}, \hat{Y}_{HT}$ under $\xi(\delta_0, \ldots, \delta_J; V(x))$ where $V(X)$ is a polynomial of degree (at most) $(J-1)$.

5.7 APPROXIMATELY BALANCED SAMPLING DESIGNS

It is obvious, balanced samples are rarely available in practice. Royall and Herson (1973), Royall and Pfeffermann (1982) recommended simple random samples as approximately balanced samples. In the case of ratio predictor, for the departure to model $\xi(1, 1; v(x))$, Godambe and Thompson (1977) suggested two types of randomized designs which would lead to balanced samples and the inference supported by randomisation. One is selection of a sample by Midzuno sampling or simple random sampling until a sample with sufficient degree of balancing is achieved. Other is stratification with respect to x-variable, also suggested by Royall and Herson (1973*b*). Royall and Cumberland (1981) proposed the following sampling design for generating $\bar{p}(1)$ approximately. Select a sample s out of $\binom{N}{n}$ possible samples at random. If $|\bar{x}_s - \bar{X}| < \delta$, a pre-assigned small positive quantity, s is considered as a sample in the proposed sampling design. Otherwise, s is rejected and a fresh random selection out of $\binom{N}{n}$ possible samples is done. For this procedure,

$$\begin{aligned} p(s) &= 0 \text{ if } |\bar{x}_s - \bar{X}| \geq \delta \\ &= c_\delta^{-1} \text{ otherwise,} \end{aligned}$$

where c_δ is the number of samples satisfying $|\bar{x}_s - \bar{X}| < \delta$.

An objection to this scheme is that $\pi_i = 0$ for a unit i so large in size that $|\bar{x}_s - \bar{X}| \geq \delta$ for all $s \ni i$, a forbidding weakness from a design-based sampling theorist's view point (Sarndal, 1981). Other objections are that δ and c_δ are arbitrary.

Iachan (1985) proposed the following modification of the above s.d. Select s with probability

$$p_1(s) = \lambda_1 \text{ if } |\bar{x}_s - \bar{X}| < \delta$$

$$= \frac{\lambda_1 \delta}{|\bar{x}_s - \bar{X}|} \text{ otherwise}$$

The constant $\lambda_1 = \lambda_1(\delta)$ is determined from the condition $\Sigma\, p_1(s) = 1$ so that

$$\lambda_1 c_\delta + \lambda_1 \delta \sum_{s \in D_1^c} |\bar{x}_s - \bar{X}|^{-1} = 1$$

where, $D_1 = \{s : |\bar{x}_s - \bar{X}| < \delta\}$.

Thus, $\lambda_1 = (c_\delta + d\delta)^{-1}$

where, $d = \sum_{s \ni D_1} c\,|\bar{x}_s - \bar{X}^{-1}|$

Hence, $$\pi_i = \lambda_1 \left\{ n(D_{1i}) + \delta \sum_{s \in D_{1i}} c\,|\bar{x}_s - \bar{X}|^{-1} \right\} > 0,\ i = 1, \ldots, N$$

when $D_{1i} = \{s \ni i : |\bar{x}_s - \bar{X}| < \delta\}$

and $n(D_{1i})$ is the number of samples in D_{1i}. Here A^c denotes the complement of the set A.

Mukhopadhyay (1985*b*) showed that simple random sampling and *pps* $\sqrt{x}$ sampling are asymptotically equivalent to balanced sampling designs for using the ratio predictor.

Like the balanced s.d. an overbalanced s.d. is rarely available in practice. Scott *et al.* (1978) showed that *pps*-selection yields an approximately over-balanced sample if the sample size is large and f is small. If the sample size is large and f also large, selection with probability $\lambda x_i(1 + \lambda x_i)^{-1}$, where λ is the solution to the equation

$$\sum \lambda x_i (1 + \lambda x_i)^{-1} = n,$$

yields approximately over-balanced samples.

5.8 A POST-SAMPLE ROBUST PREDICTOR

Balanced sampling designs were recommended for preserving unbiasedness of $\hat{T}_1^*$ under $\xi(\delta_0, \ldots, \delta_J; v(x))$ Mukhopadhyay (1985*a*) suggested a post-sample predictor which remains unbiased under alternative polynomial regression models.

Suppose that the sampler believes from his prior knowledge that the underlying superpopulation model is $\xi(0, 1; v(x))$, where $v(x) = x^g$, value of g being unknown. In this case, for unbiased prediction of Y, he can use any predictor in the class $\mathscr{I} = \{\hat{T}_g^*, g \text{ real}\}$ of optimum predictors. However, if he wants to be protected for bias of $\hat{T}_g^*$ against alternative $\xi_1 = \xi(1, 1; v(x))$, he will choose, for the given s, the one which is most robust in the class $\mathscr{I}$ with respect to ξ_1. Since $H_g(0, s)$ is a monotonically increasing function of g (Lemma 5.1), one can find a value of $g = g^*$ (say), for which $H_g(0, s)$ is as near to zero as possible. By Theorem 5.7, $\hat{T}_{g^*}$ will be the most bias-robust in $\mathscr{I}$, for given s, wrt ξ_1.

Similarly, we may be interested in $\xi_2 = \xi(0, 0, \delta_2, \ldots \delta_J; v(x))$. Since $H_g(j, s)$ is a monotonically decreasing function of g for $j \geq 2$ and since β's are assumed non-negative, $\left|\sum_2^J \delta_j \beta_j H_g(j, s)\right| = \gamma_g(s)$ (say), is a monotonically decreasing function of g. For a given s, we take that value of $g = g^{**}$ (say) for which $\gamma_g(s)$ is as near to zero as possible and construct the predictor $\hat{T}_{g^{**}}^*$ which will be most bias-robust in $\mathscr{I}$ wrt ξ_2.

5.9 ROBUSTNESS OF $\hat{T}^*(X, v)$ UNDER ALTERNATIVE MODELS

Consider $\xi(\mathbf{X}, v)$ and an alternative model ξ^* $(\mathbf{X}^*, v^*)$ as:

$$\mathscr{E}(\mathbf{Y}) = \mathbf{X}^*\beta^*, \ \mathscr{D}(\mathbf{Y}) = \sigma^2\mathbf{V}^*,$$

$$V^* = \text{Diag}\ (v_1^*, \ldots, v_N^*)$$

$$\mathbf{X}^* = [\mathbf{X}, \mathbf{Z}], \ \mathbf{Z} = \begin{bmatrix} z_{11}, \ldots, z_{1t} \\ \ldots\ldots\ldots\ldots \\ z_{N1} \ldots z_{Nt} \end{bmatrix} \tag{5.9.1}$$

$$\beta^{(*)} = \begin{bmatrix} \beta \\ \Upsilon \end{bmatrix}, \ \Upsilon = \begin{bmatrix} \gamma_1 \\ \vdots \\ \gamma_t \end{bmatrix}$$

showing that in ξ^*, additional t variables $z_1, \ldots, z_t$ have been added to the original $(r + 1)$ variables $x_0, \ldots, x_r$ in ξ, $\gamma_1, \ldots, \gamma_t$ being the corresponding regression coefficients. We consider

Theorem 5.11: (Pereire *et.al,* 1983). Royall's BLUP under $\xi(\mathbf{X}, v)$, $\hat{T}^*(\mathbf{X}, v)$ remains unbiased under $\xi^*(\mathbf{X}^*, v^*)$, if

$$\mathbf{1}'(\mathrm{I} - \Delta_s)\,\mathbf{XB}^*\mathbf{Z} = \mathbf{1}'(\mathrm{I} - \Delta_s)\mathbf{Z} \tag{5.9.2}$$

where $\mathbf{B}^*$ is defined as

$$\hat{\beta}^* = \mathbf{B}^*\Delta_s\mathbf{Y} = (\mathbf{X}'\Delta_s\mathbf{V}^{-1}\Delta_s\mathbf{X})^{-1}(\mathbf{X}'\Delta_s\mathbf{V}^{-1}\Delta_s\mathbf{Y}) \tag{5.9.3}$$

where $\Delta_s = \text{Diag}\ (\delta_k,\ k = 1, \ldots, N)$, $\delta_k = 1(0)$ if $k \in (\notin)\ s$.

Proof:

$$\hat{T}^*(\mathbf{X}, v) = \mathbf{1}_n'\mathbf{Y}_s + \mathbf{1}_{N-n}'\mathbf{X}_{\bar{s}}\hat{\beta}^*$$

$$= \mathbf{1}'\Delta_s\mathbf{Y} + \mathbf{1}'(\mathbf{I} - \Delta_s)\mathbf{X}\hat{\beta}^*.$$

Hence,

$$\hat{T}^*(\mathbf{X}, v) - \mathbf{1}'\mathbf{Y} = \mathbf{1}'(\mathbf{I} - \Delta_s)\mathbf{X}\hat{\beta}^* - \mathbf{1}^*(\mathbf{I} - \Delta_s)\mathbf{Y}$$

$$= \mathbf{1}'(\mathbf{I} - \Delta_s)(\mathbf{X}\hat{\beta}^* - \mathbf{I})\mathbf{Y} \tag{5.9.4}$$

Since, $\hat{T}^*(\mathbf{X}, v)$ is ξ-unbiased.

$$\mathbf{1}'(\mathbf{I} - \Delta_s)(\mathbf{XB}^* - \mathbf{I})\mathbf{X} = 0'$$

Hence $\hat{T}^*(\mathbf{X}, v)$ is $\xi^*(\mathbf{X}^*, v^*)$-unbiased if

$$\mathbf{1}'(\mathbf{I} - \Delta_s)(\mathbf{XB}^* - \mathbf{I})\mathbf{X}^* = \mathbf{0}'$$

i.e., $\mathbf{1}'(\mathrm{I} - \Delta_s)\,(\mathbf{XB}^* - \mathrm{I})\mathbf{Z} = \mathbf{0}$

We consider,

Lemma 5.2: (Zyskind, 1967). Consider linear model $\eta : \mathscr{E}(\mathbf{Y}) = \mathbf{X}\beta$, $\mathscr{D}(\mathbf{Y}) = \Sigma$, where Σ is any symmetric non-negative definite matrix. A linear function $\mathbf{h}'\mathbf{Y}$ is BLU-estimator of its expectation under η *iff*

$$\sum \mathbf{h} \in \mathscr{C}\ (\mathbf{X}),$$

column space of $\mathbf{X}$ *i.e.*, there exist constants $c_0, c_1, \ldots, c_r$ such that

$$\sum \mathbf{h} = \mathbf{X}c$$

where $\mathbf{c} = (c_0, c_1, \ldots, c_r)'$.

Definition 5.2: (Pereira *et al.*) The ξ-best predictor $T^+(\xi)$ is said to be robust (both wrt expectation and variance) under alternative ξ^* if $T^+(\xi) = T^+(\xi^*)$.

Theorem 5.12: (Pereira *et al.*) Royall's BLUP $\hat{T}^*(\mathbf{X}, v)$ remains $\hat{T}^*(\mathbf{X}^*, \mathbf{V}^*)$-robust *iff*

(*a*) (5.9.2) holds

(*b*) there exists a vector $\mathbf{c}' = (c_0, c_1, \ldots, c_{r+1})$ of scalars such that

$$\Delta_s\mathbf{V}^*\mathbf{B}^{*\prime}\mathbf{X}'(1 - \Delta_s)\mathbf{1} = \Delta_s\mathbf{X}^*\mathbf{c} \tag{5.9.5}$$

Proof: Condition (a) is required for unbiasedness. Again, $\hat{T}^*(\mathbf{X}, v)$ is ξ^*-best *iff* $\{\hat{T}^*(\mathbf{X}, v) - \mathbf{1}_s' \mathbf{Y}_s\}$ is BLU-estimator of its expectation under ξ^*. Now

$$\hat{\mathbf{T}}^*(\mathbf{X}, v) - \mathbf{1}_s' \mathbf{Y}_s = \mathbf{1}'(I - \Delta_s)\mathbf{X}\mathbf{B}^*\Delta_s \mathbf{Y}. \tag{5.9.6}$$

Hence, by Lemma 5.2, $\hat{T}^*(\mathbf{X}, v)$ is ξ^*-best *iff* (5.9.6) along with (a) holds.

Example 5.2: (Royall and Herson, 1973; Tallis, 1978). For $\mathbf{X} = \mathbf{1}$, $\mathbf{V} = I$,

$$\mathbf{Z} = \begin{bmatrix} x_1 & x_1^2 & \cdots & x_1^J \\ - & - & \cdots & - \\ x_N & x_N^2 & \cdots & x_N^J \end{bmatrix}$$

$$\hat{T}^*(\mathbf{X}, v) = N\sum_s Y_k / n$$

is ξ^*-BLUP *iff* $n^{-1}\sum_s x_k^j = (N-n)^{-1}\sum_{\bar{s}} x_k^j$, $j = 1, \ldots, J$ which holds for $s_{b(J)}$ and v_k^* $= \sum_0^J c_j x_k^j$ for some constants $c_0, c_1, \ldots, c_J$ ($k = 1, \ldots, N$).

The following theorem useful for prediction under $\xi(\mathbf{X}, v)$ follows readily from the above result.

Theorem 5.13: Under $\xi(\mathbf{X}, v)$, for balanced samples $s_b(x_0, x_1, \ldots, x_r)$, balanced (of order one) in each of the variables $x_0, x_1, \ldots, x_r$ for which $\bar{x}_{js} = \bar{X}_j$, $j = 0, 1, \ldots, r$, $\hat{T}^*(\mathbf{X}, v) = N\bar{Y}_s$, provided $v_k = \sum_{j=0}^r a_j x_{kj}$, $k = 1, \ldots, N$, for some arbitrary constants $a_0, a_1, \ldots, a_r$.

Example 5.3: (Rodrigues, 1983). $\mathbf{X} = \mathbf{1}$, $\mathbf{V} = \text{Diag}\,(f(x_1), \ldots, f(x_N))$, $\mathbf{Z}$ as in Example 5.2.

$$\hat{T}^*(\mathbf{X}, v) = \sum_s Y_k + (N-n)\sum_s Y_k \left(\sum_s 1/f(x_k)\right)^{-1} \Big/ f(x_k)$$

is ξ^*-unbiased if

$$\left(\sum_s \frac{1}{f(x_k)}\right)^{-1} \sum_s \frac{x_k^j}{f(x_k)} = \frac{1}{N-n}\sum_{\bar{s}} x_k^j,\ j = 1, \ldots, J \tag{5.9.7}$$

$\hat{T}^*$ is ξ^*-robust if (5.9.7) holds and if

$$f^*(x_k) = f(x_k)\sum_0^J c_j x_k^j$$

for some constants c_j, $j = 0, 1, \ldots, J$.

Example 5.4: (Scott *et al.*, 1978), $\mathbf{X} = (x_1, \ldots, x_N)'$, $\mathbf{V}$ as in Example 5.3.

$$\mathbf{Z} = \begin{bmatrix} 1 & x_1^2 & x_1^3 & \ldots & x_1^J \\ - & - & - & \ddots & - \\ 1 & x_N^2 & x_N^3 & \ldots & x_N^J \end{bmatrix}$$

Here, $$\mathbf{B}^* = \left(\sum_s \frac{x_k^2}{f(x_k)}\right)^{-1} \left(\frac{\delta_1 x_1}{f(x_1)}, \ldots, \frac{\delta_N x_N}{f(x_N)}\right)', \; \delta_k = 1(0) \text{ if}$$

$x_k \in s$ (otherwise),

when conditions (5.5.2) and (5.5.4) required for the robustness of $\hat{T}^*(0, 1; v(x))$ follows.

❑ ❑ ❑

CHAPTER

6 Predictors Based on Both Superpopulation Models and Sampling Designs

6.1 INTRODUCTION

Both Royall's class of predictors $\hat{T}^*$ and Cochran's class of projection predictors $\hat{\hat{T}}^*$ are completely dependent on the assumed superpopulation model ξ and do not depend on the sampling design used to select s. If ξ, therefore, differs markedly from the underlying true model, the optimal predictors $\hat{T}^*, \hat{\hat{T}}^*$ may fail (the predictors will be biased, in general etc.) considerably. One will, however, be satisfied if the predictors remain robust (their desirable properties are retained more or less) under departure of the assumed model from the true one.

Various authors [Cassel, Sarndal and Wretman (1976, 1977), Brewer (1979), Wright (1983)], therefore, suggested predictors which combine randomisations both due to sample selection and superpopulations, and as such are expected to retain some of the optimal properties at least approximately, under alternative models. We consider such predictors in this chapter.

6.2 DIFFERENT CHOICES OF $\hat{\beta}$ UNDER ξ(X, v)

Considering the BLUP of β under $\xi(\mathbf{X}, v)$ as given in (4.4.4) we note that a predictor $\hat{\beta}$ of β is of the form,

$$\hat{\beta} = (\mathbf{Z}'_s\mathbf{X}_s)^{-1}\mathbf{Z}'_s\mathbf{Y}_s \tag{6.2.1}$$

where $\mathbf{Z}_s$ is a $n \times (r + 1)$ matrix of weights z_{kj} to be suitably chosen such that a predictor $\hat{\beta}$ of β has some desirable properties. $\mathbf{Z}'_s\mathbf{X}_s$ is of full rank. Different choices of $\mathbf{Z}_s$ are:

(*a*) π^{-1} *weighted:* $\mathbf{Z}_s$ and corresponding $\hat{\beta}$ may be called π^{-1}-weighted if

$$\mathbf{\Pi}_s^{-1}\mathbf{1}_n \in \mathscr{C}(\mathbf{Z}_s) \tag{6.2.2}$$

i.e., $\mathbf{II}_s^{-1}\mathbf{1}_n = \mathbf{Z}_s\lambda,$

$\mathscr{C}(\mathbf{Z}_s)$ denoting the column space of $\mathbf{Z}_s$, $\lambda = (\lambda_0, \lambda_1, ..., \lambda_r)'$, a vector of constants λ_j. Here $\mathbf{II}$ = Diag (π_k; k = 1, ..., N), $\mathbf{II}_s$ = Diag (π_k; $k \in s$). If β contains only one parameter, $\mathbf{Z}_s = (c/\pi_i, i \in s)_{n\times 1}$, c an arbitrary constant, for π^{-1}-weighting. For two or more parameters β_j, if $\mathbf{Z}_s$ contains a column with elements proportional to $\pi_k^{-1}(k \in s)$, (6.2.2) is always satisfied. This concept has been generalised by Brewer *et al.* (Section 8.3).

If $\xi(\mathbf{X}, v)$ contains an intercept term, *i.e.*, if x_{k0} = 1, k = 1, ..., N, one choice for (6.2.2) is

$$\mathbf{Z}_s = \mathbf{II}_s^{-1}\,\mathbf{X}_s$$

when,

$$\hat{\beta} = (\mathbf{X}'_s\mathbf{II}^{-1}_s\mathbf{X}_s)^{-1}\,\mathbf{X}'_s\,\mathbf{II}_s^{-1}\mathbf{Y}_s = \hat{\beta}(\Pi^{-1}) \text{ (say)} \tag{6.2.3}$$

If the variance v_k satisfies

$$\mathbf{V}_s\,\mathbf{1}_n = \mathbf{X}_s c \tag{6.2.4}$$

for some arbitrary vector of constants $\mathbf{c} = (c_0, c_1, \ldots, c_r)'$ then the choice

$$\mathbf{Z}_s = \mathbf{II}_s^{-1}\mathbf{V}_s^{-1}\mathbf{X}_s$$

satisfies (6.2.2), when

$$\hat{\beta} = (\mathbf{X}'_s\mathbf{II}_s^{-1}\mathbf{V}_s^{-1}\mathbf{X}_s)^{-1}(\mathbf{X}'_s\,\mathbf{II}_s^{-1}\mathbf{V}_s^{-1}\mathbf{Y}_s)$$

$$= \hat{\beta}(\mathbf{II}^{-1}\mathbf{V}^{-1}) \text{ (say)} \tag{6.2.5}$$

A predictor of β arising from π^{-1}-weighting will be denoted by $\hat{\beta}_{\mathbf{II}} = (\hat{\beta}_{\mathbf{II}_j}, j = 0, \ldots, r)$ in general.

(*b*) *BLU-weighted:* Here $\mathbf{Z}_s = \mathbf{V}_s^{-1}\mathbf{X}_s$ when

$$\hat{\beta} = (\mathbf{X}'_s\mathbf{V}_s^{-1}\mathbf{X}_s)^{-1}(\mathbf{X}'_s\,\mathbf{V}_s^{-1}\mathbf{Y}_s)$$

$$= \hat{\beta}^* = \hat{\beta}^*(\mathbf{V}^{-1}) \text{ (say)} \tag{6.2.6}$$

If

$$\mathbf{II}_s^{-1}\mathbf{1}_n \in \mathscr{C}(\mathbf{V}_s^{-1}\,\mathbf{X}_s) \tag{6.2.7}$$

then $\hat{\beta}^*$ is also a π^{-1}-weighted $\hat{\beta}_{\mathbf{II}}$.

(*c*) *Weighted by an arbitrary matrix Q:* Here $\mathbf{Z}_s = \mathbf{Q}_s\mathbf{X}_s$, where $\mathbf{Q}$ is an arbitrary $N \times N$ diagonal matrix of weights and $\mathbf{Q}_s$ is a submatrix of $\mathbf{Q}$ corresponding to the units $k \in s$. Here

$$\hat{\beta} = (\mathbf{X}'_s\,\mathbf{Q}_s\mathbf{X}_s)^{-1}(\mathbf{X}'_s\,\mathbf{Q}_s\mathbf{Y}_s)$$

$$= \hat{\beta}\,(\mathbf{Q})\text{(say)} \tag{6.2.8}$$

6.3 GENERALISED REGRESSION PREDICTORS FOR DIFFERENT CHOICES OF $\hat{\beta}$

For $\xi(\mathbf{X}, v)$, Cassel, Sarndal and Wretman (1976), Sarndal (1980) suggested a generalised predictor of Y,

$$\sum_s \frac{Y_k}{\pi_k} + \sum_{j=0}^{r}\left(X_j - \sum_s \frac{x_{kj}}{\pi_k}\right)\hat{\beta}_j^*$$

$$= \mathbf{1}_n' \mathbf{\Pi}_s^{-1}\mathbf{Y}_s + (\mathbf{1}'\mathbf{X} - \mathbf{1}_s'\mathbf{\Pi}_s^{-1}\mathbf{X}_s)\hat{\beta}^* \tag{6.3.1}$$

$$= \hat{T}_{GR}(\mathbf{V}^{-1}) = \hat{T}_{GR}^* \text{ (say)}$$

where $\hat{\beta}_j^*$ is BLUP of β_j obtained from the jth element of $\hat{\beta}^*$ given in (6.2.6). If β is known, (6.3.1) is a generalised difference predictor, studied by Cassel *et al.* (1976), Isaki and Fuller (1982), among others.

$\hat{T}_{GR}^*$ is m-unbiased under $\xi(\mathbf{X}, v)$. It is, however, p-biased. For $\hat{\beta}_{\mathbf{II}}$, we shall denote the corresponding generalised predictor $\hat{T}_{GR}$ as $\hat{T}_{GR\mathbf{II}}$, *i.e.*, $\hat{T}_{GR\mathbf{II}}$ is the same as (6.3.1) except that $\hat{\beta}^*$ is replaced by $\hat{\beta}_{\mathbf{II}}$.

For a π^{-1}-weighted $\hat{\beta}$, $\hat{T}_{GR\mathbf{II}}$ coincides with the projection predictor $\hat{\beta}_{\mathbf{II}} X = \hat{\hat{T}}_{\mathbf{II}}$ (say). Hence, under (6.2.7), $\hat{T}_{GR}^* = \hat{\hat{T}}^*$. If $\pi_k \propto \sqrt{v_k}$, (6.2.7) holds, provided the variance function is

$$V^{1/2}\mathbf{1}_N = \mathbf{Xc}$$

where $\mathbf{V}^{1/2} = \text{Diag}\,(\sqrt{v_k}, k = 1, \ldots, N)$, $\mathbf{c} = (c_0, \ldots, c_r)$, a vector of suitable constants.

Example 6.1: Condition (6.2.7) is satisfied if $\mathbf{X}$ in $\xi(\mathbf{X}, v)$ contains $\mathbf{1}_N$, $v_k = 1$ and $\pi_k = n/N \forall k$. Under ξ with $\mathscr{E}(Y_k) = \beta x_k$, $\mathscr{V}(Y_k) = \sigma^2 x_k$ and $\pi_k = n/N \forall k$, (6.2.7) holds when $\hat{T}_{GR}^* = \hat{\hat{T}}^* = X\overline{Y}_s/\overline{x}_s$. For ξ with $\mathscr{E}(Y_k) = \beta x_k$, $\mathscr{V}(Y_k) = \sigma^2 x_k^2$ and $\pi_k = nx_k/X$ ($k = 1, \ldots, N$), $\hat{T}_{GR}^* = \hat{\hat{T}}^* = \dfrac{X}{n}\sum_s \dfrac{Y_k}{x_k}$. For arbitrary weights $\mathbf{Q}$ in $\hat{\beta}(\mathbf{Q})$, generalised regression predictor of $\mathbf{Y}$ is

$$\hat{T}_{GR}(\mathbf{Q}) = \hat{T}_{GR} \text{ (say)}$$

$$= \sum_s \frac{Y_k}{\pi_k} + \sum_{0}^{r} \hat{\beta}_j(\mathbf{Q})\left(X_j - \sum_s \frac{x_{kj}}{\pi_k}\right)$$

$$= \mathbf{1}_n'\mathbf{\Pi}_s^{-1}\mathbf{Y}_s + (\mathbf{1}'\mathbf{X} - \mathbf{1}_s'\mathbf{\Pi}_s^{-1}\mathbf{X}_s)\,\hat{\beta}(\mathbf{Q}) \tag{6.3.2}$$

As stated in Section 6.2, $\mathbf{Q}$ may be $\mathbf{V}^{-1}, \mathbf{II}^{-1}, \mathbf{II}^{-1}\mathbf{V}^{-1}$ or any arbitrary diagonal matrix. $\hat{T}_{GR}(\mathbf{Q})$ is *m*-unbiased. It is, however, *p*-biased in general.

We shall sometimes use the symbol $\hat{T}_{GR}(\mathbf{X}^*, v^*)$ to denote a generalised regression predictor based on model $\xi(X^*, V^*)$, using an arbitrary $\mathbf{Q}$. Sarndal (1980c) showed that

$$E\mathscr{E}(\hat{T}_{GR}(\mathbf{Q}) - Y)^2 = \sigma^2 E[A_s + B_s + 2D_s] \tag{6.3.3}$$

where,

$$A_s = \sum_s \frac{v_k}{\pi_k^2}(1 - 2\pi_k) + \sum v_k \tag{6.3.4}$$

$$B_s = [\{(\mathbf{1'X} + \mathbf{1'}_n \mathbf{II}_s^{-1}\mathbf{X}_s)\,\mathbf{G'}_s - 2\mathbf{1'}\}\mathbf{V}_s\mathbf{G}_s(\mathbf{1'X} - \mathbf{1'}_n\,\mathbf{II}_s^{-1}\mathbf{X}_s)'] \tag{6.3.4.1}$$

$$D_s = (\mathbf{1'X} - \mathbf{1'}_n\mathbf{II}_s^{-1}\mathbf{X}_s)\mathbf{G'}_s\mathbf{V}_s(\mathbf{I} - \mathbf{G}_s\,\mathbf{X'}_s)\mathbf{II}_s^{-1}\mathbf{1}_n \tag{6.3.4.2}$$

where $\mathbf{G}_s$ is given from

$$\hat{\beta}_s = \mathbf{G'}_s\mathbf{Y}_s.$$

For both $\mathbf{II}^{-1}$ and $\mathbf{V}^{-1}$ weighting $\mathbf{D}_s = 0$, when

$$E\mathscr{E}(\hat{T}_{GR} - Y)^2 = \sigma^2 E[A_s + B_s] \tag{6.3.5}$$

Example 6.2: $\xi : \mathscr{E}(Y_k) = \beta,\ \mathscr{V}(Y_k) = \sigma^2,\ \mathscr{C}(Y_k, Y_{k'}) = 0,\ k \neq k' = 1, \ldots, N$. For predicting $\bar{Y}$,

$$\hat{T}_{GR}(\mathbf{II}^{-1}) = \left(\sum_s \frac{1}{\pi_k}\right)^{-1}\left(\sum_s \frac{Y_k}{\pi_k}\right),$$

$$\hat{T}_{GR}^* = \sum_s \frac{Y_k}{N\pi_k} + \bar{Y}_s\left\{1 - \sum_s \frac{1}{N\pi_k}\right\}.$$

For designs with $\pi_k = n/N$ for all k, both the predictors coincide. Also

$$\mathscr{E}(\hat{T}_{GR}(\mathbf{II}^{-1}) - \bar{Y})^2 = \sigma^2\left[\left(\sum_s \frac{1}{\pi_k}\right)^{-1}\left(\sum_s \frac{1}{\pi_k^2}\right) - \frac{1}{N}\right]$$

$$\mathscr{E}(\hat{T}_{GR} - \bar{Y})^2 = \sigma^2\left[\sum_s \frac{1}{N^2\pi_k^2} - \frac{1}{n}\left(\sum_s \frac{1}{N\pi_k}\right)^2 + \frac{1}{n} - \frac{1}{N}\right]$$

For an arbitrary sampling design $p \in \rho_n$,

$$E\mathscr{E}(\hat{T}_{GR} - Y)^2 = \frac{\sigma^2}{n}\left[\left(\frac{1}{N}\sum \frac{1}{a_k} - f\right) - E\left(\frac{1}{n}\sum_s \frac{1}{a_k} - 1\right)^2\right]$$

where $a_k = (N\pi_k)/n$, $f = n/N$. A similar calculation for $\hat{T}_{GR}(\mathbf{II}^{-1})$ gives the same first term but a different second term. One may compare these average mse's with that of the Horvitz-Thompson predictor $\hat{Y}_{HT}\left(=\sum_s \frac{Y_i}{N\pi_i}\right)$ for different sampling designs. We have

$$\mathscr{E}(\hat{Y}_{HT} - \bar{Y})^2 = \sigma^2\left[\sum_s \frac{1}{N^2\pi_k^2} - 2\sum_s \frac{1}{N^2\pi_k} + \frac{1}{N}\right] + \beta^2\left(\sum \frac{1}{N\pi_k} - 1\right)^2$$

Example 6.3: $\xi : \mathscr{E}(Y_k) = \beta_0 + \beta x_k$, $\mathscr{V}(Y_k) = \sigma^2 v_k$, $\mathscr{C}(Y_k, Y_{k'}) = 0$, $k \neq k' = 1, \ldots, N$. For $\bar{Y}$

$$\hat{T}_{GR}(\mathbf{II}^{-1}) = \left(\sum_s \frac{1}{\pi_k}\right)^{-1}\left(\sum_s \frac{Y_k}{\pi_k}\right) + \hat{\beta}(\mathbf{II}^{-1})$$

$$\left(\bar{X} - \left(\sum_s x_k / \pi_k\right) \Big/ \left(\sum_s 1/\pi_k\right)\right)$$

where,

$$\hat{\beta}(\mathbf{II}^{-1}) = \left\{\sum_s \pi_k^{-1}\left(\sum_s (x_k Y_k)\pi_k^{-1}\right) - \left(\sum_s Y_k\pi_k^{-1}\right)\left(\sum_s x_k\pi_k^{-1}\right)\right\}$$

$$\left\{\sum_s \pi_k^{-1}\left(\sum_s x_k^2\pi_k^{-1}\right) - \left(\sum_s x_k\pi_k^{-1}\right)^2\right\}^{-1};$$

$$\hat{T}_{GR}^* = \sum_s \frac{Y_k}{N\pi_k} + \hat{\beta}^*\left(\bar{X} - \sum_s \frac{x_k}{N\pi_k}\right) + \hat{\beta}_0^*\left(1 - \sum_s \frac{1}{N\pi_k}\right)$$

when $\hat{\beta}^*$'s are the generalised least square predictors.

6.4 (p, Q, R)-STRATEGIES OF WRIGHT

Wright (1983) generalised the above concept further. Under $\xi(\mathbf{X}, v)$ a predictor of the residual for the ith observation $Y_i - x_i'\beta$ is

$$e_i = Y_i - \mathbf{x}_i'\hat{\beta}(\mathbf{Q}),\ i \in s \tag{6.4.1}$$

One may, therefore, predict Y_i as

$$\left.\begin{aligned}\hat{Y}_i &= \mathbf{x}_i'\,\hat{\beta}\,(\mathbf{Q}) + r_i e_i,\ i \in s\\ &= \mathbf{x}_i'\,\hat{\beta}\,(\mathbf{Q}),\ i \notin s\end{aligned}\right\} \tag{6.4.2}$$

where r_i is a suitable weight. Hence a predictor of $\theta(\mathbf{Y}) = \mathbf{a}'\mathbf{Y}$ where $\mathbf{a}' = (a_1, \ldots, a_N)$ a vector of arbitrary constants, is

$$\begin{aligned}\mathbf{a}'\hat{\mathbf{Y}} &= \mathbf{a}'\mathbf{X}\,\hat{\beta}\,(\mathbf{Q}) + \mathbf{a}_s'\mathbf{R}_s e_s\\ &= \mathbf{a}'\mathbf{X}\hat{\beta}(\mathbf{Q}) + \mathbf{a}'\Delta\,\mathbf{Re}\\ &= \mathbf{a}'[\mathbf{X}\hat{\beta}(\mathbf{Q}) + \Delta R(\mathbf{Y} - \mathbf{X}\hat{\beta}(\mathbf{Q}))]\\ &= \mathbf{a}'[\Delta R\mathbf{Y} + (I - \Delta R)\mathbf{X}\hat{\beta}(\mathbf{Q})]\\ &= \hat{\theta}(\mathbf{Q}, \mathbf{R})\ \text{(say)}\end{aligned} \tag{6.4.3}$$

where $\mathbf{R} = \text{Diag}\ (r_k;\ k = 1, \ldots, N)$, $\mathbf{R}_s = \text{Diag}\ (r_k;\ k \in s)$, $\mathbf{e} = (e_1, \ldots, e_N)'$, $(\mathbf{a}_s = (a_k;\ k \in s))$ $\mathbf{e}_s = (e_k;\ k \in s)$, $\mathbf{a}_s = (a_k;\ k \in s)$, $\Delta = \text{Diag}\ (\delta_k;\ k = 1, \ldots, N)$, $\delta_k = 1\ (0)$, if $k \in (\notin)s$.

For population total Y

$$\hat{Y}(\mathbf{Q}, \mathbf{R}) = \sum_s r_k \left\{Y_k - \sum_0^r x_{kj}\hat{\beta}_j(\mathbf{Q})\right\} + \sum_0^r X_j\beta_j(\mathbf{Q}) \tag{6.4.4}$$

For different choices of $\mathbf{Q}$ and $\mathbf{R}$ one gets different predictors. Some choices of $\mathbf{Q}$ are, as before, V^{-1}, II^{-1}, $(VII)^{-1}$ and of $\mathbf{R}$ are $\mathbf{O}$, I and $\mathbf{II}^{-1}$.

(*a*) The choice $\mathbf{R} = \mathbf{O}$ gives simple projection predictors of the type considered by Cochran (1977), Sarndal (1980*a*, *b*), Isaki and Fuller (1989).

We shall sometimes denote $\hat{Y}(\mathbf{Q}, \mathbf{O})$ as $\hat{\hat{T}}_Q$.

Example 6.4: ξ is as in Example 6.2. $\hat{Y}(\mathbf{II}^{-1}, \mathbf{O}) = N\left(\sum_s Y_k/\pi_k\right)\left(\sum_s 1/\pi_k\right)^{-1}$ is a special case of consistent ratio predictor of Brewer (1963) and Hájek (1971).

Example 6.5: $\xi : \mathscr{E}\ (Y_k) = \beta x_k$, $\mathscr{V}\ (Y_k) = \sigma^2$, $\mathscr{C}\ (Y_k, Y_{k'}) = 0$, $k \neq k' = 1, \ldots, N$,

$$\hat{Y}(\mathbf{II}^{-1}, \mathbf{0}) = X\left(\sum_s x_k Y_k/\pi_k\right)\left(\sum_s x_k^2/\pi_k\right)^{-1}$$

$$\hat{Y}(\mathbf{II}^{-1}, \mathbf{V}^{-1}, \mathbf{0}) = X\left(\sum_s x_k Y_k/\pi_k v_k\right)\left(\sum_s x_k^2/\pi_k v_k\right)^{-1}$$

which reduces to ratio type predictor $n^{-1}X\sum_s Y_k/x_k$ if $v_k = x_k$, $\pi_k = nx_k/X$.

(*b*) **R** = *I* gives linear predictors.

$$\hat{Y}(\mathbf{V}^{-1}, I) = \sum_s Y_k + \sum_0^r \hat{\beta}_j^* \left(X_j - \sum_s x_{k_j} \right)$$

considered by Royall (1970, 1976). The choices $\mathbf{Q} = \mathbf{II}^{-1}$, $\mathbf{II}^{-1}\,\mathbf{V}^{-1}$ have been considered by Sarndal (1980*c*). We shall often denote $\hat{Y}(\mathbf{Q}, I)$ as $\hat{\mathbf{T}}_{\mathbf{Q}}$.

We note that for **R** = *I* or **Q**, Wright's predictors are completely independent of the sampling design.

(*c*) $\mathbf{R} = \mathbf{II}^{-1}$ gives the class of generalised regression predictors considered by Cassel *et al.* (1976, 1977), Sarndal (1980*c*).

$$\hat{Y}(\mathbf{V}^{-1}, \mathbf{II}^{-1}) = \sum_s \frac{Y_k}{\pi_k} + \sum_0^r \hat{\beta}_j^* \left(X_j - \sum_s \frac{x_{kj}}{\pi_k} \right)$$

$$= \hat{T}_{GR}^*$$

Example 6.6: ξ as in Example 6.5 with a^2 replaced by v_k.

$$\hat{Y}(\mathbf{II}^{-1}\mathbf{V}^{-1}, \mathbf{II}^{-1}) = \sum_s \frac{Y_k}{\pi_k} + \left(\sum_s \frac{x_k Y_k}{v_k \pi_k} \right)$$

$$\left(\sum_s \frac{x_k^2}{v_k \pi_k} \right)^{-1} \left(X - \sum_s \frac{x_k}{\pi_k} \right)$$

For $v_k = x_k^g, \pi_k \propto \sqrt{v_k}, p \in \rho_n$ this reduces to

$$\frac{N m_{g/2}}{n} \sum_s \frac{Y_k}{x_k^{g/2}} + \left(\sum_s \frac{Y_k}{x_k^{3g/2-1}} \right) \left(\sum_s \frac{1}{x_k^{3g/2-2}} \right)^{-1}$$

$$\left(X - \frac{N}{n} m_{g/2} \sum_s x_k^{1-g/2} \right)$$

where $m_h = \left(\sum x_k^h \right) / N$.

Wright's predictor $T(\hat{\mathbf{Q}}, \mathbf{R})$ accompanied by a sampling design *p* will be denoted as a (*p*, **Q**, **R**)-strategy.

We note that the mean of the ratio predictor $\hat{T} = \frac{X}{n} \sum_s Y_k / x_k$ occurs in the case of model in Example 6.6 when (*i*) $\mathbf{R} = \mathbf{O}$, $\mathbf{Q} = \mathbf{V}^{-1}$, $v_i = x_i^2$ (*ii*) $\mathbf{R} = \mathbf{II}^{-1}$, $\pi_i \propto x_i$, $p \in \rho_n$, whatever **Q**. Similarly, the ratio predictor $\hat{T}_R = (\bar{Y}_s X)/\bar{x}_s$ occurs here when, (*a*) $\mathbf{R} = I$, $\mathbf{Q} = \mathbf{V}^{-1}$, $v_i = x_i$ and (*b*) $\mathbf{R} = I$, $\mathbf{Q} = \mathbf{II}^{-1}$, $\pi_i \propto x_i$, $p \in \rho_n$.

6.5 PREDICTION ERROR AND VARIANCE OF $\hat{\theta}(\mathbf{Q}, \mathbf{R})$

From (6.4.3)

$$\hat{\theta}(\mathbf{Q}, \mathbf{R}) = \mathbf{a}'[\mathbf{R} + (I - \mathbf{R}\Delta)\hat{\mathbf{C}}]\Delta\mathbf{Y} = \mathbf{a}'\hat{\mathbf{T}}\Delta\mathbf{Y} \tag{6.5.1}$$

where,

$$\hat{\mathbf{C}} = \mathbf{X}(\mathbf{X}'\mathbf{Q}\Delta\mathbf{X})^{-1}\mathbf{X}'\mathbf{Q} \tag{6.5.2}$$

$$\hat{\mathbf{T}} = \hat{\mathbf{C}} + \mathbf{R} - \mathbf{R}\Delta\hat{\mathbf{C}} \tag{6.5.3}$$

We have,

$$(I - \hat{\mathbf{T}}\Delta) = (I - \mathbf{R}\Delta)(I - \hat{\mathbf{C}}\Delta) \tag{6.5.4}$$

Note that

$$\hat{\mathbf{C}}\Delta\mathbf{Y} = \mathbf{X}\hat{\beta}(\mathbf{Q}), \hat{\mathbf{C}}\Delta\mathbf{X} = \mathbf{X} \tag{6.5.5}$$

Prediction error is

$$\begin{aligned}\theta(\mathbf{Y}) - \hat{\theta}(\mathbf{Q}, \mathbf{R}) &= \mathbf{a}'(I - \hat{\mathbf{T}}\Delta)\mathbf{Y} \\ &= \mathbf{a}'(I - \mathbf{R}\Delta)(I - \hat{C}\Delta)(\mathbf{X}\beta + \mathbf{C}) \\ &= \mathbf{a}'(I - \mathbf{R}\Delta)(I - \hat{\mathbf{C}}\Delta)\mathbf{C}\end{aligned} \tag{6.5.6}$$

Hence,

$$\mathscr{E}(\theta)(\mathbf{Y}) - \hat{\theta}(\mathbf{Q}, \mathbf{R})) = 0$$

i.e., $\hat{\theta}(\mathbf{Y})$ is model, unbiased.

Model variance of $\hat{\theta}(\mathbf{Y})$ is

$$\begin{aligned}&\mathbf{a}'(I - \hat{\mathbf{T}}\Delta)\mathbf{V}(I - \hat{\mathbf{T}}\Delta)\mathbf{a}\sigma^2 \\ &= \mathbf{a}(I - \Delta)\mathbf{V}(I - \Delta)\mathbf{a}\sigma^2 + \mathbf{a}'(I - \hat{\mathbf{T}})\Delta\mathbf{V}(I - \hat{\mathbf{T}}')\mathbf{a}\sigma^2\end{aligned} \tag{6.5.7}$$

When $\mathbf{R} = I$, $\theta(\mathbf{Y}) = Y$,

$$\begin{aligned}\mathscr{V}(\hat{Y}) &= 1'(I - \Delta)\mathbf{V}(I - \Delta)\mathbf{1} + \mathbf{1}'(\Delta - I)\hat{\mathbf{C}}\Delta\mathbf{V}\hat{\mathbf{C}}'(\Delta - I)\mathbf{1} \\ &= 1'(I - \Delta)[\mathbf{V} + \hat{\mathbf{C}}\mathbf{V}\Delta\hat{C}'](I - \Delta)\mathbf{1}\end{aligned} \tag{6.5.8}$$

Since, here, $\hat{T} - I = (I - \Delta)\hat{\mathbf{C}}$. Royall (1970, 1976) has shown that (6.5.8) is minimised for a given s, when $\mathbf{Q} = \mathbf{V}^{-1}$. Thus $\hat{Y}(V^{-1}, I)$ gives Royall's BLU-predictor.

6.6 BIAS OF $\hat{T}$ (Q, R) UNDER MODEL MIS-SPECIFICATION

Consider an alternative model $\xi' : \mathbf{Y} = \mathbf{Z}\Upsilon + \mathbf{f}$ with $\mathscr{E}(\mathbf{f}) = \mathbf{0}$. Bias of $\hat{\theta}(\mathbf{Q}, \mathbf{R})$ under ξ' is, denoting $\mathscr{E}'$ as expectation operator wrt ξ',

$$\begin{aligned}\mathscr{E}[\hat{\theta}(\mathbf{Q},\mathbf{R}) - \theta] &= \mathbf{a}'(\hat{\mathbf{T}}\Delta - I)\mathbf{Z}\Upsilon \\ &= \mathbf{a}'[(\hat{\mathbf{T}} - I)\Delta\mathbf{Z} - (I - \Delta)\mathbf{Z}]\Upsilon \qquad (6.6.1) \\ &= 0\end{aligned}$$

if
$$\mathbf{a}'(\hat{\mathbf{T}} - I)\Delta\mathbf{Z} = \mathbf{a}'(I - \Delta)\mathbf{Z} \qquad (6.6.2)$$

for every s with $p(s) > 0$. For Royall's Linear predictor, where $\mathbf{R} = I$, (6.6.2) reduces to the condition, in case $\theta = Y$,

$$\mathbf{1}'(I - \Delta)\mathbf{C}'\Delta\mathbf{Z} = \mathbf{1}'(I - \Delta)\mathbf{Z} \qquad (6.6.3)$$

for every $s : p(s) > 0$. Balancing condition (5.3.11) of Definition 5.1 for ratio predictor, conditions (5.9.2) of Pereira *et al.* (1983) occur as special cases of (6.6.3).

6.7 SARNDAL'S GENERALISED STRATEGIES

Sarndal (1980*b*) considered a more general strategy than (p, **Q**, **R**). His predictor for a sampling plan p is

$$\hat{T}(p, \mathbf{Q}, \mathbf{W}_s) = \sum_s w_{ks}Y_k + \sum_0^r \hat{\beta}_j(Q)\left(X_j - \sum_s w_{ks}x_{kj}\right) \qquad (6.7.1)$$

where $\mathbf{W}_s = ((w_{ks}))$, a $n \times n$ diagonal matrix of w_{ks}, w_{ks} being such that $\sum_s w_{ks}Y_s$ is design-unbiased for $Y\left(\sum_{s \ni k} w_{ks}p(s) = 1 \,\forall\, k\right)$.

Special cases of w_{ks} are: (i) $w_{ks} = \left[\binom{N-1}{n-1}p(s)\right]^{-1}$ (0), $k \in (\notin)s$, which if $p(s) \propto \sum_s x_k$ (Lahiri-Sen-Midzuno scheme), reduces to $\bar{X}\big/\sum_s x_k$, and (ii) $w_{ks} = w_k = 1/\pi_k(0)$, $k \in (\notin)s$.

Under ξ with $\mathscr{E}(Y_k) = \beta x_k$, $\mathscr{V}(Y_k) = \sigma^2 x_k$, $\mathscr{C}(Y_k, Y_{k'}) = 0$, $k \neq k' = 1, \ldots, N$, nine predictors have been considered by Sarndal (some of the predictors coinciding) for choices of $\mathbf{Q} = \mathbf{V}^{-1}, \mathbf{\Pi}^{-1}, \mathbf{V}^{-1}\mathbf{\Pi}^{-1}$ for weights $w_{ks} = n^{-1}$, $\bar{X}\big/\sum_s x_k$, $\bar{X}/nx_k$, $\sqrt{x}/n\sqrt{x_k}$

with matching design *srs*, $pps \sum x_k$, *ppsx*, $pps\sqrt{x}$ respectively, where $\overline{\sqrt{x}} = \sum \sqrt{x_k}/N$.

Considering $\mathbf{Q} = \mathbf{V}^{-1}$, for $\xi(\mathbf{X}, v)$,

$$\hat{T}_s^*(p, \mathbf{V}^{-1}, \mathbf{W}_s) - \hat{\hat{T}}_s^* = \mathbf{1}'_n \mathbf{W}_s (I_n - \mathbf{J}_s)\mathbf{Y}_s$$

where $\mathbf{J}_s = \mathbf{X}_s(\mathbf{X}'_s \mathbf{V}_s^{-1} \mathbf{X}_s)^{-1} \mathbf{X}'_s \mathbf{V}_s^{-1}$ as defined in (4.6.3). Hence

$$\hat{T}_s^*(p, \mathbf{V}^{-1}, \mathbf{W}_s) = \hat{\hat{T}}_s^* \ \forall\, s : p(s) > 0$$

iff

$$\mathbf{1}'_s \mathbf{W}_s = \mathbf{b}'_s \mathbf{X}'_s \mathbf{V}_s^{-1} \ \forall\ s\text{: } p(s) > 0 \tag{6.7.2}$$

for $b'_s = (b_{0s}, b_{1s}, \ldots, b_{rs})$. (6.7.2) means

$$w_{ks} = \sum_{j=0}^{r} b_{js} x_{kj} / v_k \ \forall\, k \in s \text{ and all } s : p(s) > 0 \tag{6.7.3}$$

Example 6.8: $\xi : \mathscr{E}(Y_k) = \beta x_k$, $\mathscr{V}(Y_k) = \sigma^2 x_k$, $\mathscr{C}(Y_k, Y_{k'}) = 0$, $k \neq k' = 1, \ldots, N$. Here $\hat{T}_s^* = \hat{\hat{T}}_s^* \ \forall\, s : p(s) > 0$ (Example 4.3). (6.7.3) reduces to $w_{ks} = b_s \ \forall\, s : p(s) > 0$. This is satisfied for $w_{ks} = N/n$ (to be coupled with *srs*), $(w_{ks})^{-1} \propto \sum_s x_k$ (to accompany with Lahiri-Sen-Midzuno sampling design). Hence, for these weightings $\hat{T}_s(p, \mathbf{V}^{-1}, \mathbf{W}_s) = \hat{\hat{T}}_s^* = \hat{T}_s^* \ \forall\, s$ with $p(s) > 0$.

Note 6.1: Montenari (1987) generalised Wright's results, defining an enlarged class of *QR*-predictors to include suitable predictors for linear regression models with general covariance matrix, $\xi(\mathbf{X}, \mathbf{V})$.

Brewer, Samiuddin and Asad (1991) considered an estimator of population total exclusively from the design-based approach. The estimator $\hat{Y}_E = \sum_{i \in s} c_{is} y_i$ is such that it is, (*a*) *p*-unbiased, (*b*) has the 'ratio property' (*i.e.*, $y_i \propto x_i$, value of an auxiliary variable $x \forall i$, its sampling variance is zero), (*c*) the weighted sum of the c_{is}^2, *i.e.* $\sum_{s \in \mathscr{S}} p(s) \sum_{i \in s} w_i c_{is}^2$ is as small as possible, w_i being some suitable weights. The condition (*c*) is required, because c_{is}, should be bounded (if c_{is} are allowed to take extreme values, $\hat{Y}_E$ becomes unstable), specially for more frequently occurring samples, which calls for weights $p(s)$. Again, if y_i's are roughly proportional to x_i's, c_{is} should be bounded, specially for those units with large values of *x*. This calls for weights w_i. Solution becomes very unwieldy even for $n = 2$. For $n = 2$, they obtained an estimator $\hat{Y}_E$ by appealing to a model ($\mathscr{E}(Y_i) = \beta x_i$, $\mathscr{V}(Y_i) = \sigma_i^2$, $\mathscr{C}(Y_i, Y_{i'}) = 0$,

$i \neq i'$) to obtain unique value of w_i and compared the performance of $\hat{Y}_E$ with $\hat{Y}_{HT}$, Hájek's predictor, $\hat{T}^*_{GR}$ and $\hat{\hat{T}}^*_{GR}$ (defined in Section 6.3) under Brewer's scheme by semi-empirical and empirical studies with artificial population 1–5 and natural populations 1–20 of Rao and Bayless (1969).

6.8 DISCUSSIONS

The approach in this chapter is the use of a combination of randomisations due to model as well as sampling designs for making inference. We shall consider this approach as model-based as distinct from a model-dependent philosophy of the previous two chapters. Strategies considered here have in-built robustness against possible model-failures, as opposed to specific model-failures considered in Chapter 5. A sampling design, here, is entrusted with playing a more positive role than was demanded of it in the last two chapters. A predictor is chosen which has both model-based and design-based interpretation. An optimum strategy is one whose model-expected design-mean square error is minimum in a certain class. Thus, if the model is correctly specified, the optimality is retained in the class. If the model breaks down, it is still optimum in the class of rival strategies (including itself) which generate from the assumed model. Some asymptotic properties of these model-based strategies are considered in the next chapter.

❑ ❑ ❑

CHAPTER 7

Some Asymptotically Optimum Sampling Strategies

7.1 INTRODUCTION

The model-dependent strategies considered by Royall and his co-authors are very sensitive to model mis-specifications. The BLU-predictors have been shown to be subject to serious bias when the assumed model is incorrect (vide, also Hansen *et al.* (1983)). Works of Cassel *et al.* (1976, 1977), Sarndal (1980), Wright (1983), among others, combine the randomisations both due to model (often a multiple regression model) and sampling design. In exact analysis the strategies often do not possess any desirable properties (unbiasedness, attainment of a minimum variance bound, etc.), while in asymptotic analysis most of these properties hold. The asymptotics also provide new criteria for finding optimal sampling strategies.

7.2 BREWER'S ASYMPTOTICALLY DESIGN UNBIASED SAMPLING STRATEGIES

Starting from a superpopulation model Brewer (1979) considered a class of predictors of a particular form, imposed conditions for them to be asymptotically design unbiased (ADU) and chose the strategy with the minimum value of asymptotic average variance as the optimal one. His framework for asymptotics is based on the following assumptions:

(*i*) The original population is replicated $(k - 1)$ times to give k populations of size N each.

(*ii*) A sample s_j of size n is selected from the jth population using the same sampling procedure $(j = 1, 2, \ldots)$.

(*iii*) The k populations are aggregated to yield a population of size $N(k) = kN$, having a population total $Y(k) = kY$; the k samples are aggregated to form a sample $s(k) = \bigcup_{1}^{k} s_j$ of size $n(k) = kn$.

(*iv*) The predictor $\hat{T}_s(k)$ of $Y(k)$ is formed from $s(k)$. Finally $k \to \infty$.

Definition 7.1: $\hat{T}$ is ADU for Y if

$$\lim_{k\to\infty} E\left(\frac{\hat{T}_s(k)}{N(k)}\right) = \frac{Y}{N}$$

Under model $\xi(0, 1; \sigma_k^2)$, Brewer considered the predictor

$$\hat{T}_{BR} = \sum_s Y_k + \hat{\beta}_W \sum_{\bar{s}} x_k \tag{7.2.1}$$

where
$$\hat{\beta}_W = \left(\sum_s w_k Y_k\right) \Big/ \left(\sum_s w_k x_k\right),$$

w_j's being suitable positive constants, assumed to be bounded and x_i's taken as positive quantities. $\hat{T}_{BR}$ is model-unbiased and is of the same form as Royall's BLUP with $\hat{\beta}^*$ replaced by $\hat{\beta}_W$.

For $\hat{T}_{BR}$ to be ADU,

$$\lim_{k\to\infty} E\left\{\sum_{j=1}^{k} \hat{T}_{BRs_j}/N_k\right\}$$

$$= \frac{1}{N}\left[\sum \pi_i Y_i + \left(\sum \pi_i Y_i w_i\right)\left(\sum \pi_i x_i w_i\right)^{-1} \sum (1-\pi_i) x_i\right]$$

by Slutsky's theorem, [Cramer (1946), p. 254] (7.2.2)

$$= \frac{Y}{N}$$

which holds *iff*
$$\pi_i = \frac{\alpha}{(\alpha + w_i)} \tag{7.2.3}$$

where
$$\alpha = \left(\sum \pi_j x_j w_j\right) \Big/ \sum (1-\pi_j) x_j \tag{7.2.4}$$

Subject to the requirement (7.2.3) and (7.2.4), the asymptotic model mean square error of $(\hat{T}_{BR}/\sqrt{N})$.

$$AM(\hat{T}_{BR}/\sqrt{N}) = \lim_{k\to\infty} \varepsilon\left\{\frac{(\hat{T}_{BRs(k)} - Y(k))^2}{N(k)}\right\}$$

$$= \frac{1}{N}\left[\left\{\sum (1-\pi_i) x_i\right\}^2 \left\{\sum \pi_i w_i x_i\right\}^{-2} \left\{\sum \pi_i w_i^2 \sigma_i^2\right\} + \sum (1-\pi_i)\sigma_i^2\right] \tag{7.2.5}$$

is minimised. Substituting (7.2.4) in (7.2.3) and simplifying gives

$$AM(\hat{T}_{BR}/\sqrt{N})$$

$$= \sum \frac{w_i \sigma_i^2}{N\alpha} \tag{7.2.6}$$

Applying Cauchy-Schwartz inequality to $\{\alpha/(\alpha + w_i)\}$ and $\{(\alpha + w_i)\sigma_i^2/\alpha\}$, we have (7.2.6)

$$\geq \frac{1}{Nn}\left(\sum \sigma_i\right)^2 - \frac{1}{N}\sum \sigma_i^2, \tag{7.2.7}$$

the equality is achieved when

$$w_i = \alpha\left(\frac{\sum \sigma_j}{n\sigma_i - 1}\right) \tag{7.2.8}$$

i.e.,

$$\pi_i = \frac{n\sigma_i}{\sum \sigma_j} \tag{7.2.9}$$

for which it is required that $\sigma_i \leq \sum \sigma_j / n \forall i$. Hence, the optimal predictor from (7.2.1) is

$$\hat{T}^*_{BR} = \sum_s Y_i + \left\{\sum_s Y_i\left(\frac{1}{\pi_i} - 1\right)\right\}\left\{\sum_s x_i\left(\frac{1}{\pi_i} - 1\right)\right\}^{-1} \sum_{\bar{s}} x_i \tag{7.2.10}$$

with

$$AM(\hat{T}^*_{BR}) = \sum \sigma_i^2/\pi_i - \sum \sigma_i^2 \tag{7.2.11}$$

which equals the Godambe-Joshi lower bound [GJLB] (3.7.7) of any p-unbiased strategy under the model considered. To use $\hat{T}^*_{BR}$, any $p \in \rho_n$ with $\pi_i \propto \sigma_i$ is optimum.

We note that $\hat{T}^*_{BR}$ is ADU for any $p \in \rho_n$, satisfying (7.2.3) and (7.2.4). If only ADU-ness is required of a predictor, no reference to superpopulation is necessary and one has merely to choose $w_i \propto \left(\frac{1}{\pi_i} - 1\right)$ in (7.2.1). Of course, the form of the predictor depends on the superpopulation model.

Example 7.1: In case of srs, $\hat{T}^*_{BR}$ is ordinary ratio predictor $\hat{T}_R = (\bar{Y}_s/\bar{x}_s)X$. Thus $\hat{T}_R$ is ADU and asymptotically design consistent (ADC) [ADC follows from the works of Robinson and Tsui (1979), vide note (7.1)]. Here $w_i = 1$, $i = 1, \ldots, N$.

Example 7.2: $\sigma_i^2 = \sigma^2 x_i$ in Brewer's model. Here, Royall's BLUP, $\hat{T}^* = \hat{T}^*_R$. Brewer's strategy is $(\hat{T}^*_{BR}, \pi p\sqrt{x})$ where

$$AM(\hat{T}_{BR}^{*}) = \frac{1}{N}\left[\frac{\left(\sum\sqrt{x_k}\right)^2}{n} - X\right]\sigma^2$$

$$AM(\hat{T}^{*}) = \frac{(N-n)X\sigma^2}{nN}$$

Thus $AM(\hat{T}_{BR}^{*}) < AM(\hat{T}^{*})$. If f is small, proportional reduction in AM is

$$\lambda = \frac{XN - \left(\sum\sqrt{x_i}\right)^2}{NX} = V(\sqrt{x_i})/\bar{X}$$

which increases with the relative dispersion of x_i. For many economic variables, c.v. of $x > 1$, indicating that there is considerable reduction in asymptotic mse in using $\hat{T}_{BR}^{*}$ over $\hat{T}^{*}$.

Remark 7.1: Under Brewer's model, Rao (1984) considered the class of predictors $\sum_s Y_i + \hat{\beta}_G \sum_{\bar{s}} x_i$, where

$$\hat{\beta}_G = \sum_s (\beta_{si} - 1)Y_i \Big/ \sum_s (\beta_{si} - 1)x_i, \tag{7.2.12}$$

β_{si} being some constant attached to i. (7.2.12) is, perhaps, prompted by writing

$$Y = \sum_s Y_i + \left(Y - \sum_s Y_i\right)\sum_{\bar{s}} x_i \Big/ \left(X - \sum_s x_i\right) \tag{7.2.13}$$

We note that (7.2.13) is an identity in Y_i, not dependent on any model. Following Brewer, he gives a slightly different proof of Brewer's results. Thus an ADU-estimator of Y can be obtained from (7.2.13) without appealing to any superpopulation model.

Note 7.1: Robinson and Tsui (1979) considered a different asymptotic framework and proved ADU-ness and ADC of $\hat{T}_{BR}^{*}$. They considered a sequence of populations $\mathcal{P}_k : \{Y_i, x_i, w_{ki}, i = 1, \ldots, N_{(k)}\}, k = 1, 2, \ldots$ with $N_{(k)} \to \infty$ as $k \to \infty$. A sample s_k of size $n_{(k)}$ drawn from $\mathcal{P}_k$, using the similar sampling design for all k, having inclusion probabilities $\pi_{ki}, \pi_{kii'}$. One has thus a sequence of predictors

$$\hat{T}_{BR}^{(k)} = \sum_{s_k} Y_i + \hat{\beta}^{(k)} \sum_{\bar{s}_k} x_i$$

where,
$$\hat{\beta}^{(k)} = \sum_s w_{ki} Y_i \Big/ \sum_s w_{ki} x_i.$$

They proved.

Theorem 7.1:

$$\lim_{k\to\infty}\left\{\frac{\hat{T}_{BR}^{(k)}}{N_{(k)}} - \frac{Y'(k)}{N_{(k)}}\right\}^2 = 0$$

where $Y'(k)$ is the population total for the kth population, holds under the following assumptions.

(*i*) y_i and x_i are bounded uniformly in i and w_{ki} in (k, i);

(*ii*) $\frac{1}{N_{(k)}}\sum_{s_k} w_{ki}x_i$ is bounded away from zero, uniformly in large k, for all s_k. This happens if at most $(n_{(k)} - 1)$ of $w_{ki}x_i$, $i = 1, 2, \ldots, N_{(k)}$ are zero and the remaining are greater than some positive numbers and $\lim_{k\to\infty} \frac{N_{(k)}}{n_{(k)}} < \alpha$.

(Brewer assumed x_i and w_i are positive).

(*iii*) $\lim_{k\to\infty} \sup_{i\le j\le N_{(k)}} |\pi_{kij} - \pi_{ki}\pi_{ki}| = 0$

(*iv*) For large k, $\pi_{ki} = \alpha_k/(\alpha_k + w_{ki})$, $i = 1, \ldots, N_{(k)}$

where, $$\alpha_k = \left[\sum_1^{N_{(k)}} \pi_{kj}w_{kj}x_j\right]\left[\sum_1^{N_{(k)}} (1-\pi_{kj})x_j\right]^{-1}.$$

Note 7.2: Theorem 7.1 states that $\hat{T}_{BR}^{(k)}/N_{(k)}$ is mean square consistent and $\hat{T}_{BR}^{(k)}/N_{(k)}$ is bounded, it is ADC $[P\{|\hat{T}_{BR}^{(k)}/N_{(k)} - Y'(k)/N_{(k)}| > \delta\}] \to 0$ as $k \to \infty$, probability is in the sampling design sense] and ADU

$$\left[\lim_{k\to\infty} E(\hat{T}_{BR}^{(k)}/N_{(k)} - Y'(k)/N_{(k)}) = 0\right].$$

Note 7.3: Robinson and Tsui showed that the asymptotic value of p-expectation of model mse of $\hat{T}_{BR}/\sqrt{N_{(k)}}$ as $k \to \infty$ is the same as that obtained by Brewer and the same optimum sampling design as obtained by him is also obtained under this approach.

Note 7.4: Mukhopadhyay *et al.* (1988), considering Brewer's model and the class of projection type predictors $\hat{y} = \hat{\beta}_w X$ where $\hat{\beta}_w = \left(\sum_s w_iY_i\right)\left(\sum_s w_ix_i\right)^{-1}$ showed Hájekj's (1971) predictor

$$\hat{T}_{H\alpha} = \left(\sum_s Y_i/\pi_i\right)\left(\sum_s x_i/\pi_i\right)^{-1} X$$

to be optimal ADU with $\pi p\sigma$ (*i.e.* $\pi_i \propto \sigma_i$) as the optimal design achieving the bound (8.2.11).

They also noted that Royall's predictor $\hat{T}^* = \sum_s Y_i + \hat{\beta}^* \sum_{\bar{s}} x_i$ and projection predictor $\hat{\hat{T}}^* = \hat{\beta}^* X$ are not ADU, except for some conditions on **X** and **V** (vide Example 7.6). Little (1983) identified classes of models that yield asymptotically design consistent prediction estimates.

7.3 ADU-NESS OF WRIGHT'S (p, Q, R) STRATEGIES

Wright (1983) in his asymptotics assumed both population size and sample size increasing to infinity with sampling fraction fixed. To conceptualise this he used Brewer's framework and applied it to multiple regression model ξ (**X**, v). We have k populations with $(\mathbf{Y}^{(t)}, \mathbf{X}^{(t)})$, $\mathbf{Y}^{(t)} = \mathbf{Y}$, $\mathbf{X}^{(t)} = \mathbf{X}$ for each t the amalgamated population being of size $N_k = kN$, the quantity to be predicted is

$$\frac{1}{k}\sum_{t=1}^{k} \mathbf{a}'\mathbf{Y}^{(t)} = \mathbf{a}'\mathbf{Y}(k) = \theta(\mathbf{Y}(k)) \text{ (say)} \tag{7.3.1}$$

where $\mathbf{Y}(k) = \frac{1}{k}\sum_{t=1}^{k} \mathbf{Y}^{(t)}$, by $\frac{1}{k}\sum_{t=1}^{k} \hat{\theta}_t(\mathbf{Q}, \mathbf{R})$, (7.3.2)

$\hat{\theta}_t(\mathbf{Q}, \mathbf{R})$ being Wright's predictor of $\mathbf{a}'\mathbf{Y}^{(t)}$ based on s_t. Here $\mathbf{a}' = (a_1, ..., a_N)$ and **Q** is a positive semi-definite diagonal matrix. Strategy (p, **Q**, **R**) is ADU for $\mathbf{a}'\mathbf{Y} = \theta(\mathbf{Y})$ if

$$\lim_{k\to\infty} E\left[\frac{1}{k}\sum_{t=1}^{k} \hat{\theta}_t(\mathbf{Q}, \mathbf{R})\right] = \theta(\mathbf{Y}) \tag{7.3.3}$$

(7.3.3) tacitly implies both population size and sample size tend to infinity with sampling fraction fixed. Clearly $\theta(\mathbf{Y}(k)) = \theta(\mathbf{Y})$. On the basis of k samples s_t define pooled estimate of β as

$$\hat{\beta}(\mathbf{Q}(k)) = \left\{\sum_{j=1}^{k} (\mathbf{X}'\mathbf{Q}\Delta_j \mathbf{X})\right\}^{-1} \left\{\sum_{j=1}^{k} (\mathbf{X}'\mathbf{Q}\Delta_j \mathbf{Y}^{(j)})\right\}$$

$$= (\mathbf{X}'\mathbf{Q}\bar{\Delta}\mathbf{X})^{-1}(\mathbf{X}'\mathbf{Q}\bar{\Delta}\mathbf{Y}) \tag{7.3.4}$$

where, $\bar{\Delta} = \sum_{j=1}^{k} \Delta_j / k$, $\Delta_j = \text{Diag}\,(\delta_{ij}, i = 1, ..., N)$,

$(\delta_{ij} = 1(0)$ if $i \in (\notin)\ s_j)$.

$$\hat{\theta}_t(\mathbf{Q}, \mathbf{R}) = \mathbf{a}'[\Delta_t \mathbf{R}\mathbf{Y}^{(t)} + (I - \Delta_t \mathbf{R})\mathbf{X}^{(t)}\hat{\beta}(\mathbf{Q}(k))] \tag{7.3.5}$$

$$\frac{1}{k}\sum_{t=1}^{k}\hat{\theta}_t(\mathbf{Q},\mathbf{R}) = \mathbf{a}'[\bar{\Delta}\mathbf{R}\mathbf{Y} + (I-\bar{\Delta}\mathbf{R})\mathbf{X}\hat{\beta}(\mathbf{Q}(k))] \tag{7.3.6}$$

$$= \mathbf{a}'[\mathbf{R} + (I-\bar{\Delta}\mathbf{R})\hat{\mathbf{C}}(k)]\bar{\Delta}\mathbf{Y} \tag{7.3.7}$$

$$= \mathbf{a}'\hat{T}(k)\bar{\Delta}\mathbf{Y} \tag{7.3.8}$$

where,

$$\hat{\mathbf{C}}(k) = \mathbf{X}(\mathbf{X}'\mathbf{Q}\bar{\Delta}\mathbf{X})^{-1}\mathbf{X}'\mathbf{Q} \tag{7.3.9}$$

$$\hat{\mathbf{T}}(k) = \hat{\mathbf{C}}(k) + \mathbf{R} - \mathbf{R}\bar{\Delta}\hat{\mathbf{C}}(k) \tag{7.3.10}$$

Note that

$$\hat{\mathbf{C}}(k)\bar{\Delta}\mathbf{Y} = \mathbf{X}\hat{\beta}(\mathbf{Q}(k)),\ \hat{\mathbf{C}}(k)\bar{\Delta}\mathbf{X} = \mathbf{X}, \tag{7.3.11}$$

vide (6.5.1)–(6.5.5)

Now,

$$\lim_{k\to\infty} E(\bar{\Delta}) = \mathbf{II} \tag{7.3.12}$$

$$\lim_{k\to\infty} E(\hat{\mathbf{C}}(k)) = \mathbf{X}(\mathbf{X}'\mathbf{QIIX})^{-1}\mathbf{X}'\mathbf{Q} = \mathbf{C} \text{ (say)} \tag{7.3.13}$$

$$\lim_{k\to\infty} E(\hat{\mathbf{T}}(k)) = \mathbf{C} + \mathbf{R} - \mathbf{RIIC} = \mathbf{T} \text{ (say)} \tag{7.3.14}$$

Hence,

$$\lim_{k\to\infty} E\left[\frac{1}{k}\sum_{t=1}^{k}\hat{\theta}_t(\mathbf{Q},\mathbf{R})\right] = \mathbf{a}'\mathbf{TIIY} \tag{7.3.15}$$

Thus the strategy (p, **Q**, **R**) is ADU for $\theta(Y)$ or for the characteristic **a** *iff*

$$\mathbf{a}'(\mathbf{TII} - \mathbf{I})\mathbf{Y} = 0,\ \mathbf{Y} \in R^N \tag{7.3.16}$$

or,

$$\mathbf{a}'(I - \mathbf{RII})(I - \mathbf{CII})\mathbf{Y} = 0,\ \mathbf{Y} \in R^N \tag{7.3.17}$$

(as-$\mathbf{TII} + I = (I - \mathbf{RII})(I - \mathbf{CII})$), which immediately implies that if $\pi_i = 0$ for some i, Y_i must be zero (Note that **Y** in (7.3.16) relates to a fixed set of values of y in the population).

We note that $\theta(\mathbf{Q}, \mathbf{R})$ has been obtained from the model $\xi(\mathbf{X}, v)$ but in examining ADU-ness of (p, **Q**, **R**) we had not to appeal to the model. However, conditions of ADU-ness depend on the model through **X** as shown in (7.3.18)–(7.3.20) below.

Now,

$$\mathbf{CII} = \mathbf{X}(\mathbf{X}'\mathbf{QIIX})^{-1}\mathbf{X}'\mathbf{QII}$$

being a symmetric and idempotent matrix, is an orthogonal projection operator. For any vector **z**, **CIIz** is the projection of **z** on $\mathscr{C}$ (**X**), the column space of **X** and $(I - \mathbf{CII})\mathbf{z}$ is the projection of **z** on $\perp \mathscr{C}$ (**X**). (7.3.17) implies

$$\mathbf{a}'(I - \mathbf{R\Pi})(\mathbf{Q\Pi})^{-1}(\mathbf{Q\Pi})(I - \mathbf{C\Pi})\mathbf{Y} = 0 \ \forall \ \mathbf{Y} \in R^N$$

which holds *iff*

$$(\mathbf{Q\Pi})^{-1}(I - \mathbf{R\Pi})\mathbf{Q} \in \mathscr{C}(\mathbf{X})$$

i.e., if $$(I - \mathbf{R\Pi})\ \mathbf{a} = \mathbf{Q\Pi XA} \tag{7.3.18}$$

for some vector $$\boldsymbol{\lambda}' = (\lambda_0, \lambda_1, \ldots, \lambda_r)$$

or $$(\mathbf{\Pi}^{-1} - \mathbf{R})\mathbf{a} = \mathbf{QX}\boldsymbol{\lambda} = \mathbf{Qz} \text{ (say)} \tag{7.3.19}$$

for some vector $\mathbf{z} \in \mathscr{C}(\mathbf{X})$, or (from (7.3.18))

$$(1 - r_i\pi_i)a_i = q_i\pi_i\mathbf{x}'_i\ \boldsymbol{\lambda},\ i = 1, \ldots, N \tag{7.3.20}$$

Theorem 7.2: A strategy $(p, \mathbf{Q}, \mathbf{R})$ is ADU for $\mathbf{a}'\mathbf{Y}$ or for characteristic $\mathbf{a}$ *iff* (7.3.18) or (7.3.19) or (7.3.20) holds.

Definition 7.2: Two strategies $(p, \mathbf{Q}, \mathbf{R})$ and $(p_1, \mathbf{Q}_1, \mathbf{R}_1)$ are equivalent for the characteristic $\mathbf{a}[(p, \mathbf{Q}, \mathbf{R})\ \mathbf{a}\ (p, \mathbf{Q}_1, \mathbf{R}_1)]$, if $\hat{\theta}_s(\mathbf{Q}, \mathbf{R}) = \hat{\theta}_s(\mathbf{Q}_1, \mathbf{R}_1)$ for $\theta = \mathbf{a}'\mathbf{Y}$ and for all s with $p(s) > 0$.

Hence $(p, \mathbf{Q}, \mathbf{R})\ a\ (p, \mathbf{Q}, \mathbf{R}_1)$ if

$$\mathbf{a}'(\mathbf{R} - \mathbf{R}_1)(I - \hat{\mathbf{C}}\Delta)\mathbf{Y} = 0 \forall \mathbf{Y}, \text{ (by (6.5.1))} \tag{7.3.21}$$

or, $$\mathbf{a}'(\mathbf{R} - \mathbf{R}_1)\mathbf{Q}^{-1}\mathbf{Q}(I - \hat{\mathbf{C}}\Delta)\mathbf{Y} = 0 \forall \mathbf{Y} \tag{7.3.22}$$

$\hat{\mathbf{C}}\Delta$ being a symmetric and idempotent matrix which defines an orthogonal projection operator on $\mathscr{C}(\mathbf{X})$. Thus (7.3.22) holds *iff*

$$\mathbf{Q}^{-1}(\mathbf{R} - \mathbf{R}_1)\mathbf{a} \in \mathscr{C}(X)$$

or $$(\mathbf{R} - \mathbf{R}_1)\mathbf{a} = \quad \mathbf{QX}\boldsymbol{\lambda} \tag{7.3.23}$$

for some vector λ. (7.3.19) and (7.3.23) give

Theorem 7.3: A strategy $(p, \mathbf{Q}, \mathbf{R})$ is ADU for a *iff* $(p, \mathbf{Q}, \mathbf{R})\ \mathbf{a}\ (p, \mathbf{Q}, \mathbf{\Pi}^{-1})$.

This result strengthens the desirability of the generalised regression strategy for use in survey sampling.

Example 7.3: When $\mathbf{z} = 0$ which essentially lies in column space $\mathscr{C}(\mathbf{X})$, $\mathbf{R} = \mathbf{\Pi}^{-1}$ from (7.3.19). Thus generalised regression strategies are ADU.

Example 7.4: Consider Brewer's model. For ADU-ness, (7.3.20) requires $q_i = (1 - r_i\pi_i)a_i/\pi_i x_i\lambda$, so that for a linear predictor where $r_i = 1$, $q_i = (1 - \pi_i)a_i/\pi_i x_i\lambda$ and

$$\hat{Y} = \sum_s Y_i + \left(\sum_s Y_i(1-\pi_i)/\pi_i\right)\left(\sum_s x_i(1-\pi_i)/\pi_i\right)^{-1} \sum_{\bar{s}} x_i = \hat{T}^*_{BR}$$

is ADU. When $r_i = 0,\ q_i = a_i/\pi_i x_i \lambda$

and $$\hat{Y} = \hat{Y}_{Ha} = \left(\sum_s Y_i/\pi_i\right)\left(\sum_s x_i/\pi_i\right)^{-1} X \text{ is ADU.}$$

Example 7.5: Consider $\xi(\mathbf{X}, v)$ with first column of $\mathbf{X}$ as $\mathbf{1}$. Here one can choose $\mathbf{z} = \mathbf{1}$. Hence from (7.3.19) an ADU strategy will satisfy $q_i = (1 - r_i\pi_i)a_i/\pi_i$, $i = 1,\ldots, N$. Thus here a linear predictor with $r_i = 1$, $q_i = (1 - \pi_i)a_i/\pi_i$ and a simple projection predictor with $r_i = 0$, $q_i = a_i/\pi_i$ are ADU for $\mathbf{a}$.

Definition 7.3: $\xi(\mathbf{X}, v)$ is directed towards $\mathbf{a}$, if $\mathbf{a} \in \mathscr{C}(\mathbf{X})$. If $\xi(\mathbf{X}, v)$ is directed towods $\mathbf{a}$, (7.3.19) gives $q_i = \pi_i^{-1} - r_i$. This linear prediction strategy $(p, \mathbf{\Pi}^{-1} - I, I)$ (which covers Brewer's predictor) and projection strategy $(p, \mathbf{\Pi}^{-1}, 0)$ are ADU.

Example 7.6: Consider $(p, \mathbf{V}^{-1}, \mathbf{R})$ strategy which is ADU for $\mathbf{a}$ *iff*, $\mathbf{\Pi}^{-1}\mathbf{V}(I - \mathbf{R\Pi})\mathbf{a} \in \mathscr{C}(\mathbf{X})$ (by (7.3.18)). When $R = I$, this means $\mathbf{\Pi}^{-1}\mathbf{V}(I - \mathbf{\Pi})a \in \mathscr{C}(\mathbf{X})$. For $\mathbf{R} = 0$, requirement is $\mathbf{\Pi}^{-1}\mathbf{Va} \in \mathscr{C}(\mathbf{X})$.

7.4 ASYMPTOTIC MSE OF WRIGHT'S STRATEGIES

We have,

$$\sum_{j=1}^{k} (\mathbf{Y}^{(j)} - \hat{\mathbf{Y}}^{(j)}) = \sum_{j=1}^{k} (\mathbf{Y}^{(j)} - \hat{\mathbf{T}}(k)\Delta_j\mathbf{Y}^{(j)}) \text{ by (6.5.1)}$$

$$= \sum_{j=1}^{k} (I - \hat{\mathbf{T}}(k)\Delta_j)\mathbf{e}^{(j)}$$

where $$\mathbf{e}^{(j)} = \mathbf{Y}^{(j)} - \mathbf{X}\hat{\beta}(\mathbf{Q}(k)) \tag{7.4.1}$$

Now,

$$\mathscr{E}(\mathbf{a}'\mathbf{Y}(k) - \mathbf{a}'\hat{\mathbf{Y}}(k))^2 = \frac{1}{k^2}\mathscr{E}\left[\sum_{j=1}^{k} \mathbf{a}'(\mathbf{Y}^{(j)} - \hat{\mathbf{Y}}^{(j)})\right]^2 \Big/ k^2$$

$$= \frac{\sigma^2}{k^2}\sum_{j=1}^{k} \mathbf{a}'(I - \hat{\mathbf{T}}(k)\Delta_j)\,\mathbf{V}(I - \hat{\mathbf{T}}'(k)\Delta_j)\,\mathbf{a}$$

$$= \frac{\sigma^2}{k^2} \sum_{j=1}^{k} [\mathbf{a}'(I - \Delta_j)\mathbf{V}(I - \Delta_j)\mathbf{a} + \mathbf{a}'(I - \hat{\mathbf{T}}(k))\Delta_j \mathbf{V}(I - \mathbf{T}'(k))\mathbf{a}]$$

$$= \frac{\sigma^2}{k}[\mathbf{a}'(I - \bar{\Delta})\mathbf{V}\mathbf{a} + \mathbf{a}'(I - \hat{T}(k))\bar{\Delta}\mathbf{V}(I - \mathbf{T}'(k))\mathbf{a}] \quad (7.4.2)$$

Hence,

$$\lim_{k \to \infty} E\mathscr{E}(\mathbf{a}'\mathbf{Y}(k) - \mathbf{a}'\hat{\mathbf{Y}}(k))^2$$

$$= [\mathbf{a}'(I - \mathbf{II})\mathbf{V}\mathbf{a} + \mathbf{a}'(I - \mathbf{T})\mathbf{II}\,\mathbf{V}(I - \mathbf{T}')\mathbf{a}]\sigma^2 \quad (7.4.3)$$

If $(p, \mathbf{Q}, \mathbf{R})$ is ADU for $\mathbf{a}$, $\mathbf{a}'\mathbf{T}\mathbf{II}\mathbf{Y} = \mathbf{a}'\mathbf{Y} \forall \mathbf{Y} \in R^N$, when (7.4.3) simplifies to

$$\sigma^2 \mathbf{a}'(\mathbf{II}^{-1} - I)\mathbf{V}\mathbf{a} = \sigma^2 \sum_{1}^{N} a_i^2 (\pi_i^{-1} - 1) v_i = v_0 \text{ (say)} \quad (7.4.4)$$

Theorem 7.4: If $(p, \mathbf{Q}, \mathbf{R})$ is ADU for $\mathbf{a}$, its asymptotic variance is given by (7.4.4). For $\pi_i = N^{-1} n \forall i$, this reduces to $(N - n)\Sigma a_i^2 v_i / n = v_0'$ (say) (assuming $\sigma^2 = 1$).

Applying Cauchy-Schwartz inequality,

$$\left(\sum |a_i| \sqrt{v_i}\right)^2 \le \left(\sum \pi_i\right)\left(\sum a_i^2 v_i \pi_i^{-1}\right).$$

Hence,

Theorem 7.5: Within the class of strategies $(p, \mathbf{Q}, \mathbf{R})$ that are ADU for $\mathbf{a}$ and have fixed sample size n, the minimum asymptotic variance is

$$\left(\sum |a_i| \sqrt{v_i}\right)^2 \Big/ n - \sum a_i^2 v_i = v^* \text{ (say)} \quad (7.4.5)$$

which is achieved *iff*,

$$\pi_i \propto \sqrt{v_i}\, |a_i|. \quad (7.4.6)$$

Note 7.5: The result on lower bound of variance by Godambe and Joshi, therefore, also holds for (p, e) where e is any ADU predictor of $\mathbf{1}'\mathbf{Y}$. Since $\hat{T}_{GR}$ is always (for any $p \in \rho_n$) ADU, it follows that asymptotically,

$$E\mathscr{E}\ \{(e - \mathbf{1}'\mathbf{Y})^2\} \ge E\mathscr{E}\{(\hat{T}_{GR} - \mathbf{1}'Y)^2\} = \sum (\pi_i^{-1} - 1) v_i$$

for any p-unbiased predictor e. We also note that asymptotic average variance of $(p, \hat{T}_{GR})$ under $\xi(\mathbf{X}, v)$ depends only on π_i's involved in p and is independent of $\mathbf{Q}$ in $\hat{\beta}(\mathbf{Q})$ so long as $\mathbf{Q}$ is a positive definite matrix.

Note 7.6: Sarndal (1980) considered asymptotic mse of $\hat{T}_{GR}$ for π^{-1}-weighting and BLU-weighting in the sense of Brewer. Considering (6.3.5), $E(A_s) = \Sigma v_k(\pi_k^{-1} - 1)$ which is minimised for the choice $\pi_k \propto \sqrt{v_k}$, when it attains the minimum value of Godambe-Joshi lower bound. $E(B_s)$ is of the order of n^{-2} under certain mild assumptions on sampling design. It has, therefore, been shown that for a given p, asymptotic mse of $\hat{T}_{GR}$ is given by (7.4.4), which for an optimal sampling design $p \in \rho_n$, attains (7.4.5).

A strategy may be called (robust) r-best if it is ADU and attains v^*. If $|a_i|\sqrt{v_i}$ is called the 'relevance' of unit i, **a** $(p, \mathbf{Q}, \mathbf{R})$ strategy is r-best *iff* it is ADU with $\pi_i \propto$ relevance of i.

The same strategy can be r-best for **a** and **a*** *iff* | **a** | and | **a*** | are proportional. Thus a strategy which is r-best for Y is also r-best for all the contrasts among the subclass totals.

Example 7.7: Under $\xi(\mathbf{X}, v)$, simple BLU-projection predictor strategy $(p, \mathbf{V}^{-1}, \mathbf{0})$ is r-best for **a** *iff* $a_i v_i/\pi_i = x_i'\boldsymbol{\lambda}$ for some $\boldsymbol{\lambda} \in R^{r+1}$ (ADU; Example 7.6) and $\pi_i \propto |a_i|\sqrt{v_i}$. Hence we require (sign a_i) $\sqrt{v_i} = x_i'\boldsymbol{\lambda}$. In particular a best BLU-projection strategy exists for Y *iff* $\left(\sqrt{v_1}, \ldots, \sqrt{v_N}\right)' \in \mathscr{C}(\mathbf{X})$.

7.5 ASYMPTOTIC EFFICIENCY OF AN ADU STRATEGY

Let $(p, \mathbf{Q}, \mathbf{R})$ be an ADU strategy for **a** with sample size n and inclusion probability π_i and asymptotic variance $v(n)$. Let $(p^0, \mathbf{Q}^0, \mathbf{R}^0)$ be the best strategy, if it exists, for **a** such that for sample size n^*, its asymptotic variance $v^*(n^*) = v(n)$. The asymptotic efficiency of $(p, \mathbf{Q}, \mathbf{R})$ with respect to $(p^0, \mathbf{Q}^0, \mathbf{R}^0)$ is

$$\frac{n^*}{n} = \frac{\left(\sum |a_i|\sqrt{v_i}\right)^2}{n\sum a_i^2 v_i/\pi_i} = e \text{ (say)} \tag{7.5.1}$$

$$\text{If } \pi_i \propto a_i^2 v_i, \qquad e = (1 + CV_{a\sqrt{v}}^2)^{-1} \tag{7.5.2}$$

where $CV_{a\sqrt{v}}$ is the coefficient of variation of the relevance of $|a_i|\sqrt{v_i}$.

For designs with $\pi_i = n/N$, efficiency is also given by (7.5.2). Thus an equal probability ADU strategy is reasonably efficient if $CV_{a\sqrt{v}}^2 \simeq 0$, *i.e.* if all the units are more or less equally relevant. Such a situation is expected to hold if ξ $(\mathbf{X}, v)$ is approximately homoscedastic.

In case $CV_{a\sqrt{v}} > 1$ so that asymptotic efficiency of an equal probability sampling ADU strategy is less than 50%, one should use an optimal design with $\pi_i \propto |a_i|\sqrt{v_i}$.

When efficiency of an equal probability ADU strategy is poor, one can use stratification to improve upon the efficiency. Let $\{\mathscr{P}_h, h = 1, \ldots, L\}$ be a stratification of $\mathscr{P}$ and let CV_h be the coefficient of variation of $|a_i|\sqrt{v_i}$ within $\mathscr{P}_h$. Consider stratification such that $CV_h \leq \in$, $h = 1, \ldots, L$, for some positive small $\in$; sample allocation

$$n_h = n\left(\sum_{i \in \mathscr{P}_h} |a_i|\sqrt{v_i}\right) \Big/ \sum |a_i|\sqrt{v_i} \tag{7.5.3}$$

and stratified random sampling with $\pi_i = n_h/N_h \forall i \in \mathscr{P}_h$. The strategy (*p*, **Q**, **R**) with these values of π_i are used in each stratum. Such a strategy may be called a 'strongly stratified strategy' with sample size *n*. For this strategy the asymptotic variance

$$\dot{\mathscr{V}}(\mathbf{a}'\mathbf{Y}) = \sum_1^L \dot{\mathscr{V}}(\mathbf{a}_h'\hat{\mathbf{Y}}_h) = \sum_1^L \sum_{i \in \mathscr{P}_h} a_i^2 (N_h/n_h - 1)v_i$$

$$= \frac{1}{n}\left(\sum_1^N |a_i|\sqrt{v_i}\right)\sum_{h=1}^L (1 + CV_h^2)\sum_{i \in \mathscr{P}_h} |a_i|\sqrt{v_i} - \sum a_i^2 v_i \tag{7.5.4}$$

$$\left(\text{since } 1 + CV_h^2 = N_h \sum_{i \in \mathscr{P}_h} a_i^2 v_i \Big/ \left(\sum_{i \in \mathscr{P}_h} |a_i|\sqrt{v_i}\right)^2\right)$$

$$\leq \frac{1}{n}(1 + \in^2)\left(\sum |a_i|\sqrt{v_i}\right)^2 - \sum a_i^2 v_i \tag{7.5.5}$$

$\mathbf{a}'_h$ is the row-vector consisting of a_i, $i \in \mathscr{P}_h$, $\dot{\mathscr{V}}$ denoting asymptotic variance, so that for $\in$ sufficiently small, the minimum asymptotic variance (7.4.5) is approximately achieved. (Such a technique has also been exploited by Godambe (1982) for finding near-optimal strategies (Section 9.4). Hence the asymptotic efficiency of a strongly stratified strategy is at least $(1 + \in^2)^{-1}$.

Note 7.7: The allocation rule (7.5.3) is not actually optimal in terms of minimising $\dot{\mathscr{V}}(\mathbf{a}'\mathbf{Y})$ for a strongly stratified strategy. The optimal allocation rule is

$$n_h \propto N_h \left(\sum_{i \in \mathscr{P}_h} a_i^2 v_i\right)^{1/2}. \tag{7.5.6}$$

for which $$\dot{\mathscr{V}}(\mathbf{a}'\mathbf{Y}) = \frac{1}{n}\left[\sum_1^L (1 + CV_h^2)^{1/2} \sum_{1 \in \mathscr{P}_h} |a_i|\sqrt{v_i}\right]^2 - \sum v_i \tag{7.5.7}$$

However, for small $\in$, all strongly stratified strategies are almost equivalent in terms of asymptotic efficiency. A strongly stratified strategy with equal sample size for all strata may be of advantage. By (7.5.3), one should then construct strata such that

$\sum_{i \in \mathscr{P}_h} |a_i|\sqrt{v_i}$ is constant for all *h*.

7.6 ADU-NESS UNDER THE SET-UP OF CONTINUOUSLY INCREASING NESTED POPULATIONS

Robinson and Sarndal (1983) examined the ADU-ness and ADC of $\hat{T}_{GR}$ under the following framework of asymptotic analysis.

(*i*) There are M nested populations $\mathcal{P}_1, \subseteq \mathcal{P}_2, \ldots, \subseteq \mathcal{P}_M$ with $\mathcal{P}_1 \subseteq \mathcal{P}_2 \subseteq \ldots \subseteq \mathcal{P}_M$, $\mathcal{P}_t$ being of size N_t, $N_1 < N_2 < \ldots$

(*ii*) A sample s_t of size n_t is selected from $\mathcal{P}_t$ following the sampling design p_t with inclusion probability $\pi_{kt}, \pi_{kk't}, \pi_{kt} = \sum_{\substack{s \ni k \\ s \in \mathcal{S}_t}} p(s)$

$$\pi_{kk't} = \sum_{\substack{s \ni k \\ s \in \mathcal{S}_t}} p(s), \text{ etc,}$$

$\mathcal{S}_t$ being the sample space corresponding to $\mathcal{P}_t$. Define $\delta_{kt} = 1(0)$ according as $k \in (\notin) s_t$; $E(\delta_{kt}) = \pi_{kt}$, $E(\delta_{kt}\delta_{k't}) = \pi_{kk't}$, $\sum \pi_{kt} = n_t$, $\sum \pi_{kk't} = n_t(n_{t-1})$; $n_1 < n_2 < \ldots$; however, samples are not nested, *i.e.*, s_t is not necessarily a subset of s_{t+1}.

(*iii*) The population $(\mathbf{Y}^{(t)}, \mathbf{X}^{(t)})$, follows the model $\mathbf{Y}^{(t)} = \mathbf{X}^{(t)}\beta + \mathbf{e}^{(t)}$. $\mathscr{E}(\mathbf{e}^{(t)}) = \mathbf{0}$, $\mathscr{D}(\mathbf{Y}^{(t)}) = \sigma^2\mathbf{V}^{(t)}$, $\mathbf{V}^{(t)} = \text{Diag}\,(v_i, i = 1, \ldots, N_t)$,

$$\mathbf{X}^{(t)} = \begin{bmatrix} x_{10} & \cdots & x_{1r} \\ - & \cdots & - \\ x_{N_t0} & \cdots & x_{N_tr} \end{bmatrix}, \quad \mathbf{e}^{(t)} = \begin{bmatrix} e_1^t \\ \vdots \\ e_{N_t}^t \end{bmatrix}$$

β, σ^2 being unknown constants; *i.e.*, $\mathbf{Y}^{(t)}$ follows the model $\xi^{(t)}(\mathbf{X}^{(t)}, v^{(t)})$.

(*iv*) $M \to \infty$.

As a predictor of the population parameter $\theta_t(\mathbf{Y}^{(t)}) = \mathbf{a}'_t\mathbf{Y}^{(t)}$, $\mathbf{a}'_t = (a_1, \ldots, a_{Nt})$, consider Wright's predictor

$$\hat{\theta}_t(\mathbf{Y}^{(t)}) = \mathbf{a}_t'[\mathbf{X}^{(t)}\hat{\beta}(\mathbf{Q}_t) + \Delta_t\mathbf{R}_t\hat{\mathbf{e}}^{(t)}]$$

where $\Delta_t = \text{Diag}\,(\delta_{it}, i = 1, \ldots, N_t)$, $\mathbf{R}_t = \text{Diag}\,(r_{it}, i = 1, \ldots, N_t)$, r_{it} being the weight attached to i in $\mathcal{P}_t$ in strategy $(p_t, \mathbf{Q}_t, \mathbf{R}_t)$ and

$$\hat{\beta}(\mathbf{Q}_t) = (\mathbf{X}_t'\mathbf{Q}_t\Delta_t\mathbf{X}_t)^{-1}(\mathbf{X}_t'\mathbf{Q}_t\Delta_t\mathbf{Y}_t),$$

a generalised predictor of β based on s_t, $\hat{\mathbf{e}}^{(t)} = \mathbf{Y}^{(t)} - \mathbf{X}^{(t)}\hat{\beta}(\mathbf{Q}_t)$. Denote by η the probability distribution of infinite dimensional random variable $(Y_1, Y_2, \ldots)$. Under this set-up,

Definition 7.4: A predictor $\hat{\theta}(\mathbf{Q}, \mathbf{R})$ is ADU for θ if

$$\lim_{k \to \alpha} E_t[(\hat{\theta}_t(\mathbf{Q}_t, \mathbf{R}_t) \mid \mathbf{Y}^{(t)}) - \theta_t(\mathbf{Y}^{(t)})] = 0 \tag{7.6.1}$$

with η probability one.

Definition 7.5: $\hat{\theta}(\mathbf{Q}, \mathbf{R})$ is ADC for θ if

$$\lim_{t \to \infty} P_t[(\hat{\theta}_t(\mathbf{Q}_t, \mathbf{R}_t) - \theta_t | > \in | \mathbf{Y}^{(t)}] = 0 \tag{7.6.2}$$

for all $\in$, with η-probability one. Here P_t, E_t are respectively probability and expectation wrt p_t.

By Markov's inequality

$$P\{|\hat{\theta}_t - \theta_t| > \in\} < \frac{E|\hat{\theta}_t - \theta_t|}{\in}$$

Hence taking limit of both sides as $t \to \infty$, ADU-ness of $\hat{\theta}$ implies ADC of $\hat{\theta}$.

Under the above framework Robinson and Sarndal proved.

Theorem 7.6: The generalised regression strategy (p, $\mathbf{Q}$, $\mathbf{II}^{-1}$) is ADU and ADC for $\bar{Y}$ under assumptions

$$\underline{\lim}_{t \to \infty} (N_t \min_i \pi_{it}) = \alpha \tag{a.1}$$

$$\lim_{t \to \infty} N_t^{-1} \sum Y_i^2 < \alpha \text{ with } \eta\text{-probability one} \tag{a.2}$$

$$\overline{\lim}_{t \to \infty} \max_{1 \neq i'} |\pi_{ii't} / \pi_{it}\pi_{it} - 1| = 0 \tag{a.3}$$

$$\overline{\lim}_{t \to \infty} N_t^{-1} \sum_i x_{ij}^2 < \alpha,\ j = 0, 1, \ldots, r \tag{a.4}$$

$$\overline{\lim}_{t \to \infty} E\left(\sum_{j=0}^{r} \hat{\beta}_j^2(\mathbf{Q}_t)\right) < \alpha \text{ with } \eta\text{-probability one} \tag{a.5}$$

The assumption (a.1), (a.3) are satisfied by srs.

Note that the form of the generalised regression predictor $\hat{T}_{GR}$ has been suggested on the basis of the model $\xi(\mathbf{X}, v)$. The predictor is model-unbiased if $\hat{\beta}_j$ is so of β_j and is in general design-unbiased. However, $\hat{T}_{GR}$ is ADU and ADC whether the model $\xi(\mathbf{X}, v)$ holds or not.

Robinson and Sarndal also found the asymptotic mse of $\hat{T}_{GR}(Q)$ for arbitrary $\mathbf{Q}$ under a very general superpopulation model τ that obeys

$$\mathscr{E}(Y_i) = \mu_i,\ \mathscr{V}(Y_i) = \sigma_i^2, \tag{b.1}$$

the Y_i's are uncorrelated and

$$\overline{\lim}_{t \to \infty} \sum \mu_k^2 / N_t < \alpha,\ \sigma_k^2 < C < \alpha \tag{b.2}$$

C is independent of k; assume there exist constants β_j, $J = 0, \ldots, r$ such that

$$n_t \sum_{j=0}^{r} \mathscr{E}\{(\hat{\beta}_{jt} - \beta_j)^2 | s_t\} < H < \alpha \tag{b.3}$$

for all samples s_t. If ξ (**X**, v) is the true model,

$$\mu_i = \sum_0^r \beta_j x_{ij}\,,\ i = 1, 2, \ldots$$

Additional assumptions on sampling designs are:

$$\lim_{t\to\infty} N_t \min_i \pi_{it} / n_t > 0 \qquad (b.4)$$

$$\lim_{t\to\infty} N_t^2 \max_{i\neq i'} | \pi_{ii't} - \pi_{it}\pi_{i't} | / n_t < \alpha \qquad (b.5)$$

apart from (a.l) and (a.3). They proved:

Theorem 7.7: Under assumptions (a.1), (a.3), (a.4) and (b.1)–(b.5),

$$n_t \mathscr{E}(\hat{\bar{Y}}_t(\mathbf{Q}, \mathbf{II}^{-1}) - \bar{Y})^2 = A_t + B_t + \epsilon_t \qquad (7.6.3)$$

where,
$$A_t = \frac{n_t}{N_t^2}\sum_i \sigma_i^2(\pi_{it}^{-1} - 1)$$

$$B_t = \frac{n_t}{N_t^2}\sum\sum_{i<i'} (\pi_{it}\pi_{i't} - \pi_{ii't})\left(\frac{g_i}{\pi_{it}} - \frac{g_i'}{\pi_{i't}}\right)^2 \qquad (7.6.4)$$

$$g_i = \mu_i - \sum_{j=0}^r \beta_j x_{ij}$$

and $\epsilon_t \to 0$ as $t \to \infty$, $\mathscr{E}$ denoting expectation wrt τ.

Note 7.8: The term A_t is independent of μ_i. β_t vanishes if the assumed model τ is ξ(**X**, v). If $\mu_i = \sum_{j=0}^r \beta_j x_{ij} + \theta\pi_{it}$ (i = 1, ..., N_t) then also B_t vanishes. In these cases the minimum value of the expected variance $n_t^{-1}A_t$ is minimised wrt π_{it} by choosing $\pi_{it} = n_t\sigma_i \Big/ \sum_1^{N_t} \sigma_k$ when minimum average variance is

$$\frac{\left(\sum \sigma_i\right)^2}{N_t n_t} - \frac{\sum \sigma_i^2}{N_t}$$

which is the same as Godambe-Joshi lower bound of average variance of any p-unbiased estimator of $\bar{Y}_t$.

Thus if the supposed multiple regression model is true, at least up to its first moment (covariance between Y_i, Y_j have to be zero, of course, ($i \neq j$)), asymptotic mse of $\hat{T}_{GR}$ attains Godambe-Joshi lower bound for a specified design under some mild restrictions on the higher order moments of the distributions in the model and the class of sampling designs considered. If, however, the supposed model is not ξ(**X**, v) (at least up to its first order moment), B_t does not vanish and no simple result for the asymptotically optimum inclusion probabilities follows.

Note 7.9: Under the same framework Fuller and Isaki (1981) computed the expected mse of the projection predictor

$$\hat{T}_{\mathbf{II}^{-2}} = \mathbf{1'X}\hat{\beta}(\mathbf{II}^{-2})$$

$$\hat{\beta}(\mathbf{II}^{-2}) = (\mathbf{X'}_s\mathbf{II}_s^{-2}\mathbf{X}_s)^{-1}\mathbf{X'}_s\mathbf{II}_s^{-1}\mathbf{Y}_s$$

under $\xi(\mathbf{X}, V, \rho)$. It is shown that the average mse of the strategy that couples the above predictor with a design with $\pi_i \propto \sqrt{v_i}$ is asymptotically minimum in the class of mse's of linear predictors provided

$$\mathbf{II'1} \in \mathscr{C}(\mathbf{X}) \tag{7.6.5}$$

The condition (7.6.5) is a special case of the condition (7.3.20).

Under a finite population model which mimics the superpopulation model $\xi(\mathbf{X}, v)$, Deng and Chikkara (1991) examined the ADU-ness and ADC of $\hat{T}_{\mathbf{Q}}, \hat{\hat{T}}_{\mathbf{Q}}$ and $\hat{T}_{GR}(\mathbf{Q})$ when the sample is drawn by srswor.

7.7 DISCUSSIONS

Three types of framework for asymptotics have been considered in the analysis: (*i*) replicating the original population an infinite number of times with sampling fraction fixed [Brewer (1979), Wright (1983)], (*ii*) an infinite number of populations of increasing sizes drawn with samples of increasing sizes drawn from them, observing $\lim_{t\to\infty} N_t / n_t < \infty$ [Robinson and Tsui (1979)], and (*iii*) an infinite sequence of nested populations of increasing sizes with samples, not necessarily nested, of increasing sizes drawn from them [Isaki and Fuller (1982), Robinson and Sarndal (1983)].

Wright extended Brewer's work to multiple regression model, thus obtaining more detailed conditions (7.3.18)–(7.3.20) than (7.2.3), (7.2.4) for ADU-ness. ADU-ness of generalised regression strategies, Royal's BLU-strategies, BLU-projection strategies, Brewer's strategies have been examined and established.

In the third framework, ADU-ness and ADC of generalised regression strategy have been established. It is not difficult to prove the same properties for Wright's strategies under this set-up.

For a given values of π_i's, all the ADU strategies have the same asymptotic variance under the supposed model [(7.2.11), (7.4.4), A_t in (7.6.3)]. For optimal choice of π_i's, they attain the Godambe-Joshi lower bound. It would be of interest to differentiate between ADU-strategies, using the same set of π_i-values, by studying, for example, their second order efficiency in some sense.

❑ ❑ ❑

CHAPTER

8

Optimality of a Linear Predictor Under Multiple Regression Models

8.1 INTRODUCTION

Pfefferman (1984), Brewer, Tam and Hanif (1988), Tam (1986, 1988), among others, addressed the question when a linear predictor becomes best under multiple regression models $\xi(\mathbf{X}, \mathbf{V})$. Clearly, this would entail conditions some of which relate to the sampling design and some to the model. They also considered conditions on $\mathbf{Q}$ when $\hat{T}_{\mathbf{Q}}, \hat{\hat{T}}_{\mathbf{Q}}$ and $\hat{T}_{GR}(\mathbf{Q})$ coincide under $\xi(\mathbf{X}, \mathbf{V})$ so that $\hat{T}_{\mathbf{Q}}$ and $\hat{\hat{T}}_{\mathbf{Q}}$ may share the ADU and ADC properties of $\hat{T}_{GR}(\mathbf{Q})$ under these circumstances. The conditions under which $\hat{T}^*$ and $\hat{\hat{T}}^*$ coincide have been considered by Sarndal (1980 *a,b*), discussed in Section 4.6. Sampling designs for which $\hat{T}^*, \hat{\hat{T}}^*, \hat{T}^*_{GR}$ become identical have also been investigated.

Tam (1988) confronted a predictor to $\hat{T}_{GR}(\mathbf{Q})$ and investigated the conditions for a linear prediction strategy to be asymptotically best under $\xi(\mathbf{X}, v)$. Sufficient conditions were found to be model-unbiasedness and a condition relating the coefficient vector of the predictor, dispersion matrix of $\mathbf{Y}$ and weight matrix in $\hat{\beta}$ to the design matrix of the multiple regression model. A predictor satisfying the second condition has been termed weakly robust or robust under covariance matrix misspecification. Relevance of these two conditions for robust prediction under $\xi(\mathbf{X}, v)$ has been shown. These results are discussed in this chapter.

8.2 BLU-NESS OF A LINEAR PREDICTOR UNDER ξ(X, V)

We first consider:

Lemma 8.1: (Tam, 1986). The necessary and sufficient conditions for linear predictor $\mathbf{h}'\mathbf{Y}_{\mathbf{s}}$ where $\mathbf{h}' = (h_{s1}, \ldots, h_{ns})$ to be BLUP of $\mathbf{1}'\mathbf{Y}$ under $\xi(\mathbf{X}, \mathbf{V})$ are: for all $s : p(s) > 0$, (*i*) $\mathbf{h'X}_s = \mathbf{1'X}$ (8.2.1)

(*ii*) $\bar{V}_s\mathbf{h} - \mathbf{K1} \in \mathscr{C}(\mathbf{X}_s)$ (8.2.2)

i.e., there exist constants $\boldsymbol{\lambda} = (\lambda_0, ..., \lambda_r)$ such that $\bar{\mathbf{V}}_s\mathbf{h} - \mathbf{K1} = \mathbf{X}_s\,\boldsymbol{\lambda}$. Here $\mathbf{K} = (\bar{\mathbf{V}}_s\mathbf{V}_{sr})$. We shall write here (up to the end of note 8.1) $\bar{V}_s$ as the $n \times n$ submatrix corresponding to $k \in s$ of $\mathbf{V}$ in the model $\xi(\mathbf{X}, \mathbf{V})$.

Proof: (8.2.1) is a condition on sampling design and is required for $\xi(\mathbf{X}, \mathbf{V})$-unbiasedness of $\mathbf{h'Y}_s$. From (4.4.9), $\mathbf{h'Y}_s$ is BLUP of $\mathbf{1'Y}$ *iff* $(\mathbf{h'} - \mathbf{1'K'V}^{-1}_s)\mathbf{Y}_s$ is BLUE of its expectation. This holds by Lemma 5.2 *iff*

$$\bar{\mathbf{V}}_s(\mathbf{h} - \bar{\mathbf{V}}_s^{-1}\mathbf{K1}) \in \mathscr{C}(\mathbf{X}_s)$$

i.e.,

$$\bar{\mathbf{V}}_s\mathbf{h} - \mathbf{K1} \in \mathscr{C}(\mathbf{X}_s)$$

If $\mathbf{V}$ is a diagonal matrix, (8.2.2) reduces to

$$\mathbf{V}_s(\mathbf{h} - \mathbf{1}) \in \mathscr{C}(\mathbf{X}_s),$$

as obtained by Pereira *et al.* (1983).

Example 8.1: (Tallis, 1978). The necessary and sufficient conditions for $\hat{T}_s^*(\mathbf{X}, v) = N\bar{Y}_s\ \forall\, s: p(s) > 0$ are: (*i*) sample is balanced on $\mathbf{X}$ and (*ii*) $\mathbf{V}_s\mathbf{1} \in \mathscr{C}(\mathbf{X}_s)$.

Example 8.2: (Pfefferman, 1984). The expansion predictor $\hat{T} = N\bar{Y}_s$ is BLUP under $\xi(X, \mathbf{V})$ *iff* (*i*) $(N/n)\ \mathbf{1'}_n\mathbf{X}_s = \mathbf{1'X}$ *i.e.*, the sample is balanced.

(*ii*) $(N - n)[\bar{\mathbf{V}}_s\mathbf{1}_n - n\mathbf{V}_{sr}\mathbf{1}_{N-n}] \in \mathscr{C}(\mathbf{X}_s)$.

Example 8.3: Consider Royall's predictor,

$$\hat{T}^*(\mathbf{X}, v) = [\mathbf{1'}_\mathbf{n} + (\mathbf{1'}_{N-n}\mathbf{X}_s)(\mathbf{X'}_s\mathbf{V}_s^{-1}\mathbf{X}_s)^{-1}{}_s\,\mathbf{X'}_s\mathbf{V}_s^{-1}]\mathbf{Y}_s; \tag{8.2.3}$$

Generalised classical regression predictor of Cochran (1977)

$$\hat{T}_{GR}^*(\mathbf{X}, v) = N\bar{Y}_s + N\sum_{j=0}^{r}(\bar{X}_j - \bar{x}_{js})\hat{\beta}_j^*$$

$$= \left[\frac{N}{n}\mathbf{1}'_n + \left\{\mathbf{1'X} - \frac{N}{n}\mathbf{1}'_n\mathbf{X}_s\right\}(\mathbf{X}'_s\mathbf{V}_s^{-1}\mathbf{X}_s)^{-1}\mathbf{X}'_s\mathbf{V}_s^{-1}\right]\mathbf{Y}_s \tag{8.2.4}$$

which is a special case of Sarndal's generalised regression predictor (with $\pi_i = n/N$ $\forall\ i$).

Projection-type predictor of Cochran-Sarndal (1980), Isaki-Fuller (1982),

$$\hat{\hat{T}}^*(\mathbf{X}, v) = [\mathbf{1'X}(\mathbf{X}'_s\mathbf{V}_s^{-1}\mathbf{X}_s)^{-1}\mathbf{X}'_s\mathbf{V}_s^{-1}]\mathbf{Y}_s \tag{8.2.5}$$

Each of (8.2.3)–(8.2.5) is $\xi(X, \mathbf{V})$-unbiased.

For $\hat{T}^*(\mathbf{X}, v)$,

$$\bar{\mathbf{V}}_s\mathbf{h} - \mathbf{K1} = \bar{\mathbf{V}}_s\mathbf{V}_s^{-1}\mathbf{X}_s(\mathbf{X}_s'\mathbf{V}_s^{-1}\mathbf{X}_s)^{-1}(\mathbf{X}_{\bar{s}}'\mathbf{1}_{N-n}) - \mathbf{V}_{sr}\mathbf{1}_{N-n} \tag{8.2.6}$$

For (8.2.4), this is

$$\bar{\mathbf{V}}_s\mathbf{V}_s^{-1}\mathbf{X}_s(\mathbf{X}_s'\mathbf{V}_s^{-1}\mathbf{X}_s)^{-1}\left\{\mathbf{X}'\mathbf{1} - \frac{N}{n}\mathbf{X}_s'\mathbf{1}_n\right\}$$

$$+\frac{(N-n)}{n}\bar{\mathbf{V}}_s\mathbf{1}_n - \mathbf{V}_{sr}\mathbf{1}_{N-n} \tag{8.2.7}$$

and for (8.2.5),

$$\bar{\mathbf{V}}_s\mathbf{V}_s^{-1}\mathbf{X}_s(\mathbf{X}_s'\mathbf{V}_s^{-1}\mathbf{X}_s)^{-1}\mathbf{X}'\mathbf{1} - \bar{\mathbf{V}}_s\mathbf{1}_n - \mathbf{V}_{sr}\mathbf{1}_{N-n} \tag{8.2.8}$$

Hence $\hat{T}^*$ is $\xi(\mathbf{X}, \mathbf{V})$-best *iff*

$$\mathbf{u} \in \mathscr{C}\,(\mathbf{X}_s) \tag{8.2.9}$$

$\hat{T}^*_{GR}$ is BLUP under $\xi(\mathbf{X}, \mathbf{V})$ *iff*

$$(F-1)\mathbf{Z}_s - \mathbf{u} \in \mathscr{C}\,(\mathbf{X}_s) \tag{8.2.10}$$

$\hat{\hat{T}}^*$ is ξ $(\mathbf{X}, \mathbf{V})$-best *iff*

$$\mathbf{Z}_s + \mathbf{u} \in \mathscr{C}\,(\mathbf{X}_s) \tag{8.2.11}$$

where

$$\mathbf{u} = \mathbf{V}_{sr}\mathbf{1}_{N-n},\ \mathbf{Z}_s = \bar{\mathbf{V}}_s 1_n,\ F = \frac{N}{n}.$$

If, in particular, $\mathbf{V}$ has diagonal elements $v_1, \ldots, v_2, \ldots, v_N$ with constant non-zero correlation ρ, then

$$\mathbf{u} \in \mathscr{C}\,(\mathbf{X}_s)\ \textit{iff}\ \sqrt{\mathbf{v}} \in \mathscr{C}\,(\mathbf{X}_s) \tag{8.2.12}$$

and

$$\mathbf{Z}_s \in \mathscr{C}\,(\mathbf{X}_s)\ \textit{iff}\ \mathbf{v} \in \mathscr{C}\,(\mathbf{X}_s) \text{ and (8.2.12) holds} \tag{8.2.13}$$

where $\sqrt{\mathbf{v}} = \mathrm{v}\left(\sqrt{v_1}, \ldots, \sqrt{v_n}\right)'$, $\mathbf{v} = (v_1, \ldots, v_n)'$, assuming without loss of generality that the sample is (1, ..., n). Hence, if (8.2.13) holds, all these predictors are identical and $\xi(X, \mathbf{V})$-best for this particular form of $\mathbf{V}$.

Note 8.1: Bolfarine and Rodrigues (1988) showed that under $\xi(\mathbf{X}, \mathbf{V})$, $\hat{T}^*(\mathbf{X}, \mathbf{V}) = \hat{\hat{T}}^*(\mathbf{X}, \mathbf{V})$ *iff* (8.2.11) holds. Thus for $\mathbf{X} = \mathbf{1}, \mathbf{V} = (1 - \rho)I_N + \rho\mathbf{11}'$,

$\hat{T}^* = \hat{\hat{T}}^* = N\sum_s Y_k / n$. If, however, $\mathscr{C}(Y_i, Y_{i+h}) = \sigma^2{}_{\rho h}$, $h > 0$ and $\rho h = \rho^h$ or $q(1 - h/L)$ for some $L > 0$ and $0 < q < 1$, condition (8.2.11) is not satisfied. For $\mathbf{X} = (x_1, \ldots, x_N)'$ and $\mathbf{V}$ diagonal with $v_i \propto x_i$, $\hat{T}^* = \hat{\hat{T}}^* = \hat{T}_R$, the usual ratio predictor.

Note 8.2: Consider two models $\xi(\mathbf{X}, \mathbf{V}_1)$ and $\xi(\mathbf{X}, \mathbf{V}_2)$ and let $\hat{\hat{T}}_1^*$ and $\hat{\hat{T}}_2^*$ be the corresponding projection predictor of Y. It follows that $\hat{\hat{T}}_1^* = \hat{\hat{T}}_2^*$ *iff*

$$\mathbf{1}'\mathbf{X}\hat{\beta}_1^* = \mathbf{1}'\mathbf{X}\hat{\beta}_2^*$$

which is equivalent to

$$\mathbf{1}'\mathbf{X}(\mathbf{X}_s'\mathbf{V}_s^{*-1}\mathbf{X}_s)^{-1}\mathbf{X}_s'\mathbf{V}_s^{*-1}(I_n - \mathbf{X}_s'(\mathbf{X}_s'\mathbf{V}_s^{-1}\mathbf{X}_s)^{-1}\mathbf{X}_s'\mathbf{V}_s^{-1})\mathbf{Y}_s = 0$$

where $\mathbf{V}_s^*$, $\mathbf{V}_{\bar{s}}^*$ etc. are the elements of V_2. A sufficient condition for (8.2.14) to hold is

$$\mathbf{X}_s'\mathbf{V}_s^{*-1}(I_n - \mathbf{X}_s'(\mathbf{X}_s'\mathbf{V}_s^{-1}\mathbf{X}_s)^{-1}\mathbf{X}_s'\mathbf{V}_s^{-1}) = 0$$

As an example, if $\mathbf{X} = \mathbf{1}$, $\mathbf{V}_1 = \sigma^2 I_N$, (8.2.14) holds *iff* for all i, $\sum_{j=1}^{n} w_{ij} = \sum_{i=1}^{n}\sum_{j=1}^{n} w_{ij}/n$ where w_{ij} is the $(i, j)^{\text{th}}$ element of V_s^{*-1}. The condition are satisfied if $\mathbf{V}_2 = \rho I_N + (1 - \rho)11'$.

Note 8.3: Suppose $\mathbf{X} = (\mathbf{X}^*, \bar{\mathbf{X}})$. If due to inadvertent omission of certain explanatory variables corresponding to $\tilde{\mathbf{X}}$, the model ξ $(\mathbf{X}^*, v)$ in place of $\xi(\mathbf{X}, v)$ is used, the $\xi(\mathbf{X}^*, v)$-best predictor $\mathbf{h}'\mathbf{Y}_s$ will be $\xi(\mathbf{X}, v)$-best if $\mathbf{h}'\tilde{\mathbf{X}}_s = \mathbf{1}'\tilde{\mathbf{X}}$ (condition 8.2.1) holds, where $\tilde{\mathbf{X}}' = (\tilde{\mathbf{X}}_s, \tilde{\mathbf{X}}'_{\bar{s}})$. If mean-per-unit predictor $N\bar{Y}_s$ is used for $\xi(\mathbf{X}^*, v)$, this condition reduces to the requirement of balanced sampling designs in respect of omitted variables.

If due to mis-specification, some non-explanatory variables are included in the model, *i.e.*, if $\mathbf{W}^* = (\mathbf{X}, \mathbf{W})$, in place of $\mathbf{X}$, is used or if there is a mis-specification of the dispersion matrix in the multiple regression model, the resulting best predictor under the working model remains best under $\xi(\mathbf{X}, \mathbf{V})$ *iff* (8.2.2) holds. These results have been generalised by Tam (1988) in the case of asymptotic analysis (Section 8.5).

8.3 CHOOSING WEIGHTS IN $\hat{\beta}$ TO MAKE $\hat{T}^*, \hat{\hat{T}}^*, \hat{T}_{GR}$ IDENTICAL

Under $\xi(\mathbf{X}, v)$ for an arbitrarily weighted predictor $\hat{\beta}(\mathbf{Q})$ of $\hat{\beta}$ consider three predictors of $\mathbf{1}'\mathbf{Y}$:

$$T_{\mathbf{Q}} = \sum_{s} Y_k + \sum_{\bar{s}} \mathbf{x}_k' \hat{\beta}(\mathbf{Q}),$$

$$\hat{\hat{T}}_{\mathbf{Q}} = \sum \mathbf{x}_k' \hat{\beta}(\mathbf{Q})$$

$$\hat{T}_{GR} = \sum_{s} (Y_k - \mathbf{x}_k' \hat{\beta}(\mathbf{Q}))/\pi_k + \sum \mathbf{x}_k' \hat{\beta}(\mathbf{Q})$$

We have,
$$\hat{T}_{RC}(\mathbf{Q}) = \delta \sum_{s} (Y_k - \hat{\beta}'(\mathbf{Q}) x_k) + \sum x_k' \hat{\beta}(\mathbf{Q}) \tag{8.3.1}$$

reduces to $\hat{T}_{\mathbf{Q}}$ for $\delta = 1$ and $\hat{\hat{T}}_{\mathbf{Q}}$ for $\delta = 0$.

Now, $\hat{T}_{RC} = \hat{T}_{GR}$ if for all $s : p(s) > 0$,

$$\mathbf{1}_n' \mathbf{W}_s \mathbf{Y}_s = \mathbf{1}_n' \mathbf{W}_s \mathbf{X}_s \hat{\beta}(\mathbf{Q}) \tag{8.3.2}$$

where $\mathbf{W}_s = \text{Diag}\,(\pi_k^{-1} - \delta I_n,\ k \in s)$. Under (8.3.2), model-based predictors $\hat{T}_{\mathbf{Q}}$ and $\hat{\hat{T}}_{\mathbf{Q}}$ become identical with the generalised regression predictor $\hat{T}_{GR}$ and thus share the later's property of unbiasedness; the leading term in their expected variances would be the Godambe-Joshi lower bound which again is minimised for sampling designs $p \in \rho_n$ with $\pi_i \propto \sqrt{v_i}$.

Example 8.4: $\xi : \mathscr{E}(Y_k) = \beta x_k,\ \mathscr{V}(Y_k) = \sigma^2 v_k,\ \mathscr{C}(Y_k, Y_{k'}) = 0,\ k \neq k'$. Here (8.3.2) is satisfied *iff*

$$\hat{\beta} = \sum_{s} Y_k (\pi_k^{-1} - \delta) \Big/ \sum_{s} x_k (\pi_k^{-1} - \delta)$$

where $\hat{T}_{RC}$ reduces to Hájek's predictor (1971), $\hat{Y}_{Ha}$ for $\delta = 0$ and to Brewer's predictor $\hat{T}_{BR}^{*}$ for $\delta = 1$. A sufficient condition for (8.3.2) to hold is obtained from:

Theorem 8.1: (Brewer *et al.*, 1988). If there exist a $n \times (r + 1)$ matrix $\mathbf{Z}_s$ such that (*i*) $\mathbf{Z}_s' \mathbf{X}_s$ is of full rank (*ii*) $\mathbf{W}_s \mathbf{1}_n$ is in the column space of $\mathbf{Z}_s$, then, provided $\mathbf{1}_n' \mathbf{W}_s \mathbf{X}_s \neq 0$, a sufficient condition for (8.3.2) is

$$\hat{\beta} = (\mathbf{Z}_s' \mathbf{X}_s)^{-1} \mathbf{Z}_s' \mathbf{Y}_s \tag{8.3.3}$$

Proof: $\mathbf{1}_n' \mathbf{W}_s \mathbf{Y}_s - \mathbf{1}_n' \mathbf{W}_s \mathbf{X}_s (\mathbf{Z}_s' \mathbf{X}_s)^{-1} \mathbf{Z}_s' \mathbf{Y}_s$

$$= \boldsymbol{\lambda}' \mathbf{Z}_s' \mathbf{Y}_s - \boldsymbol{\lambda}' \mathbf{Z}_s' \mathbf{X}_s (\mathbf{Z}_s' \mathbf{X}_s)^{-1} \mathbf{Z}_s \mathbf{Y}_s$$

(since $\mathbf{W}_s \mathbf{1}_n = \mathbf{Z}_s \boldsymbol{\lambda}$ for some $\boldsymbol{\lambda}$)

$$= 0$$

If $\mathbf{X}_s$ includes $\mathbf{1}_n$, (8.3.2) is always satisfied for the choice $\mathbf{Z}_s = \mathbf{W}_s\mathbf{X}_s$ and hence $\hat{\beta} = (\mathbf{X}_s'\mathbf{W}_s\mathbf{X}_s)^{-1}\mathbf{X}_s'\mathbf{W}_s\mathbf{Y}_s$.

Example 8.5: For $\hat{\beta}$, $\mathbf{Z}_s' = \mathbf{X}_s'\mathbf{V}_s^{-1}$, (8.3.2) therefore reduces to $\mathbf{W}_s\mathbf{1}_s = \mathbf{V}_s^{-1}\mathbf{X}_s\boldsymbol{\lambda}$ for some $\boldsymbol{\lambda}$. An equivalent condition with $\delta = 0$ was considered by Isaki and Fuller (1982) as a condition for $\hat{\hat{T}}^*$ to be ADU.

Note 8.4: For π^{-1} weighting (Section 6.2), $\mathbf{W}_s = \mathbf{II}^{-1}$.

Note 8.5: For minimisation of Godambe-Joshi lower bound, one takes $\pi_k \propto \sqrt{v_k}$. Again for $\hat{\beta}^*$, $\mathbf{Z}_s' = \mathbf{X}_s'\mathbf{V}_s^{-1}$. Isaki and Fuller (1982), therefore, suggested β to be estimated by $(\mathbf{X}_s'\mathbf{II}_s^{-2}\mathbf{X}_s)^{-1}\mathbf{X}_s'\mathbf{II}_s^{-2}\mathbf{Y}_s$. $\hat{T}_{RC}$ is, therefore, ADU in this situation, being identical with $\hat{T}_{GR}$ if $(\mathbf{II}_s^{-1} - \delta\mathbf{I}_n)1_s \in \mathscr{C}(\mathbf{II}_s^{-2}\mathbf{X}_s)$ *i.e.*, $(\mathbf{II}_s - \delta\mathbf{II}_s^2)\mathbf{1}_s \in \mathscr{C}(\mathbf{X}_s)$. It has, therefore, been suggested to include π_k and π_k^2 into the set of auxiliary variables as a condition for $\hat{T}_{RC}$ to be ADU.

Note 8.6: Wright (1983) showed that $\hat{T}_{GR}(\mathbf{Q})$ is ADU for all positive definite diagonal matrices $\mathbf{Q}$ subject to certain restrictions on the sampling design ((7.3.18)–(7.3.20)). Furthermore, the strategy $(\hat{T}_{GR}, p_0)$ where $p_0 \in \rho_n$ with $\pi_i \propto \sqrt{v_i}$ attains the minimum value of GJLB. Brewer *et al.*, concluded, these results also hold for any consistent estimator $\hat{\beta}$ of β including that owing to choosing $\mathbf{Z}_s$ as an instrumental variable satisfying the conditions of Theorem 8.1, provided that the sampling designs ensures $n\{(\hat{\beta}' - \beta')(\mathbf{X}'\mathbf{1} - \mathbf{X}_s'\mathbf{II}_s^{-1}\mathbf{1}_n)\}^2 \to 0$ asymptotically. The method of construction of Z_s is given by:

Theorem 8.2: (Brewer *et al.* 1988). If $\mathbf{1}_N'\mathbf{W}_s\mathbf{X}_s \neq 0$, there exists an instrumental variable $\mathbf{Z}_s = (\mathbf{V}_s^{-1}\mathbf{X}_s\mathbf{L}_s, \mathbf{W}_s\mathbf{1}_n)$, where $\mathbf{L}_s$ is a $(r + 1) \times r$ matrix obtained by deleting from I_{r+1} a suitable column such that $\mathbf{Z}_s'\mathbf{X}_s$ is of full rank, when choice of $\hat{\beta}$ as in (8.3.3) satisfies (8.3.2).

$\mathbf{L}_s$ may be constructed by deleting the jth column of I_{r+1} if $\mathbf{1}_n'\mathbf{W}_s\mathbf{X}_s(\mathbf{X}_s'\mathbf{V}_s^{-1}\mathbf{X}_s)^{-1}\mathbf{e}_j \neq 0$ where e_j is the $(r + 1) \times 1$ vector $(0, \ldots, 0, 1, 0, \ldots, 0)'$ with 1 in the jth place.

Note 8.7: In the choice of $\mathbf{Z}_s$ as stated in Theorem 8.2, $\mathbf{Z}_s$ is the same as $\mathbf{V}_s^{-1}X_s$ occurring in $\hat{\beta}^*$ except one column which is to be replaced by $\mathbf{W}_s\mathbf{1}_n$. The choice of the column to be replaced is given by the last part of Theorem 8.2.

If the entries in a column of $\mathbf{V}_s^{-1}\mathbf{X}_s$ are proportional to $\mathbf{W}_s\mathbf{1}_n$ then replacing that column by $\mathbf{W}_s\mathbf{1}$ make $\hat{\beta}$ identical with $\hat{\beta}^*$. A suitable choice, therefore, is to replace that column of $\mathbf{V}_s^{-1}\mathbf{X}_s$ the entries of which are most nearly proportional to the corresponding elements of $\mathbf{W}_s\mathbf{1}_s$.

Example 8.6: $\xi : \mathscr{E}(Y_k) = \beta_1 x_{k1} + \beta_2 x_{k2}$, $\mathscr{V}(Y_k) = \sigma^2 v_k$, $\mathscr{E}(Y_k, Y_{k'}) = 0$, $k \neq k'$. Consider

$$\mathbf{Z}_s = [\mathbf{V}_s^{-1}\mathbf{X}_s\mathbf{L}_s, \mathbf{W}_s\mathbf{1}_n]$$

where $L_s = (\mathbf{1}_1, \mathbf{1}_2)'$, $\mathbf{1}_1$, $\mathbf{1}_2$ left arbitrary for the time being. Then

$$\mathbf{Z}'_s\mathbf{X}_s = \begin{bmatrix} \sum_{k=1}^{n} \frac{(x_{k1}1_1 + x_{k2}1_2)x_{k1}}{v_k} & \sum_{k=1}^{n} \frac{(x_{k1}1_1 + x_{k2}1_2)x_{k2}}{v_k} \\ \sum\sqrt{v_k}\sum\frac{x_{k1}}{n\sqrt{v_k}} - \delta\sum x_{k1} & \sum\sqrt{v_k}\sum\frac{x_{k2}}{n\sqrt{v_k}} - \delta\sum x_{k1} \end{bmatrix}$$

where $$\pi_k = \frac{n\sqrt{v_k}}{\sum\sqrt{v_k}}.$$

If the second row of $\mathbf{Z}'_s\mathbf{X}_s$ viz. $\mathbf{1}'_n\mathbf{W}_s\mathbf{X}_s$ is null, then $|\mathbf{Z}'_s\mathbf{X}_s| = 0$. Otherwise, $|\mathbf{Z}'_s\mathbf{X}_s| = 0$ for choices $1_1 = g_s 1_2$, g_s being some scalar function of v_k, x_{k1}, x_{k2}, δ. It follows that for all choices of 1_1 and 1_2 such that $1_1 \neq g_s 1_2$, $\mathbf{Z}'_s\mathbf{X}_s$ is of full rank, when one can use $\hat{\beta}$ as in (8.3.3).

However, the choice $L_s = (1, 0)'$ or $(0, 1)'$ are particularly simple as then $\mathbf{V}_s^{-1}\mathbf{X}_s L_s$ is the same as $\mathbf{V}_s^{-1}\mathbf{X}_s$ with one column excluded. The column to be excluded is given by Note 8.7.

8.4 SAMPLING DESIGNS FOR WHICH $\hat{T}^*, \hat{\hat{T}}^*, \hat{T}_{GR}$ COINCIDE

We have seen in Example 8.5 that under conditions

$$\mathbf{V}_s\mathbf{W}_s\mathbf{1}_s = \mathbf{X}_s\lambda, \tag{8.4.1}$$

where $\mathbf{W}_s = \text{Diag}(\mathbf{\Pi}^{-1} - I_n, k \in s)$, $\boldsymbol{\lambda}$ a $(r + 1) \times 1$ vector of suitably chosen constants, $\hat{T}^*_{GR} = T^*$, under $\xi(\mathbf{X}, v)$. Sarndal (1980*c*) showed that under certain restrictions on the second order inclusion-probabilities $\hat{T}^*_{GR}$ is ADU (vide Note 7.6). Hence under sampling designs p satisfying (8.4.1), $\hat{T}^*$ is ADU.

Solving for π_i from (8.4.1)

$$\pi_i(\boldsymbol{\lambda}) = \frac{v_i}{(v_i + \boldsymbol{\lambda}'\mathbf{x}_i)} \tag{8.4.2}$$

Assuming that $\mathbf{X}\boldsymbol{\lambda}$ has positive elements for some $\boldsymbol{\lambda}$, one can use this to make $0 \leq \pi_i(\boldsymbol{\lambda}) \leq 1$ for $i = 1, ..., N$ for which designs $\hat{T}^* = \hat{T}^*_{GR}$. However, such a design may be a variable size design.

To find a fixed size sampling design satisfying (8.4.2) for which it is required that $\pi_i \in [0, 1]$ for all i and $\sum \pi_i = n$, we note that as $\pi_i(\boldsymbol{\lambda})$ is a continuous function of $\boldsymbol{\lambda}$ and $\sum \pi_i(0) = N$ and $\lim_{\lambda \to \infty} \sum \pi_i(\boldsymbol{\lambda}) = 0$, there exists a $\boldsymbol{\lambda}^0$ with non-negative members for which $\sum \pi_i(\boldsymbol{\lambda}^0) = n$ and $0 < \pi_i(\boldsymbol{\lambda}^0) \le 1 \ \forall \ i = 1, \ldots, N$.

Example 8.7: If $\mathbf{V1}_N = \mathbf{X}\boldsymbol{\Upsilon}$ for some $1 \times (r+1)$ vector $\boldsymbol{\Upsilon}$, then taking $\boldsymbol{\lambda} = (N-n)\Upsilon/n$ one gets $\pi_i(\boldsymbol{\lambda})$ with $\pi_i = n/N \ \forall \ i = 1, \ldots, N$.

Example 8.8: Let $k = \sum \sqrt{v_i}/n$, $\sqrt{\mathbf{V}} = \text{Diag}\,(\sqrt{v_1}, \ldots, \sqrt{v_N})$. Suppose $k\sqrt{\mathbf{V1}_N} - \mathbf{V1}_N = \mathbf{X}\boldsymbol{\lambda}$ for some $\boldsymbol{\lambda}$. The design p with $\pi_i = \sqrt{v_i}/k$ satisfies (8.4.1). Note that under above conditions, and for the above sampling design, $\hat{T}^*$ is ADU and attains minimum value of GJLB to the expected variance of a sampling strategy (Example 7.6).

Considering $\hat{\hat{T}}^*$, it has been noted (Section 6.3) that under conditions

$$\mathbf{II}_s^{-1}\mathbf{1}_n = \mathbf{V}_s^{-1}\mathbf{X}_s\boldsymbol{\lambda}_s \tag{8.4.3}$$

where $\boldsymbol{\lambda}_s = (\lambda_{0s}, \ldots, \lambda_{rs})'$, $a(r+1) \times 1$ vector of suitable constants, $\hat{T}^*_{GR} = \hat{\hat{T}}^*$.

Example 8.9: If $\mathbf{V}_s^{1/2}\mathbf{1}_s = \mathbf{X}_s\boldsymbol{\lambda}_s$ then (8.4.4) is satisfied for $\mathbf{II}_s = k^{-1}\,\mathbf{V}_s^{1/2}$, where $k = \sum \sqrt{v_i}/n$.

Example 8.10: For $\mathbf{V}_s\mathbf{1}_s = \mathbf{X}_s\mu$ and $\pi_k = n/N$ for all k, where μ is $(r+1) \times 1$ vector of suitable constants, (8.4.3) is satisfied and $\hat{T}^* = \hat{\hat{T}}^* = \hat{\hat{T}}^*_{GR}$ and each is ADU and ADC for $\mathbf{1}'Y$.

Note 8.8: Wright (1983) has shown that $\hat{T}_{GR}(\mathbf{Q})$ is ADU under some mild restrictions on the second order inclusion probabilities of the sampling design where $\mathbf{Q}$ is any positive definite diagonal matrix and $(\hat{T}_{GR}(\mathbf{Q}), p_0)$ attains minimum value of GJLB asymptotically. Here $\hat{\beta}^*$ is retained in each of $\hat{T}^*, \hat{\hat{T}}^*$ and $\hat{\hat{T}}^*_{GR}$ and sampling designs are identified for which $\hat{T}^*$ or $\hat{\hat{T}}^*$ or both equal $\hat{\hat{T}}^*_{GR} \ \forall\, s : p(s) > 0$. Since such designs are not often optimal designs p_0 (as in Example 8.10), the asymptotic mse of the resulting strategy tends to be higher than the minimum value of GJLB. p_0 has been defined in Note 8.6.

8.5 BESTNESS OF A LINEAR STRATEGY UNDER ξ(X, v)

By a linear strategy we shall mean a linear predictor $\mathbf{h}'\mathbf{Y}_s$ where $\mathbf{h}' = (h_{1s}, \ldots, h_{rs})$, coupled with a sampling design p.

Definition 8.1: (Wright, 1983; Tam, 1988). A strategy (e, p) is best in the asymptotic sense or 'asymp. best' or (simply) best if it is ADU and attains $(\text{GJLB})_0$ (minimum value of GJLB) $\left[=\left(\sum \sqrt{v_i}\sigma\right)^2/n-\sum v_i\sigma^2\right]$ asymptotically.

From Wright it is known that $\hat{T}_{GR}(\mathbf{Q})$ is ADU for any p (with some mild restrictions on the second order inclusion probabilities) for any positive definite diagonal matrix $\mathbf{Q}$. Further, $H(\mathbf{Q}) = (p_0, \hat{T}_{GR}(\mathbf{Q}))$ is 'asymp. best' under $\xi(\mathbf{X}, v)$.

Theorem 8.3: (Tam, 1988). The sufficient conditions for a strategy $(\mathbf{h}'\mathbf{Y}_s, p_0)$ to be best under $\xi(\mathbf{X}, v)$ are, for all $s : p(s) > 0$,

$$(i)\ \mathbf{h}'\mathbf{X}_s = \mathbf{1}'\mathbf{X} \text{ (model-unbiasedness)} \tag{8.5.1}$$

$$(ii)\ \mathbf{Q}_s^{-1}[\mathbf{h} - k\mathbf{V}_s^{-1/2}\mathbf{1}_s] \in \mathscr{C}(\mathbf{x}_s) \tag{8.5.2}$$

for some positive definite diagonal matrix $\mathbf{Q}_s$, where $k = \sum \sqrt{v_i}/n$.

Proof: (8.5.1) is required for ξ-unbiasedness of $\mathbf{h}'\mathbf{Y}_s$.

From (8.5.2)

$$\mathbf{h}' = \boldsymbol{\lambda}_s'\mathbf{X}_s'\mathbf{Q}_s + \mathbf{1}_n'\mathbf{II}_s^{-1} \quad (\text{because } kV_s^{-1/2} = \mathbf{II}_s^{-1})$$

for some $(r + 1) \times 1$ vector $\boldsymbol{\lambda}_s' = (\boldsymbol{\lambda}_{0s}, \ldots, \lambda_{rs})$.

Using (i) to solve for $\boldsymbol{\lambda}_s$,

$$\boldsymbol{\lambda}_s' = \mathbf{1}'\mathbf{X}(\mathbf{X}_s'\mathbf{Q}_s\mathbf{X}_s)^{-1} - \mathbf{1}'\mathbf{II}_s^{-1}\mathbf{X}_s'(\mathbf{X}_s'\mathbf{Q}_s\mathbf{X}_s)^{-1} \tag{8.5.3}$$

Using (8.5.3) in (8.5.2) one gets $\mathbf{h}'\mathbf{Y}_s = \hat{T}_{GR}(\mathbf{Q}_s)$.

Note 8.9: If only (8.5.2) holds $\mathbf{h}'\mathbf{Y}_s - \hat{T}_{GR}(\mathbf{Q}_s) = (\mathbf{h}'\mathbf{X}_s - \mathbf{1}'\mathbf{X})\hat{\beta}(\mathbf{Q}_s)$ [= 0 *iff* (8.5.1) also holds]. Thus, condition (8.5.2) links up $\mathbf{h}'\mathbf{X}_s, \hat{T}_{GR}(\mathbf{Q}_s)$ with an estimate of model-bias.

Suppose the true model is $\xi(\mathbf{X}, v)$, but due to ignorance the statistician uses a working model ξ' which may differ either in design matrix or in dispersion matrix or both. Again mis-specification in design matrix may be either due to, (i) omission of explanatory variables from those in $\mathbf{X}$ or (ii) inclusion of non-explanatory variables along with those in $\mathbf{X}$. One is, therefore, interested in finding conditions under which a strategy based on a working model remains 'asymp. best' under $\xi(\mathbf{X}, v)$.

(a) Mis-specification of design matrix: Let $\mathbf{X} = (\mathbf{P}, \widetilde{X})$ where $\mathbf{P}$ is $N \times m$, $p + m = r + 1$.

and,

$$\mathbf{P} = \begin{bmatrix}\mathbf{P}_s \\ \mathbf{P}_{\bar{s}}\end{bmatrix}, \quad \widetilde{\mathbf{X}} = \begin{bmatrix}\widetilde{\mathbf{X}}_s \\ \widetilde{\mathbf{X}}_{\bar{s}}\end{bmatrix}$$

Let also $\mathbf{W} = (\mathbf{X}, \tilde{\mathbf{Q}})$ where $\mathbf{Q}$ is $N \times q$,

$$\mathbf{W} = \begin{bmatrix} \mathbf{W}_s \\ \mathbf{W}_{\bar{s}} \end{bmatrix}, \ \tilde{\mathbf{Q}} = \begin{bmatrix} \tilde{\mathbf{Q}}_s \\ \tilde{\mathbf{Q}}_{\bar{s}} \end{bmatrix}.$$

Lemma 8.2: $(\hat{T}^*_{GR}(\mathbf{P}_s, v), p_0)$ is best under $\xi(\mathbf{X}, v)$ if $\hat{T}^*_{GR}(P_s, v)$ satisfies (8.5.1) *i.e.*, if

$$\mathbf{h}'(\mathbf{P}_s)\tilde{\mathbf{X}}_s = \mathbf{1}'\tilde{\mathbf{X}} \ \forall \ s : p(s) > 0.$$

Proof: We have to show $\hat{T}^*_{GR}(\mathbf{P}_s, v)$ satisfies (8.5.1) and (8.5.2).

$$\mathbf{h}(\mathbf{P_s}) - \mathbf{II}_s^{-1}\mathbf{1_n} = (\mathbf{V_s^{-1}P_s})(\mathbf{P_s'V_s^{-1}P_s})(\mathbf{P'1} - \mathbf{P_s'II_s^{-1}1_n}).$$

Hence

$$\mathbf{V_s}[\mathbf{h}(\mathbf{P_s}) - \mathbf{II_s}^{-1}\mathbf{1_n}] = \mathbf{P_s}(\mathbf{P'_sV_s}^{-1}\mathbf{P_s})(\mathbf{P'1} - \mathbf{P'_sII_s}^{-1}\mathbf{1_n}) \in \mathscr{C}(\mathbf{P_s}) \subset \mathscr{C}(\mathbf{X_s})$$

so that (8.5.2) holds.

By the condition of the Lemma, (8.5.1) holds.

Lemma 8.3: $(\hat{T}^*_{GR}(\mathbf{W}_s, v), p_0)$ is best under $\xi(\mathbf{X}, v)$ *iff* $\hat{T}^*_{GR}(\mathbf{W_s}, v)$ satisfies (8.5.2).

Proof: Here (8.5.1) is automatically satisfied. The Lemma only requires satisfaction of (8.5.2). It is shown in Theorem 8.4, and (8.5.6) that (8.5.2) gives conditions for robustness wrt mis-specification of covariance matrix.

Since (8.5.1) and (8.5.2) give conditions for robustness of predictors under model mis-specifications we consider.

Definition 8.2: *Weak Robustness:* A predictor $\mathbf{h}'\mathbf{Y}_s$ is said to be weakly robust or robust under covariance matrix mis-specification *iff* for all s with $p(s) > 0$ (8.5.2) holds.

Definition 8.3: *Strong Robustness:* A predictor $\mathbf{h}'\mathbf{Y}_s$ is said to be strongly robust or robust against design matrix and covariance matrix mis-specifications *iff* for all s with $p(s) > 0$, both (8.5.1) and (8.5.2) hold.

Clearly, a strongly robust predictor is ADU and coupled with p_0 gives the best sampling strategy under $\xi(\mathbf{X}, v)$.

The following theorem gives the conditions for $\hat{T}_{GR}$ based on multiple regression model with one explanatory variable u and dispersion matrix $\mathbf{G}$ to be weakly robust under $\xi(\mathbf{X}, v)$.

Theorem 8.4: The predictor,

$$\mathbf{h}'\mathbf{Y}_s = \mathbf{1}_s'\mathbf{II}_s^{-1}\mathbf{Y}_s + (\mathbf{1}'\mathbf{U} - \mathbf{1}_s'\mathbf{II}_s^{-1}\mathbf{U}_s)\hat{\beta}(\mathbf{G}_s)$$

is weakly robust for positive definite diagonal matrix $\mathbf{G}_s$ if there exists a positive definite diagonal matrix $\mathbf{Q}_s$ such that $\mathbf{Q}_s^{-1}\mathbf{G}_s\mathbf{U}_s \in \mathscr{C}(\mathbf{X}_s) \forall s : p(s) > 0$.

Here $\mathbf{U} = (\mathbf{U}'_s \mathbf{U}'_{\bar{s}})$ is a $N \times 1$ vector,

$$\hat{\beta}(\mathbf{G}_s) = (\mathbf{U}'_s\mathbf{G}_s\mathbf{U}_s)^{-1}(\mathbf{U}'_s\mathbf{G}_s\mathbf{Y}_s) \text{ and } \mathbf{II}_s = k^{-1}V_s^{1/2}.$$

Proof: Clearly, $Q_s^{-1}[\mathbf{h} - k\overline{\mathbf{V}}_s^{1/2}\mathbf{1}_s] = c\mathbf{Q}_s^{-1}\mathbf{G}_s\mathbf{U}_s$, c being some scalar. Hence by the assumption of the theorem $\mathbf{h}'\mathbf{Y}_s$ satisfies (8.5.2).

In particular if $\mathbf{U}_s \in \mathscr{C}$ $(\mathbf{X}_s)$, $\mathbf{h}'\mathbf{Y}_s$ always robust. We note that HT-predictor with $\pi_i \propto \sqrt{v_i}$ is weakly robust under ξ $(\mathbf{X}, v)$.

Note 8.10: If $\mathbf{U}_s = \mathbf{X}_s$, (with number of explanatory variables ≥ 1), weakly robust predictor of Theorem 8.4 becomes strongly robust under $\xi(\mathbf{X}, v)$.

We also note that in Theorem 8.4 though dispersion matrix $\mathbf{G}_s$ has been considered in the working model, sampling design has been related to $\mathbf{V}_s$ of the true model $\xi(\mathbf{X}, v)$. The sampling design related to the alternative dispersion matrix has been considered in (8.5.6).

If there are many weakly robust predictors of $\mathbf{1}'\mathbf{Y}$ one should choose the one with the least absolute value of model-bias. This follows from

Lemma 8.4: In a weakly robust ADU predictor $\mathbf{h}'\mathbf{Y}_s$ of $\mathbf{1}'\mathbf{Y}$

$$E\mathscr{E}\ (\mathbf{h}'\mathbf{Y}_s - \mathbf{1}'\mathbf{Y})^2$$

$$= E\mathscr{E}\ (\hat{T}_{GD} - \mathbf{1}'\mathbf{Y})^2 + E\{(\mathbf{h}'\mathbf{X}_s - \mathbf{1}'\mathbf{X})\beta\}^2 \tag{8.5.4}$$

asymptotically for all p, where $\hat{T}_{GD}$ is the generalised difference predictor.

Proof: Since $\mathbf{h}'\mathbf{Y}_s$ is weakly robust it follows by Note 8.9,

$$\mathbf{h}'\mathbf{Y}_s - \hat{T}_{GR}(\mathbf{Q}_s) = (\mathbf{h}'\mathbf{X}_s - \mathbf{1}'\mathbf{X})\hat{\beta}(\mathbf{Q}_s)$$

for some positive definite diagonal matrix $\mathbf{Q}_s$. Now

$$E\ \mathscr{E}\,(\mathbf{h}'\mathbf{Y}_s - \mathbf{1}'\mathbf{Y})^2 = E\mathscr{E}\ (\hat{T}_{GD} - \mathbf{1}'\mathbf{Y})^2 + \{(\mathbf{h}'\mathbf{X}_s - \mathbf{1}'\mathbf{X})\beta\}^2$$

$$+ (\mathbf{h}'\mathbf{X}_s - \mathbf{1}_s'\mathbf{\Pi}_s^{-1}\mathbf{X}_s)(\hat{\beta}(\mathbf{Q}_s) - \beta)]$$

Hence the average mse of $\mathbf{h}'\mathbf{Y}_s$ attains (8.5.4) asymptotically, from Wright (1983).

The following lemma compares between a ADU predictor $\mathbf{h}_1'\mathbf{Y}_s$ and a ADU weakly robust predictor $\mathbf{h}'\mathbf{Y}_s$ under $\xi(\mathbf{X}, v)$.

Lemma 8.5: $E\mathscr{E}\ (\mathbf{h}_1'\mathbf{Y}_s - \mathbf{1}'\mathbf{Y})^2 - E\mathscr{E}\ (\mathbf{h}'\mathbf{Y}_s - \mathbf{1}'\mathbf{Y})^2$

$$= E\,\mathscr{V}\,(\mathbf{h}_1'\mathbf{Y}_s) - E\,\mathscr{E}\,(\hat{T}^*_{GR}(\mathbf{Q}_s) - \mathbf{1}'\mathbf{Y})^2 + E\{\gamma(\mathbf{h}_1'\mathbf{Y}_s)\}^2$$

$$- E\{\gamma(\mathbf{h}'(\mathbf{Y}_s)\}^2 - \mathscr{V}\,(\mathbf{1}'\mathbf{Y})$$

since $\mathbf{h}'\mathbf{Y}_s$ satisfies (8.5.2) and by (4.2.5), where $\gamma(e)$ is the model-bias in the predictor e.

$$= E\left[\sum_j I_{sj}h_{1sj}^2 v_j\sigma^2\right] - \sum v_j\sigma^2/\pi_j + E\{\gamma(\mathbf{h}_1'\mathbf{Y}_s)\}^2 - E\{\gamma(\mathbf{h}'\mathbf{Y}_s)\}^2 \tag{8.5.5}$$

where $\mathbf{h}'_1 Y_s = \sum I_{sj} h_{1sj} Y_j$, $I_{sj} = 1(0)$ if $j \in (\notin) s$. Since $\mathbf{h}'_1 \mathbf{Y}_s$ is ADU for $\mathbf{1}'\mathbf{Y}$, $E(I_{sj} h_{1sj}) = 1$, $E(I_{sj}) = \pi_j$, $E(h_{1sj} \mid I_{sj} = 1) = \pi_j^{-1}$.

Hence $\qquad E(I_{sj} h_{1sj}^2) = E(I_{sj}) E(h_{1sj}^2 \mid I_j) \geq \pi_j^{-1}$.

Hence (8.5.5) $\geq E\,\{\Upsilon\,(\mathbf{h}'_1 \mathbf{Y}_s)\}^2 - E\,\{\Upsilon\,(\mathbf{h}'\mathbf{Y}_s)\}^2$

Hence, a weakly robust ADU-predictor is preferable to a ADU predictor provided it has a smaller model bias. If both are model-unbiased, a weakly robust ADU predictor is preferable to a ADU predictor.

We now consider the situation when the model with mis-specified covariance matrix is used to provide a sampling strategy. Suppose that due to ignorance the statistician uses the model $\xi_1(\mathbf{X}, g)$ with dispersion matrix $G = \text{Diag}\,(g_1, \ldots, g_N)$ in place of $\xi(\mathbf{X}, v)$ and hence uses the optimum sampling design $p_1 \in \rho_n$ having $\pi_i(p_1) = \sqrt{g_i}/k_1$ where $k_1 = \left(\sum \sqrt{g_i}\right)/n$ together with the generalised regression predictor $\hat{T}^*_{GR}(\mathbf{X}, g) = \hat{Y}(\mathbf{G}_s^{-1}, \mathbf{II}_1^{-1})$ where $\mathbf{II}_1 = \text{Diag}\,(\pi_i(p_1), i = 1, \ldots, N)$. This strategy will not be optimal under $\xi(\mathbf{X}, v)$. However, this is ADU under $\xi(\mathbf{X}, v)$ and it can be shown that asymptotically (vide Note 7.5)

$$E_{p1}\mathscr{E}(e - \mathbf{1}'\mathbf{Y})^2 \geq E_{p1}\mathscr{E}(\hat{T}^*_{GR}(\mathbf{X}, g) - \mathbf{1}'\mathbf{Y})^2$$

$$= \sum (k/\sqrt{g_i} - 1) v_i \sigma^2 = v_0 \text{ (say)} \tag{8.5.6}$$

for any ADU predictor e.

More generally for any linear predictor $\mathbf{h}'\mathbf{Y}_s$ which is strongly robust under $\xi(\mathbf{X}, v)$, *i.e.*, which satisfies (8.5.1) and (8.5.2), asymptotically,

$$E_{p1}\mathscr{E}(e - \mathbf{1}'\mathbf{Y})^2 \geq E_{p1}\mathscr{E}\;(\mathbf{h}'\mathbf{Y}_s - \mathbf{1}'\mathbf{Y})^2 = v'_0$$

for any ADU e. The proof follows from Theorem (8.3) and is as in the proof of (8.5.6).

❑ ❑ ❑

CHAPTER

9 Sampling Designs Making Predictors Robust and Near-Optimal

9.1 INTRODUCTION

The approach in this chapter is to find sampling designs within a class of optimal designs which endow robustness and near-optimality to some simple predictors. The asymptotic approaches of chapters 7 and 8 are not the concern of this chapter. Optimality here is in the sense of attaining GJLB. It has been shown that such sampling designs, often stratified sampling designs, exist for some models.

Godambe (1983) observed that under a usual superpopulation model, optimality of a well-known predictor, which is exactly optimal for a subclass of priors, can be extended to the other subclasses of the model by choosing a suitable sampling design so that the strategy becomes optimal or near-optimal under these subclasses. The design so chosen belongs to the class of designs providing optimality to the strategy under the original subclass. In some situations, the procedure gives an alternative method of prediction by doing away with the necessity of estimating unknown parameters involved in the model, as opposed to model-based approaches of Cassel *et al.* (1976) and others. These results have been discussed in Section 9.4.

9.2 SAMPLING DESIGNS MAKING OPTIMAL STRATEGIES ROBUST UNDER ALTERNATIVE MODELS

Consider the model $\xi_\beta(\mathbf{a})$ such that

$$\begin{aligned} &(i)\ \mathscr{E}(Y_i) = a_i + \beta x_i \\ &(ii)\ \mathscr{V}(Y_i) = \sigma^2 f(x_i) = \sigma^2 f_i \\ &(iii)\ \mathscr{C}(Y_i, Y_{i'}) = 0,\ i \neq i' = 1, \ldots, N \end{aligned} \tag{9.2.1}$$

$\mathbf{a} = (a_1, \ldots, a_N)$, $f(x)$ a function of x. It is well-known

$$E_p\mathscr{E}(\hat{T} - Y)^2 \geq E_{p0}\mathscr{E}(\hat{T}_{GD} - Y)^2 = \left(\sigma\sum\sqrt{f_i}\right)^2/n - \sum_i f_i\sigma^2$$

where, $$\hat{T}_{GD} = \sum_s (Y_i - a_i - \beta x_i)/\pi_i + a + \beta X,$$

$a = \Sigma a_i$ and $p_0 \in \rho_\pi = \{p : \pi_i(p) \propto \sqrt{f(x_i)}\} \in \rho_n$ and $\hat{T}$ is any p-unbiased predictor.

Thus any design $p \in \rho_\pi$ preserves the optimality of $\hat{T}_{GD}$ under ξ.

To choose a suitable design in the class ρ_π consider an alternative model ξ' such that $\mathscr{E}(Y_i) \neq a_i + \beta x_i$ whereas (*ii*) and (*iii*) in (9.2.1) still hold. From (4.2.5), for any p-unbiased predictor,

$$\mathscr{E}V(\hat{T}) = \{E\mathscr{V}(\hat{T}) - \mathscr{V}(Y)\} + E(\mathscr{E}(\hat{T} - Y))^2 \tag{9.2.2}$$

Defining a measure of robustness of strategy H under alternative model ξ' as $|\,\mathscr{E}V(H) - \mathscr{E}'V(H)\,|$, $\mathscr{E}'$ denoting expectation under ξ', so that among a class of strategies one with the least change in average variance is the most robust in the class, we have

$$\begin{aligned}\mathscr{E}'V'(\hat{T}_{GD}) - \mathscr{E}V(\hat{T}_{GD}) &= E[\mathscr{E}'(\hat{T}_{GD} - Y)]^2 \\ &= \Delta_p(\hat{T}_{GD}\,|\,\xi,\xi') \text{ (say)}\end{aligned} \tag{9.2.3}$$

as the term in the curly bracket of (9.2.2) remains the same for both ξ and ξ' and $\hat{T}_{GD}$ is p-unbiased under ξ'. Hence one should choose a p in ρ_π such that the right hand side of (9.2.3) is as small as possible. It has been shown (Godambe and Thompson, 1977) that

$$\begin{aligned}\Delta_{p0}(\hat{T}_{GD}\,|\,\xi,\xi') &= \frac{1}{2}\sum{}'(\pi_i\pi_j - \pi_{ij})(D_i/\pi_i - D_j/\pi_j)^2 \\ &= V_{p0}(\hat{T}_{GD})\Big]_{\mathbf{y}=(D_1,\ldots,D_N)}\end{aligned}$$

where $D_i = \mathscr{E}'(Y_i) - a_i - \beta x_i$. Now it is very likely that $(D_i/\pi_i - D_j/\pi_j)^2$ is large for dissimilar units i and j, *i.e.*, for units with greater difference between x_i and x_j. Hence one should choose $\rho'_0 \in \rho_\pi$ such that $\pi_i\pi_j - \pi_{ij}$ is zero or even negative for units i and j with large $|\,x_i - x_j\,|$. This condition can be approached by stratifying the population into strata which are homogeneous in x-values and choosing from each stratum an independent sample such that the π_i's are realised. Thus the appropriate randomisation provided through homogeneous stratification render the optimality of $\hat{T}_{GD}$ robust against possible departure of the superpopulation model $\xi_\beta(\mathbf{a})$ from the condition (*i*) of (9.2.1).

9.3 SAMPLING DESIGNS TO MAKE RATIO PREDICTOR NEARLY OPTIMAL

Consider model $\xi_\beta(\mathbf{0})$ as in (9.2.1) with $f(x_i) = x_i$ for which ratio predictor $\hat{T}_R = \bar{Y}_s X/\bar{x}_s$ is BLUP. For any sampling design for which $\hat{T}_R$ is p-unbiased (*e.g.*, Sen-Midzuno-Lahiri scheme).

$$\mathscr{E}\,V(\hat{T}_R) = \frac{N\sigma^2 X}{n - \mathscr{V}(Y)} \tag{9.3.1}$$

Now GJLB of average variance is

$$N\sigma^2 X/n - N^2\sigma^2\, V(\sqrt{x})/n - \mathscr{V}(Y) \tag{9.3.2}$$

which is attained by (p_0, T_{HT}) with $p_0 \in \rho_n$ and $\pi_i(p_0) \propto \sqrt{x_i}$, $i = 1, ..., N$. Here $V(\sqrt{x})$ denotes variance of $\sqrt{x_1}, \ldots, \sqrt{x_N}$. To make (9.3.1) nearly equal to (9.3.2) consider the following stratified sampling design. Divide the population into L strata $\mathscr{P}_j$, homogeneous wrt x values and draw samples of size $n_j = nN_j\sqrt{\bar{X}_j} \Big/ \sum_1^L N_k\sqrt{\bar{X}_k}$, $\bar{X}_j = \sum_{\mathscr{P}_j} x_i / N_j$ by srs. Consider the predictors, both of which are approximately p-unbiased for large or moderate n (Cochran, 1977, pp. 164–65).

$$\hat{T}_R' = \sum (\bar{y}_j/\bar{x}_j) N_j \bar{X}_j$$

$$\hat{T}_R'' = \sum (N_j \bar{y}_i) X \Big/ \sum (N_j \bar{x}_j)$$

$\bar{y}_j, \bar{x}_j$ denoting sample means for the jth stratum. Now

$$\mathscr{E}\,V(T_R') \simeq \sum N_j^2 \bar{X}_j \sigma^2/n_j - \mathscr{V}(Y)$$

$$= NX\sigma^2/n - N^2\sigma^2 V(\sqrt{z})/n - \mathscr{V}(Y)$$

$$\simeq \mathscr{E}\,V(\hat{T}_R'')$$

Here $V(\sqrt{z})$ denotes variance of $\sqrt{z_1}, \ldots, \sqrt{z_N}$ where $z_i = \bar{X}_j$ for $i \in \mathscr{P}_j$. Clearly $V(\sqrt{z}) = V(\sqrt{x})$ if in each $\mathscr{P}_j$, all values of x_i are exactly equal. Thus if the strata $\mathscr{P}_j$ are homogeneous in x-values, the above stratified sampling scheme together with $\hat{T}_R'$ or $\hat{T}_R''$ provide a nearly optimal predictor under the model considered. Optimality here is in the sense of attaining GJLB. A choice between $\hat{T}_R$ or $\hat{T}_R''$ or can be made by considering what happens under departure of model ξ in respect of $\mathscr{E}(Y_i)$, say $\mathscr{E}(Y_i) = \gamma + \beta x_i$. Here

$$\mathscr{E}'(\hat{T}_R' - Y) = \sum_j N_j \bar{X}_j (\bar{x}_j^{-1} - \bar{X}_j^{-1})$$

$$\mathscr{E}'(\hat{T}_R'' - Y) = NX\gamma\left\{\left(\sum_j N_j \bar{x}_j\right)^{-1} - X^{-1}\right\}$$

so that $E\{\mathscr{E}'(\hat{T}_R' - Y)\}^2$ is likely to be larger than $E\{\mathscr{E}'(\hat{T}_R'' - Y)\}^2$.

9.4 ROBUST NEAR-OPTIMAL PREDICTION UNDER A SUPERPOPULATION MODEL

Consider a superpopulation model $C = \{\xi\}$, a class of prior distributions ξ. Let e_ξ be the Bayes estimator corresponding to ξ and $H_C : \{e_\xi\}$ class of Bayes estimators. We have

Definition 9.1: Let $U_C = \{e : \mathscr{E}\{e - Y) = 0 \ \forall\ s : p(s) > 0; \xi \in C\}$, the class of C-unbiased predictors of Y.

Definition 9.2: A predictor $\bar{e}$ is said to be C-UMVU (or C-best) predictor of Y if

$$\mathscr{E}(\bar{e} - Y)^2 \leq \mathscr{E}(e - Y)^2 \ \forall\ s : p(s) > 0; \xi \in C, e \in U_C$$

Definition 9.3: Let $\bar{U}_C = \{e : \mathscr{E}(e - e_\xi) = 0 \ \forall\ s : p(s) > 0, \xi \in C\}$ since $\mathscr{E}(e_\xi)$ $= \mathscr{E}(Y)$, $\bar{U}_C \equiv U_C$.

Definition 9.4: A predictor $\bar{e}$ is said to be H_C-closest if

$$\mathscr{E}(\tilde{e} - e_\xi)^2 \leq \mathscr{E}(e - e_\xi)^2, \xi \in C, e \in \bar{U}_C.$$

Thus, a H_C-closest predictor has the minimum average distance from e_ξ under prior ξ, for all $\xi \in C$, among all $e \in \bar{U}_C$.

Now, $$\mathscr{E}(e - e_\xi)^2 = \mathscr{E}(e - Y)^2 + \mathscr{E}(Y - e_\xi)^2 - 2\mathscr{E}(e - Y)(Y - e_\xi) \quad (9.4.1)$$

where the product term

$$\begin{aligned} &= -2\mathscr{E}_{\xi \cdot d}[e\mathscr{E}_{\xi/d}(Y - e_\xi)] + 2\mathscr{E}_\xi[Y(Y - e_\xi)] \\ &= 2\mathscr{E}_\xi[Y(Y - e_\xi)], \end{aligned} \quad (9.4.2)$$

$\mathscr{E}_{\xi|d}$ denoting expectation for the given data d and $\mathscr{E}_{\xi \cdot d}$ expectation wrt d, considering d as a random variable. Hence from (9.4.1) minimisation of $\mathscr{E}(e - e_\xi)^2$ wrt e is the same as minimising $\mathscr{E}(e - Y)^2$. Thus $\tilde{e} \equiv \tilde{e}$, if they exist.

Lemma 9.1: If C is a complete class of distribution in R^N, no C-best (H_C-closest) estimator exists except for a census design.

Proof: Since C is complete, $[\mathscr{E}(e - Y) = 0, \xi \in C] \Rightarrow [e - Y = 0$, except for sets of Lebesgue measures zero]. Now $e = Y$ only for a census design.

Consider now the model $C(\mathscr{A}) = U_{\mathbf{a} \in \mathscr{A}} C(\mathbf{a})$, where $C(\mathbf{a})$ is such that

(*i*) $\mathbf{Y}$ when distributed as ξ in $C(\mathbf{a})$ involves $Y_1, \ldots, Y_N$ are independent;

(*ii*) $\mathscr{E}(Y_i) = a_i + \beta x_i$, $\mathbf{a} = (a_1, \ldots, a_N)$

(*iii*) $\mathscr{V}(Y_i) = \sigma^2 f(x_i)$

$f(x)$, a known function of x; β, $\sigma^2 (> 0)$ belongs to some specified intervals.

Noting that if $\mathscr{A}$ is an N-dim. interval, $C(\mathscr{A})$ is a complete class of distributions so that no $C(\mathscr{A})$-best predictor of Y exists, Godambe (1982) considered a different concept of optimality by bringing in a non-informative sampling design p.

Definition 9.5: For a given p, let

$$U_{pC} = \{e : E\mathscr{E}(e - Y) = 0, \xi \in C\}$$

$$\bar{U}_{pC} = \{e : E\mathscr{E}(e_\xi - e) = 0, \xi \in C\}$$

$$B_p = \{e : E(e - Y) = 0\}$$

where E denotes expectation wrt a given sampling design p. Since,

$$E\mathscr{E}(e_\xi) = E\mathscr{E}(Y), \bar{U}_{pC} = U_{pC}.$$

Definition 9.6: A predictor $\bar{e}_p$ is said to be $p - C$ best if

$$E\mathscr{E}(\bar{e}_p - Y)^2 \le E\mathscr{E}(e - Y)^2, \xi \in C, e \in B_p.$$

Definition 9.7: A predictor $\bar{e}_p$ is said to be $p - H_C$ closest if

$$E\mathscr{E}(\tilde{e}_p - Y)^2 \le E\mathscr{E}(e - Y)^2, \xi \in C, e \in \bar{U}_{pC}.$$

Lemma 9.2: In $C(\mathscr{A})$, if $\mathscr{A}$ is an N-dimensional interval, $\bar{U}_{pC}(\mathscr{A}) = B_p$.

Proof: If $\mathscr{A}$ is an N-dim. interval, $C(\mathscr{A})$ is a complete class of distributions and hence

$$[\mathscr{E}(e - Y) = 0, \xi \in C] \Rightarrow [e = Y \text{ almost everywhere}] \tag{9.4.3}$$

Now, since p is non-informative,

$$[E\mathscr{E}(e - Y) = 0, \xi \in C] \Leftrightarrow [\mathscr{E}E(e - Y) = 0, \xi \in C]$$

$$\Leftrightarrow [E(e - Y) = 0] \text{ by (9.4.3)}$$

Lemma 9.3: In $C(\mathscr{A})$, if $\mathscr{A}$ is an N-dim. interval, $\bar{e}_p = \tilde{e}_p$, if they exist.

Proof: We have $U_{pC(\mathscr{A})} = \bar{U}_{pC(\mathscr{A})} = B_p$ by Lemma 9.2. That $e_p = \tilde{e}_p$ follows from (9.4.1) and (9.4.2) by operating E on both sides. To sum up

(a) for a complete class of distributions in C, no C-best (H_C-closest) predictor of Y exists. For most models C, no C-best predictor exists.

(b) for a complete class of priors $C(\mathscr{A})$, $U_{pC(\mathscr{A})} = \bar{U}_{pC(\mathscr{A})} = B_p$ and $p - C(\mathscr{A})$ best predictor $\bar{e}_p = p - H_{C(\mathscr{A})}$ closest predictor $\tilde{e}_p$.

One may, therefore, find an alternative criterion of optimality by considering randomisation both wrt sampling design and superpopulation model.

Definition 9.8: A strategy $(\bar{p}, \bar{\bar{e}})$ is optimal wrt (C, D), a model $C = \{\xi\}$ and a class of sampling designs $D = \{p\}$ if

$$E\mathscr{E}\,(\bar{\bar{e}}_p - Y)^2 \le E_{\bar{p}}\mathscr{E}\,(e - Y)^2, \xi \in C, p \in D, e \in B_p.$$

The above consideration gives us a justification of finding strategies (p^*, e^*), whose ξ-expected p-variance is minimum in the class of all p-unbiased strategies, as first considered by Godambe and Joshi (1965) in finding the lower bound of average variance of a p-unbiased strategy under a particular prior.

We now proceed to find optimal strategies wrt

$$C_\beta(\mathbf{a}) = \{\xi : \mathbf{a}(\xi) = \mathbf{a},\ \beta(\xi) = \beta\} \quad \text{and} \quad \rho_n.$$

Theorem 9.1: The pair $(e(\mathbf{a} + \beta\mathbf{x}), p')$ is $(C_\beta(\mathbf{a}), \rho_n)$-optimal where p' is any design in

$$\rho_\pi = \left\{p : \pi_i(p) = n\sqrt{f(x_i)}\Big/\sum \sqrt{f(x_k)}, i = 1, \ldots, N\right\}(\in \rho_n)$$

and
$$e(\delta) = \sum_s n(y_i - \delta_i)\sqrt{f(x_i)}\Big/\sum \sqrt{f(x_k)} + \sum \delta_i.$$

Proof: We have, from (4.2.5), for any p-unbiased strategy,

$$E\mathscr{E}\ (e - Y)^2 = E\mathscr{E}\mathscr{V}\,(e) + E[\mathscr{E}\ (e - Y)]^2 - \mathscr{V}\,(Y) \tag{9.4.4}$$

Now $e(\mathbf{a} + \beta\mathbf{x})$ is $C_\beta(\mathbf{a})$-unbiased so that the middle term in (9.4.4) vanishes. It follows that

$$E_{\bar{p}}\mathscr{E}\,[e(\mathbf{a} + \beta\mathbf{x}) - Y]^2 = \sigma^2\left(\sum \sqrt{f(x_i)}\right)^2\Big/n - \sigma^2\sum \sqrt{f(x_i)} \tag{9.4.5}$$

which is the minimum value of lower bound of any fixed-size p-unbiased strategy for Y under priors with

$$\mathscr{E}\ (Y_i) = \mu_i,\ \mathscr{V}\ (Y_i) = \sigma^2 f(x_i),\ \mathscr{E}\ (Y_i, Y_{i'}) = 0,\ i \ne i'.$$

Theorem 9.2: The necessary and sufficient conditions for a p-unbiased strategy (p, e) to be $(C(\mathscr{A}), \rho_n)$-optimal are

(*i*) $\mathscr{E}\ (e - Y) = 0, \xi \in C(\mathscr{A})$ *i.e.*, e is $C(\mathscr{A})$-optimal

(*ii*) $p \in \rho_n$

(*iii*) $e \in F = \{e(\boldsymbol{\delta}), \boldsymbol{\delta} \in R^N\}$.

Proof: Necessity. (*i*) Suppose (p, e_0) is $(C(\mathscr{A}), \rho_n)$-optimal, while for some $\xi \in C_\beta(\mathbf{a})$, $E[\mathscr{E}\,(e_0 - Y)]^2 > 0$. Then from (9.4.4) and (9.4.5) (since $e(\mathbf{a} + \beta\mathbf{x})$ is $C_\beta(\mathbf{a})$-unbiased) $E_p\mathscr{E}\,(e_0 - Y)^2$ is greater than the right hand side of (9.4.5), contradicting optimality of (p, e_0). (*ii*) If $p \notin \rho_\pi$, the lower bound in (9.4.5) cannot be attained.

Suppose both (p, e_0), $(p, e(\mathbf{a} + \beta\mathbf{x}))$, $p \in \rho_\pi$ are $(C_\beta(\mathbf{a}), \rho_n)$-optimal. Hence, for all $\xi \in C_\beta(\mathbf{a})$,

$$0 = E_p\mathscr{E}\ \{e_0 - Y)^2 - E\mathscr{E}\,(e(\mathbf{a} + \beta\mathbf{x}) - Y)^2 \tag{9.4.6}$$

$$= E_p\mathscr{E}\ (e_0 - e(\mathbf{a} + \beta\mathbf{x}))^2 + 2E_p\mathscr{E}\ (e_0 - e(\mathbf{a} + \beta\mathbf{x}))(e(\mathbf{a} + \beta\mathbf{x}) - Y)$$

Now $(C_\beta(\mathbf{a}), \rho_n)$-optimality of $(p, e(\mathbf{a} + \beta\mathbf{x})$ implies covariance term is zero. Hence $e_0 = e(\mathbf{a} + \beta\mathbf{x})$ with probability one, probability being wrt joint distribution of p, ξ, $p \in \rho_\pi$, $\xi \in C(\mathscr{A})$. The sufficiency can be checked directly.

Therefore, any strategy (p_0, e_0) satisfying all the conditions of Theorem 9.2 is $(C(\mathscr{A}), \rho_n)$-optimal. Assuming $C_0(\mathbf{0}) \in C(\mathscr{A})$ we take $e^0 = e(\mathbf{0})$. Hence (c^0, p), $p \in \rho_\pi$ is $(C(\mathscr{A}), \rho_n)$-optimal if

$$(a)\ e(\mathbf{0}) \in B_p \tag{9.4.7a}$$

$$(b)\ E_p[\mathscr{E}\ (e(\mathbf{0}) - Y)]^2 = 0,\ \xi \in C(\mathscr{A}) \tag{9.4.7b}$$

$$(c)\ E\mathscr{E}\ [(e(\mathbf{0}) - e(\mathbf{a} + \beta\mathbf{x})]^2 = 0,\ \xi \in C_\beta(\mathbf{a}) \subset C(\mathscr{A}) \tag{9.4.7c}$$

Sampling designs $p(\in \rho_\pi)$ satisfying conditions (b) and (c) above extends the $(C_0, (\mathbf{0}), \rho_n)$-optimality of $(e(\mathbf{0}), p)$ to $(C(\mathscr{A}), \rho_n)$-optimality. Since $e(\mathbf{0})$ has a very simple form we now examine the designs $p \in \rho_\pi$ which make $(e(\mathbf{0}), p)$, $(C(\mathscr{A}), \rho_n)$-optimal or near optimal.

Now, $\qquad e(\mathbf{a} + \beta\mathbf{x}) - e(\mathbf{0}) = -\mathscr{E}\ (e(\mathbf{0}) - Y)$, when $\xi \in C_\beta(\mathbf{a})$.

Hence in this case both the conditions (9.4.7b) and (9.4.7c) are identical and reduce to

$$E_p[e(\mathbf{a} + \beta\mathbf{x}) - e(\mathbf{0})]^2 = \gamma_p \text{ (say)}$$

$$= 0,\ \xi \in C_\beta(\mathbf{a}) \subset C(\mathscr{A}) \tag{9.4.8}$$

Suppose no sampling design $p(\in \rho_\pi)$ satisfying (9.4.8) exists; however, a design p'_0 exists for which

$$\gamma p'_0 \leq \in, \tag{9.4.9}$$

$\in$, a pre-assigned small quantity. The use of $(p'_0, e(\mathbf{0}))$ provides robust estimation in the following two senses:

(i) Suppose the underlying true model is $C_\beta(\mathbf{a})$ so that one should use $(p, e(\mathbf{a} + \beta\mathbf{x}))$, $p \in \rho_\pi$; however, by mis-specification, the statistician considered the model as $C_0(\mathbf{0})$ and hence used $e(\mathbf{0})$. In this situation the use of $(p'_0, e(\mathbf{0}))$ would ensure that samples for which $|\ e(\mathbf{a} + \beta\mathbf{x}) - e(\mathbf{0})\ |$ is small are obtained with high frequency. This endows inference (criterion)-robustness to $(p'_0, e(\mathbf{0}))$ under $C_\beta(\mathbf{a})$.

(ii) (9.4.9) implies that

$$| E_{p_0'} \mathscr{E} (e(\mathbf{0}) - Y)^2 - E_{p_0'} \mathscr{E} (e(\mathbf{a} + \beta(\mathbf{x})) - Y)^2 | \le \in$$

so that the difference between the efficiency of optimal strategy $(p, e(\mathbf{a} + \beta\mathbf{x})), p \in \rho_\pi$ and $(p'_0, e(\mathbf{0}))$ under $\xi \in C_\beta(\mathbf{a})$ is small. This gives efficiency-robustness to $(p'_0, e(\mathbf{0}))$ under alternative $C_\beta(\mathbf{a})$.

Definition 9.9: A pair (p, e) is near optimal and robust wrt $(C(\mathscr{A}), \rho_n)$ for Y if, (*i*) $e \in B_p$, (*ii*) $e \in F$, (*iii*) $p \in \rho_\pi$ and (*iv*) $E_p[\mathscr{E}(e - Y)\}^2$ is very small compared to the right hand side of (9.4.5) for all $\xi \in C(\mathscr{A})$.

Obviously, $(p_0, e(\mathbf{0}))$ is near optimal and robust wrt $(C(\mathscr{A}), \rho_n)$. The design p'_0 may sometimes be attained through stratified sampling.

Stratification I: We have for any $\xi \in C_\beta(\mathbf{a})$,

$$E[\mathscr{E}(e(\mathbf{0}) - Y]^2$$

$$= E\left[\left\{\sum_s (a_i + \beta x_i) \Big/ n\sqrt{f(x_i)}\right\} \sum \sqrt{f(x_k)} - \sum a_i - \beta X\right]^2 \tag{9.4.10}$$

Stratify the population $\mathscr{P}$ into L strata $\mathscr{P}_j$ $(j = 1, \ldots, L)$ as follows and draw srs of size $n_j \cdot C(\mathscr{A})$ should also satisfy the required conditions:

$$(i)\ \alpha_j \le x_i / \sqrt{f(x_i)} \le \alpha'_j, i \in \mathscr{P}, j = 1, \ldots, L$$

with max $(\alpha_j - \alpha'_j) = \in_1$, a pre-assigned small quantity;

$$(ii)\ \gamma_j \le a_i / \sqrt{f(x_i)} \le \gamma'_j, i \in \mathscr{P}_j, j = 1, \ldots, L$$

with max $(\gamma'_j - \gamma_j) = \in_2$. (9.4.11)

$$(iii)\ n_j = n \sum_{\mathscr{P}_j} \sqrt{f(x_i)} \Big/ \sum \sqrt{f(x_k)}$$

Under these conditions,

$$(9.4.10) \le \left[\sum_1^L \{(\gamma'_j - \gamma_j) + |\beta| (\alpha'_j - \alpha_j)\} \sum_{\mathscr{P}_j} \sqrt{f(x_i)}\right]^2$$

$$\le \left[(\in_2 + |\beta| \in_1) \sum \sqrt{f(x)}\right]^2 \tag{9.4.12}$$

Hence, $(p'_0, e(\mathbf{0}))$ is near-optimal and robust wrt $(C(\mathscr{A}), \rho_n)$ if (9.4.12) is small compared to the right hand side of (9.4.5) *i.e.*, if

$$n\left[(\in_2 + |\beta| \in_1)\sum \sqrt{f(x_i)}\right]^2 \Big/ \sigma^2\left[\left(\sum \sqrt{f(x_i)}\right)^2 - n\sum f(x_i)\right] \qquad (9.4.13)$$

is small for possible values of β, σ^2. If $\in_1 = \in_2 = 0$, $(p'_0, e(\mathbf{0}))$ is exactly optimal.

In particular, suppose $a_i = a^{(j)}x_i$ for all $i \in \mathscr{P}_j$. Here $\mathscr{E}(Y_i) = \beta^{(j)}x_i$ where $\beta^{(j)} = a^{(j)} + \beta$, $i \in \mathscr{P}_j$. This is equivalent to the model $C(\mathbf{a})$ with $\mathbf{a} = \mathbf{0}$ and β varying from stratum to stratum. Here $\in_2 = 0$ and (9.4.13) reduces to

$$n\left[\in_1 \sum_{1}^{L} |\beta^{(j)}| \sum_{\mathscr{P}_j} \sqrt{f(x_i)}\right]^2 \Big/ \sigma^2\left[\left(\sum \sqrt{f(x_i)}\right)^2 - n\sum f(x_i)\right]$$

Special cases:

Case (1): $f(x) = x^2$. Here $E[\mathscr{E}(e(\mathbf{0}) - Y)]^2$ is independent of β and $\in_1 = 0$ in (9.4.13). Also $e(\mathbf{0}) = X\sum_{s} Y_i / x_i n = e_1$(say); p'_0 is the stratified simple random design with $n_j = nX_j/nX$, $X_j = \sum_{\mathscr{P}_j} x_i$, p_1 (say). If further, $\in_2 = 0$ (which is equivalent to the condition $\mathbf{a} = \mathbf{0}$ and β varying from stratum to stratum), (p_1, e_1) becomes $(C(\mathscr{A}), \mathscr{P}_n)$-optimal.

Example 9.1: $f(x) = x^2$ · $\mathscr{P}$ is divided into four strata with $X_j = X/4$, $j = 1, ..., 4$. Suppose $C(\mathscr{A})$ satisfies $\in_2 = 0$ with $\gamma_1 \le \ldots \le \gamma_4$. Consider strategies (p', e_1), $(\bar{p}, e_1)$ where $p', \bar{p} \in \rho_\pi \in \rho_n$ and

p' : (*i*) draw one unit from $\mathscr{P}_1 \cup \mathscr{P}_2$ with inclusion probability $2x_i / X$;
(*ii*) draw another unit from $\mathscr{P}_3 \cup \mathscr{P}_4$ with inclusion probability $2x_i / X$.

$\bar{p}$: (*a*) same as in (*i*) for p'; (*b*) if $i \in (\notin)\ \mathscr{P}_2$, a unit i' is drawn from $\mathscr{P}_3(\mathscr{P}_4)$ with probability $4x_{i'} / X$. Here

$$E_{\bar{p}}[\mathscr{E}(e_1 - Y)]^2 \le E_{p'}[\mathscr{E}(e_1 - Y)]^2,\ \xi \in C(\mathscr{A})$$

with **a** satisfying the required conditions. Hence, considering near-optimality $(\bar{p}, e_1)$ is preferable to (p', e_1).

Case (2): $f(x) = x$. Let $(e(\mathbf{0}), p_0) = (p_2, e_2)$ where $e_2 = \sum_{s} Y_i\left(\sum \sqrt{x_i}\right) \Big/ n\sqrt{x_i}$, p_2 is a stratified random design (9.4.11) with $n_j = n\sum_{\mathscr{P}_j} \sqrt{x_i} \Big/ \sum \sqrt{x_k}$. When $\mathbf{a} = \mathbf{0}$, Cassel *et al.* (1976) suggested use of $(p, \hat{T}^*_{GR})$, $p \in \rho_\pi$,

$$\hat{T}^*_{GR} = \sum_{s} (Y_i - \hat{\beta}^* x_i)\sum \sqrt{x_i} / n\sqrt{x_i} + \hat{\beta}^* X$$

where $\hat{\beta}^*$ is the least square predictor of β. The present theory suggests elimination of β in contrast to its estimation, through choice of a suitable sampling plan p_2.

Stratification II: Sometimes (9.4.13) cannot be made sufficiently small if $\mathscr{A}$, range of values of β and σ^2 are wide. In such cases, appropriate stratification as follows can sometimes be made to achieve near-optimal estimation.

Let $f(x) = x$. Consider the pair (p_r, e_r) where $e_r = \sum_{j=1}^{L} \bar{y}_j X_j / \bar{x}_j$, $\bar{y}_j \sum_{i \in s_j} y_i / n_j$, p_r stratified design as in p_2 but with $n_j = n\sqrt{N_j \bar{X}_j} \Big/ \sum \sqrt{N_k \bar{X}_k}$ and probability of selecting s_j, $p_r(s_j) = \sum_{s_j} x_i \Big/ \binom{N_{j-1}}{n_{j-1}} X_j$ (Midzuno sampling scheme), $j = 1, \ldots, L$.

Suppose also $\mathscr{A}$ is such that $a_j = a^{(j)} x_i$, $i \in \mathscr{P}_j$. Here e_r is p-unbiased as well as $C(\mathscr{A})$-unbiased.

$$E_{pr}\, \mathscr{E}\, (e_r - Y)^2 = \sigma^2 (N - n) X/n - \sigma^2 N^2 V\left(\sqrt{z}\right)/n, \ \xi \in C(\mathscr{A}) \tag{9.4.14}$$

$V\left(\sqrt{z}\right)$ = variance of $\left(\sqrt{z_1}, \ldots, \sqrt{z_N}\right)$, $z_i = X_j / N_j$, $i \in \mathscr{P}_j$. Again for all $e \in B_p$, $p \in \rho_n$,

$$E\mathscr{E}\, (e - Y)^2 \geq \sigma^2 (N - n) X/n - \sigma^2 N^2 V\left(\sqrt{x}\right)/n \tag{9.4.15}$$

Hence, if the stratification in p_r is sharp enough to make the relative excess in (9.4.14) over (9.4.15),

$$\frac{N^2\left[V\left(\sqrt{x}\right) - V\left(\sqrt{z}\right)\right]}{(N - n)X - N^2 V\left(\sqrt{x}\right)}$$

sufficiently small, (p_r, e_r) is close to $(C(\mathscr{A}), \rho_n)$-optimal.

Consider again, (p_R, e_R) where $e_R = X \sum_1^L N_j \bar{Y}_j \Big/ \sum_1^L N_j \bar{X}_j$, p_R a stratified sampling design as in p_r but with $p_R(s_j) = [\binom{N_j}{n_j}]^{-1}$. The estimator e_R is neither p_R-unbiased nor $C(\mathscr{A})$-unbiased. If, however, $\mathscr{A} = \{\mathbf{a} : a_i = \gamma, i = 1, \ldots, N\}$,

$$E_{pr}\{\mathscr{E}\, (e_r - Y)\}^2 = \gamma^2 E_{pr}\left[\sum_1^L N_j (\bar{X}_j / \bar{x}_j - 1)\right]^2$$

$$> E_{pr}\{\mathscr{E}\, (e_R - Y)\}^2 = \gamma^2 N^2 E_{pr}\left[X \Big/ \left(\sum_1^L N_j \bar{x}_j\right) - 1\right]^2$$

the later being generally much smaller than the former. Thus if the stratification underlying p_R is sharp enough in x-values and if $\mathscr{A} = \{\mathbf{a} : a_i = \gamma, i = 1, \ldots, N\}$, (p_R, e_R) is close to $(C(\mathscr{A}), \rho_n)$ optimality. Particularly if $\gamma \gg \sigma$, (p_R, e_R) is preferable to (p_r, e_r) or (p_2, e_2).

❑ ❑ ❑

CHAPTER

10

ϕ-Optimality and Some Related Criteria

10.1 INTRODUCTION

Looking back we note that through chapters one to three we considered models under fixed population set-up where probability sampling was extensively used for drawing inference from a sample survey. The superpopulation model distribution was used there in order to assess the performance of several strategies through the values of their average mean square errors. The Godambe-Joshi lower bound (and similar bounds) was considered as a milestone to the results in this area to which the strategies looked for their optimality. Works by Brewer (1963), Royall (1970, 1976), Royall and Herson (1973*a*, *b*) thrived from the postulation of a superpopulation model of which the finite population at hand is a sample and ushered in the purposive design as an optimal sampling design to base their optimal predictors. Researches by Cassel *et al.* (1976), Brewer (1979), Sarndal (1980), Godambe (1982), Wright (1983), and others, as discussed through chapters four to nine, salvaged the use of probability sampling on the grounds of robustness under model-failures and hence created the realm of strategies which are based on both models and random designs. We are of the opinion that the available auxiliary and prior information should first be formulated in suitable superpopulation models and design-model based strategies should be used for achieving (near-) optimal results under the assumed model and robustness under departure from the supposed model in some suspected direction.

Wynn (1977*a, b*), Cheng and Li (1983, 1987), among others, attempted to combine these diverse approaches (probability sampling and purposive sampling along with the corresponding optimal estimators/predictors) in an unified set-up by bringing in the concept of ϕ-optimality from the theory of optimal designs. These results will be discussed in this chapter. In Section 10.2, we consider the minimaxity of a strategy under a class of finite population models which mimic some superpopulation models. The minimaxity of some well-known strategies have also been considered. Section 10.3 considers ϕ-optimality and shows the relation between model-optimality and minimaxity. The concept of universal optimality has also been considered. Lastly we consider some optimality criteria for choosing sampling designs, namely, Wynn's minimax purposive sampling, *D*-optimality and *G*-optimality.

10.2 MINIMAX STRATEGIES UNDER A CLASS OF FINITE POPULATION MODELS

Consider the following finite population model equivalent to the superpopulation model $\xi(0, 1; v(x))$.

$$y_i = \beta x_i + g(x_i)\,\delta_i,\ i = 1, ..., N,\ \beta \in \mathscr{B} \subset R \tag{10.2.1}$$

where β is an unknown parameter, x_i a known value (assumed positive) of an auxiliary variable, $g(x)$ a known function of x, often of the form x^{α}, $(\alpha \geq 0)$ and $\boldsymbol{\delta} = (\delta_1, ..., \delta_N)$ lies in a neighbourhood L of $\mathbf{0} = (0, ..., 0)$. For example,

$$L = L_2(M) = \left\{\boldsymbol{\delta} : \sum_{i=1}^{N} \delta_i^2 \leq M\right\} \tag{10.2.2}$$

$$L = L_{\alpha}(M) = \{\boldsymbol{\delta} : |\,\delta_i\,| \leq M\} \tag{10.2.3}$$

for some $M(> 0)$. In (10.2.1), $g(x)$ is the counterpart of $\sqrt{v(x)}$ and δ_i is akin to $\in_i$ in the expression $Y_i = \beta x_i + \in_i \sqrt{v(x_i)}$ in the model $\xi(0, 1; v(x))$. However, unlike $\in = (\in_1, ..., \in_N)$, $\boldsymbol{\delta}$ is restricted to lie in some restricted set $L_2(M)$ or $L_{\alpha}(M)$. The restriction $\sum \delta_i^2 \leq M$ in (10.2.2) is akin to restricting $\sigma^2 \leq M/N$ in $\xi(0, 1; v(x))$. We shall assume, without loss of generality, that

$$x_1 \leq x_2 \ldots \leq x_N \tag{10.2.4}$$

In (10.2.1) the parameter space is $(\mathscr{B}, L)$. The set $\mathscr{B}$ may be bounded or unbounded.

Definition 10.1: A strategy H^* is minimax in a class of strategies $\mathscr{H} = \{H : (p, e)\}$ under the model (10.2.1) if it minimises the supremum risk

$$\sup_{(\mathscr{B}, L)} R(H; \beta, \delta) \tag{10.2.5}$$

in the class $\mathscr{H}$.

We shall write the parameter spaces

$$(\mathscr{B},\ L_2(M)) = \Theta_1,\ (\mathscr{B},\ L_{\infty}(M)) = \Theta_2 \tag{10.2.6}$$

Considering the squared error loss function and confining to the class of non-randomized linear strategies $\mathscr{H} = \{H : (p, e)\}$, $p \in \rho_n$, $e = \sum_{i=1}^{N} b_{si} y_i$ (with $b_{si} = 0$ for $i \notin s$) and writing $\mathbf{G} = \text{Diag}\,(g(x_1), ..., g(x_N))$, $\mathbf{b}_s = (b_s1, ..., b_{sN})'$, the risk of a strategy H in estimating the total Y for a specified (β, δ) is

$$R(p, e; \beta, \delta) = \sum_{s \in \mathscr{S}} p(s)\left(\sum_{i=1}^{N} b_{si}(y_i - 1)\right)^2$$

$$= \sum_{s \in \mathcal{S}} p(s)\mathbf{y}'(b_s - \mathbf{1})(b_s - \mathbf{1})'\mathbf{y}$$

$$= \sum_{s \in \mathcal{S}} p(s)[\beta^2 \mathbf{x}'(\mathbf{b}_s - \mathbf{1})(\mathbf{b}_s - \mathbf{1})'\mathbf{x}$$

$$+ \beta\boldsymbol{\delta}'\mathbf{G}(\mathbf{b}_s - \mathbf{1})(\mathbf{b}_s - \mathbf{1})'\mathbf{x}$$

$$+ \beta\mathbf{x}'(\mathbf{b}_s - \mathbf{1})(\mathbf{b}_s - \mathbf{1})'G\delta$$

$$+ \boldsymbol{\delta}'\mathbf{G}(\mathbf{b}_s - \mathbf{1})(b_s - \mathbf{1})'G\delta] \tag{10.2.7}$$

This is a quadratic function in β if $(\mathbf{b}_s - \mathbf{1})'\mathbf{x} \neq 0$ for some s with $p(s) > 0$. To keep the risk (10.2.7) finite over $(\mathcal{B}, L)$ whenever $\mathcal{B}$ is unbounded, we assume that H satisfies

$$(\mathbf{b}_s - \mathbf{1})'\ \mathbf{x} = 0\ \forall\ s : p(s) > 0 \tag{10.2.8}$$

This is the condition of model-unbiasedness of the predictor $\sum_{i=1}^{N} b_{si}Y_i$ under $\xi(0, 1; v(x))$, and has been termed 'representative condition' by Hájek (1959). Even when $\mathcal{B}$ is bounded we shall restrict ourselves to the class H of strategies $\mathcal{H}_r$ in $\mathcal{H}$ which satisfies (10.2.8). In this case

$$R(H; \beta, \delta) = \boldsymbol{\delta}'\mathbf{G} \sum_{s \in \mathcal{S}} p(s)(\mathbf{b}_s - \mathbf{1})(\mathbf{b}_s - \mathbf{1})'\mathbf{G}\boldsymbol{\delta} \tag{10.2.9}$$

is independent of β and is a quadratic form in δ. The matrix

$$\sum_{s \in \mathcal{S}} p(s)(\mathbf{b}_s - \mathbf{1})(\mathbf{b}_s - \mathbf{1})' = \mathbf{Q}(H) \text{ (say)} \tag{10.2.10}$$

has been termed the 'risk generating matrix' (*rgm*) and the matrix

$$\mathbf{GQ}(H)\mathbf{G} = \mathbf{R}(H) \text{ (say)} \tag{10.2.11}$$

the adjusted risk generating matrix (*argm*) of the strategy H for estimating Y under the model (10.2.1). Clearly,

Proposition 10.1: $\mathbf{R}(H)$ is singular and satisfies $\mathbf{R}(H)\mathbf{G}^{-1}\mathbf{x} = 0$.

Proposition 10.2: If $L = L_2(M)$, a minimax strategy is the one which minimises the maximum eigenvalue of $\mathbf{R}(H)$.

A lower bound to minimax strategy is given in

Proposition 10.3: Let $g(x) = \sqrt{x}$. Then

$$\min_{H \in \mathcal{H}_r} \sup_{\Theta_1} R(H;\beta,\delta) \geq \frac{QX}{M\left(X^2 - \sum_{i=1}^{N} x_i^2\right)}$$

where
$$Q = M^2\left\{\frac{1}{n}\left(\sum_{i=1}^{N-n} x_i\right)^2 + \sum_{i=1}^{N-n} x_i^2\right\} \tag{10.2.12}$$

Consider Rao-Hartley Cochran (1962) strategy, H_{RHC} discussed in Chapter 2 and Hansen-Hurwitz strategy H_{HH}. Note that H_{RHC} is a randomized strategy depending on the random stratification of the population and as such does not belong to $\mathcal{H}_r$. Cheng and Li (1983) proved,

Proposition 10.4: Let $g(x) = \sqrt{x}$. Then

$$\mathbf{R}(H_{RHC};\beta,\boldsymbol{\delta}) = \mu\mathbf{R}(H_{HH};\beta,\boldsymbol{\delta}) = \mu XM/n \tag{10.2.13}$$

where
$$\mu = (N-n)/(N-1) + k(n-k)/N(N-1)$$

and
$$N = nr + k.$$

Moreover,
$$\mathbf{R}(H_{RHC}) = aI_N + b\mathbf{G}\mathbf{J}_N\mathbf{G} \tag{10.2.14}$$

where,
$$a = \mu X/N^2 n,\ b = -\mu/N^2 n,$$

J_N is a $N \times N$ matrix of ones. Each of the $(N-1)$ non-zero eigenvalues of $\mathbf{R}(H_{RHC})$ equals $X\mu/n$.

The following theorem arising from Propositions 10.3 and 10.4 states that H_{RHC} is approximately minimax.

Theorem 10.1: Let $g(x) = \sqrt{x}$. If

$$\mu\left(X^2 - \sum_{i=1}^{N} x_i^2\right)M^2/nQ \leq 1 + \epsilon \tag{10.2.15}$$

then *RHC*-strategy is approximately minimax in the sense that for any $H \in \mathcal{H}_r$,

$$\sup \mathbf{R}(H_{RHC},\beta,\boldsymbol{\delta}) \leq (1+\epsilon)\sup_{\Theta_1}\mathbf{R}(H;\beta,\boldsymbol{\delta}).$$

The *HH*-strategy is approximately minimax if the sampling fraction $f \simeq 0$.

Corollary 10.1: If $n|N$ and x_i is a constant $\forall i$, then H_{RHC} reduces to (srswor, $N\bar{y}$) $= H_0$ (say). The minimaxity of H_0 follows.

Note 10.1: The minimaxity of H_{HH} derived by Scott and Smith (1975) in the class of all strategies using e_{HH} was discussed in Section 3.6. A similar result applies here also. When $n = 1$, the estimator $e_{HH} = y_i X/x_i$ is the only representative estimator. The problem then becomes one of choosing a sampling design. It has been shown that if $g(x) = \sqrt{x}$ and if there exists a set s_0 such that $\sum_{i \in s_0} x_i = X/2$ then *ppx*-selection minimises the maximum risk $R(p, e_{HH}; \beta, \boldsymbol{\delta})$ overall sampling designs p for any non-empty $\mathcal{B}$ and $\boldsymbol{\delta} \in L_2(M)$.

Consider now the case $g(x) = x$ and $L = L_\alpha(M)$. The following proposition gives a lower bound to the minimax risk.

Proposition 10.5: Let $g(x) = x$. Then

$$\min_{H \in \mathcal{H}_r} \sup_{\Theta_2} R(H; \beta, \boldsymbol{\delta}) \begin{cases} \geq QN/(N-1) \ (N \text{ even}) \\ \geq Q(N+1)/N \ (N \text{ odd}) \end{cases} \tag{10.2.16}$$

where Q is given in (10.2.12).

Proposition 10.6: Let $g(x) = x$. Then

(*a*) $\sup_{\Theta_2} \mathbf{R}(H_{HH}; \beta, \boldsymbol{\delta}) \leq X^2M^2/n$

(*b*) $\sup_{\Theta_2} \mathbf{R}(H_{RHC}; \beta, \boldsymbol{\delta}) \leq \mu X^2M^2/n$ (10.2.17)

hence,

Theorem 10.2: If $\mu X^2M^2(N-1)/nQN \leq 1 + \in$ when N is even and $\mu X^2M^2N/nQ(N+1)$ when N is odd, then H_{RHC} is $(1 + \in)$-approximately minimax in the sense that for any $H \in \mathcal{H}_r$

$$\sup_{\Theta_2} R(H_{RHC}; \beta, \boldsymbol{\delta}) \leq (1+\in) \sup_{\Theta_2} \mathbf{R}(H; \beta, \delta) \tag{10.2.18}$$

H_{HH} is approximately minimax when $f \simeq 0$.

A finite population model equivalent to regression superpopulation model $\xi(\mathbf{X}, v)$ is

$$y_i = \mathbf{x}'_i\boldsymbol{\beta} + g(x_i)\,\delta_i,\ i = 1, \ldots, N \tag{10.2.19}$$

where $\mathbf{x}_i = (x_{i0}, \ldots, x_{ir})'$, the value of the auxiliaries $x_0, \ldots, x_r$ on unit i and $\boldsymbol{\beta} = (\beta_0, \beta_1, \ldots, \beta_r)'$, $\boldsymbol{\delta} \in L$, some neighbourhood of $\mathbf{0}$. Assuming square error loss function, the risk of $H = (p, e) \in \mathcal{H}$, $p \in \rho_n$, $e = \sum b_{si} y_i$ is

$$R(p, e; \boldsymbol{\beta}, \boldsymbol{\delta}) = \sum_{s \in \mathcal{P}} p(s) \left[\left\{ \sum_{i=1}^{N} (b_{si} - 1) x_i' \right\} \boldsymbol{\beta} \right.$$

$$\left. + \sum_{i=1}^{N} (b_{si} - 1) \delta_i g(\mathbf{x}_i) \right]^2 \tag{10.2.20}$$

A strategy is representative if

$$\sum_{i=1}^{N} (1 - b_{si}) \mathbf{x}_i = \mathbf{0} \; \forall \; s : p(s) > 0 \tag{10.2.21}$$

which are the model-unbiasedness condition of $\sum_{i=1}^{N} b_{si} Y_i$ under $\xi(\mathbf{X}, v)$. As before, the risk of a representative strategy is given by (10.2.9) and its adjusted risk generating matrix by (10.2.11).

Writing $\lambda_1 \geq \lambda_2 \ldots \geq \lambda'_{N-q}$ as the $(N - q)$ largest non-zero eigenvalues of $\mathbf{R}$ where q is the rank of $\mathbf{X}$, a minimax strategy is one which minimises the value of λ_1.

The maximum risk of a strategy H over a parameter space of interest may, therefore, be used as a criterion for comparison among several strategies.

Definition 10.2: A strategy H_1 will be better than H_2 in the maximum risk sense wrt parameter space $(\mathcal{B}, L)$ $(H_1 \; \mathcal{B}, L \succ H_2)$ if

$$\sup_{\beta \in \mathcal{B}, \boldsymbol{\delta} \in L} R(H_1; \beta, \boldsymbol{\delta}) < \sup_{\beta \in \mathcal{B}, \boldsymbol{\delta} \in L} R(H_2; \beta, \boldsymbol{\delta}) \tag{10.2.22}$$

For the model $\xi(1, 1; x)$ Brewer (1963), Royall (1970) suggested the ratio predictor and purposive design p^* (Section 4.3). Denoting this strategy as H_{BR}, it follows that

$$\lambda_1(H_{BR}) - X \sum_{i=1}^{N-n} x_i \Big/ \sum_{i=N-n+1}^{N} x_i,$$

$$\lambda_i = 0 (i = 2, \ldots, N).$$

Proposition 10.7: Let $g(x) = \sqrt{x}$. If $\sum_{i=1}^{N-n+1} x_i \geq \sum_{i=N-n+1}^{N} x_i / n$ then for any $M\ (> 0)$

$$H_{RHC} \; \mathcal{B}, L_2(M) \succ H_{HH} \; \mathcal{B}, L_2(M) \succ H_{HBR}$$

in the sense of Definition 10.2.

Thus if the n largest values of x are not too extreme, H_{RHC} and H_{HH} will be better than H_{BR} in the minimax sense.

Proposition 10.8: Let $g(x) = \sqrt{x}$ and $n \mid N$. Then for any $M\ (> 0)$,

(*i*) $H_{RHC}\ \mathscr{B},\ L_2(M) \succsim H_R$

(*ii*) $H_{RHC}\ \mathscr{B},\ L_2(M) \succsim H_M$

where H_R = (srswor, $\hat{Y}_R$), H_M = Midzuno's strategy.

10.3 ϕ-OPTIMALITY AND RECONCILIATING DIFFERENT OPTIMALITY CRITERIA

It is evident from (10.2.9) and (10.2.11) that the comparison between two strategies H_1 and H_2 reduces to the comparison between their adjusted risk generating matries $\mathbf{R}(H_1)$, $\mathbf{R}(H_2)$.

Definition 10.3: A strategy H_1 is said to be at least as good as $H_2 (H_1 \succsim H_2)$ *iff* the matrix $\mathbf{R}(H_2) - \mathbf{R}(H_1)$ is non-negative definite (nnd).

The adjusted *rgm* $\mathbf{R}(H)$ plays the same role as the information matrix M in the theory of optimal designs as developed by Kiefer and Wolfwitz (1959), Kiefer (1959, 1970, 1975) and their followers. An excellent review of the theory of optimal designs can be found in Shah and Sinha (1989). In the theory of optimal designs a design with information matrix M_1 is said to be at least as good as another with information matrix M_2 if $M_1 - M_2$ is nnd. However, an adjusted *rgm* should be thought of as a covariance matrix and, therefore, corresponds to the inverse or generalised inverse of an information matrix.

Definition 10.4: Let $\mathscr{R}$ be the set of all adjusted risk generating matrices and Ψ be a real-valued function defined on $\mathscr{R}$. A strategy (p, e) is called Ψ-optimal in a certain class $\mathscr{H}$ if $\Psi\{\mathbf{R}(p, e)\} \le \psi\ \{\mathbf{R}(p', e')\} \forall (p', e')$ in $\mathscr{H}$.

Borrowing the ϕ_m-criteria from the theory of optimal designs which is deifined in terms of the eigenvalues of information matrix and using it in the theory of survey sampling, we have

Definition 10.5: For any $m > 0$, define

$$\phi_m\{\mathbf{R}(p, e)\} = \left\{\sum_{i=1}^{N-q} \lambda_i^m / (N - m)\right\}^{1/m} \tag{10.3.1}$$

A strategy, (p, e) is ϕ_m-optimal in $\mathscr{H}$ if

$$\phi_m\{\mathbf{R}(p, e)\} \le \phi_m\{\mathbf{R}(p', e')\}\ \forall\ (p', e') \text{ in } \mathscr{H}.$$

Definition 10.6: If $m = 1$, $\phi_1 \{\mathbf{R}(p, e)\} = \sum_{i=1}^{N-q} \lambda_i /(N-q)$

Thus ϕ_1-criterion, also called A-criterion, is the minimisation of trace of $\mathbf{R}(p, e)$. A strategy H is termed ϕ_1-optimal (or A-optimal) in $\mathscr{H}$ if

$$\text{tr } \mathbf{R}(H) = \min_{H' \in \mathscr{H}} tr \{\mathbf{R}(H')\} \tag{10.3.2}$$

It can be shown that

$$\lim_{m \to \infty} \phi_m \{\mathbf{R}(p, e)\} = \lambda_1, \tag{10.3.3}$$

the maximum eigenvalue of $\mathbf{R}(p, e)$. Hence

Definition 10.7: The ϕ_∞-criterion, also called E-criterion, is the minimisation of maximum eigenvalue of $R(H)$. A strategy H is termed ϕ_∞-optimal or E-optimal in $\mathscr{H}$ if

$$\lambda_1(H) = \min_{H' \in \mathscr{H}} \lambda_1(H) \tag{10.3.4}$$

Note 10.2: In Section 10.2, we have noted that under model (10.2.1) or (10.2.19) a minimax strategy in the class of all representative strategies is the one for which the maximum eigenvalue of the adjusted *rgm* is minimum in the class of all such strategies provided $\delta \in L_2(M)$ for any M (> 0). Hence, in such situations a minimax strategy is a E-optimal strategy.

Definition 10.8: A strategy H is L-optimal in a class $\mathscr{H}$ if

$$\text{tr}\{\mathbf{CR}(H)\} = \min_{H' \in \mathscr{H}} tr\{\mathbf{CR}(H')\}$$

where $\mathbf{C}$ is a given nnd symmetric matrix. For $C = I$, L-optimality reduces to A-optimality.

Definition 10.9: A strategy H is D-optimal in a class $\mathscr{H}$ if

$$|\mathbf{R}(H)| = \min_{H \in \mathscr{H}} |\mathbf{R}(H)|$$

where $|\mathbf{B}|$ denotes determinant of the matrix $\mathbf{B}$.

In Section 3.7 while examining the performance of several design-based strategies we compared their average mean square errors under assumed superpopulation models and a strategy with the smallest value of average mean square error in a class was considered as optimal in that class. Under Brewer-Royall prediction theoretic approach also a strategy with the smallest value of the average mean square error was considered as optimal. Denoting this optimality as model-optimality we shall in the following theorem discuss the connection between model-optimality and A-optimality.

Theorem 10.3: A representative strategy (p, e) is model-optimal under superpopulation model $\xi(\mathbf{X}, v)$ *iff* it is A-optimal under finite population model (10.2.19).

Proof: For the model $\xi(\mathbf{X}, v)$ write $Y_i = \mathbf{x}'_i\boldsymbol{\beta} + \in_i$ where $\in_i$ is independent with $(0, \sigma^2 g^2(x_i))$ which is now comparable with model (10.2.19). For a representative strategy (p, e)

$$\mathscr{E} E(e - Y)^2 = \mathscr{E} \sum_{s \in \mathscr{S}} \left\{ \sum_{i=1}^{N} (b_{si} - 1) \in_i \right\}^2 p(s)$$

$$= \sigma^2 \sum_{s \in \mathscr{S}} \left\{ \sum_{i=1}^{N} (b_{si} - 1)^2 g^2(x_i) \right\} p(s)$$

$$= \sigma^2 \text{tr} \left\{ \mathbf{G} \sum_{s \in \mathscr{S}} p(s)(\mathbf{b}_s - \mathbf{1})(\mathbf{b}_s - \mathbf{1}')\mathbf{G} \right\}$$

$$= \sigma^2 \text{tr}\mathbf{R}(p, e).$$

Hence the proof.

Remark 10.1: The *argm* reflects the performance of a strategy for a squared error loss function. For minimising the average variance one has to minimise the trace or average of λ-values, while for minimising the maximum risk one minimises the maximum eigenvalue.

We now consider the universal optimality of the strategy $(srswor, \bar{y})$. Kiefer (1975) defined the universal optimality of a design as follows:

Let,

$\mathscr{M}_N$ = {all $N \times N$ symmetric matrices of zero row-sums and zero column-sums} and $\mathscr{C} \subseteq \mathscr{M}_N$. Kiefer showed that if there exists a matrix $\mathbf{C}^* \in \mathscr{C}$ such that $\mathbf{C}^*$ is completely symmetric, trace $(\mathbf{C}^*) = \max_{C \in \mathscr{C}} \text{tr}(\mathbf{C})$ then $\mathbf{C}^*$ minmised $\Phi(\mathbf{C})$ over $\mathscr{C}$ for all real valued functions Φ defined over $\mathscr{M}_N$ satisfying (*i*) Φ is convex (*ii*) $\Phi(b\mathbf{C})$ is non-increasing in scalar $b(\geq 0)$ (*iii*) Φ is invariant under simultaneous permutations of rows and columns.

Such a matrix $\mathbf{C}^*$ is called universally optimal. Kiefer considered each matrix $\mathbf{C} \in \mathscr{E}$ as the information matrix of a block design. To bring this idea into survey sampling one has to consider the inverse or g-inverse of adjusted *rgm* of a strategy, which is a formidable task. Following Kiefer and Wynn (1981), Cheng and Li (1987), therefore, considered universal optimality in survey sampling as follows:

Definition 10.10: Let $\bar{\mathscr{C}} = \{\bar{\mathbf{C}}\} \subset \mathscr{M}_N$. The matrix in $\bar{\mathscr{C}}$ can be thought of as the adjusted risk generating matrices (or the g-inverses of information matrices). It is shown that if there exists a matrix $\bar{\mathbf{C}}^* \in \bar{\mathscr{C}}$ such that, (*a*) $\bar{C}$ is completely symmetric, (*b*) $\bar{\mathbf{C}}^*$ minimises trace ($\bar{\mathbf{C}}$) over $\bar{\mathscr{C}}$ then $\bar{\mathscr{C}}$ minimises $\Psi(\bar{\mathscr{C}})$ over $\bar{\mathscr{C}}$ for any real-valued function defined on $\mathscr{M}_N$ satisfying (*i*), (*iii*) and (*ii*′) $\psi(b\mathbf{C})$ is non-decreasing in scalar $b(\geq 0)$. Such a matrix $\bar{\mathbf{C}}^*$ is called 'weakly universally optimal'.

This is a weaker form of optimality because the class of convex decreasing functionals Φ of matrices in $\mathscr{M}_N$ is more general than the class of convex increasing functionals Ψ of the generalised inverses of matrices in $\mathscr{M}_N$. However, the class of criteria satisfying (*i*), (*ii*′) and (*iii*) are still quite large and it contains all the ϕ_m-criteria with $m \geq 1$.

Condition (*b*) is *A*-optimality. Therefore, for a strategy with symmetric adjusted *rgm*, optimality under superpopulation $\xi(\mathbf{X}, v)$ implies its weakly universal optimality under finite model (10.2.19) in the class of all strategies with symmetric adjusted *rgm*'s.

Consider model (10.2.1) with $x_1 = ... = x_N$. The representative conditions (10.2.8), here, reduces to $\sum_{i=1}^{N} b_{si} = N \,\forall\, s : p(s) > 0$. Here the adjusted *rgm* of the strategy H_0 = (srswor, $N\bar{y}$) is $aI_N + bJ_N$ where $a = (N - n)/nN(N - 1)$, $b = -a/N$ and hence is a symmetric matrix. It can be shown that H_0 is *A*-optimal under $\xi(0, 1; 1)$ (In fact it can be shown that every strategy using $N\bar{y}$ as an estimator is *A*-optimal under $\xi(0, 1; 1)$). Hence the strategy H_0 is weakly universally optimal. In particular, it is ϕ_m-optimal for $m \geq 1$.

Following Kiefer (1975), Chen and Li (1987) established the following:

Definition 10.11: Let $\mathscr{M}_{N'}$ be the set of all $N \times N$ and symmetric matrices $\mathbf{C}$ such that $\mathbf{Ca} = \mathbf{0}$ for some non-zero vector $\mathbf{a}$ and let $\bar{\mathscr{C}}' \subset \mathscr{M}_{N'}$. If there exists a matrix $\bar{\mathbf{C}}^* \in \bar{\mathscr{C}}'$ such that all the $(N - 1)$ largest eigenvalues of $\bar{\mathbf{C}}^*$ are equal and $\bar{\mathbf{C}}^*$ is *A*-optimal over $\bar{\mathscr{C}}'$, then $\bar{\mathbf{C}}^*$ minimises $\Psi(\mathbf{C})$ over $\bar{\mathscr{C}}'$ for all real-valued functions Ψ defined on $\mathscr{M}_{N'}$ satisfying (*i*), (*ii*)′ and (*iii*)′ $\Psi(\mathbf{0C0}) = \Psi(\mathbf{C})$ for any orthogonal matrix $\mathbf{0}$ such that $\mathbf{0a} = \mathbf{a}$. Such a matrix $\bar{\mathbf{C}}^*$ is again called weakly universally optimal.

Note 10.3: The consideration of the above definition comes from the fact that the representation condition (10.2.8) requires $\mathbf{Ra} = \mathbf{0}$ for some $\mathbf{a}$ for any $\mathbf{R} \in \mathscr{R}$. For example, for $g(x) = \sqrt{x}$, the requirement is $\mathbf{R}(\sqrt{x_1}, \ldots, \sqrt{x_N})' = \mathbf{0}$.

We note that Rao-Hartley-Cochran strategy is not *A*-optimal and hence is not weakly universally optimal. Under $\xi(0, 1; x)$, the strategy H_{BR} (defined in Section

10.2) is model-optimal and hence is A-optimal under the corresponding finite population model (10.2.1). Now

$$\text{tr}\{\mathbf{R}(H_{BR})\} \leq \text{tr}\{\mathbf{R}(H_{RHC})\} = \mu(N-1)XM/n.$$

For $m > 1$,

$$\phi_m\{\mathbf{R}(H_{RHC})\} = \mu X/n \text{ and}$$

and,
$$\phi_m\{\mathbf{R}(H_{BR})\} = X\sum_{\bar{s}^*} x_i \Big/ \left\{\sum_{\bar{s}^*} x_i(N-1)^{1/m}\right\}$$

Hence, H_{RHC} is $\phi_m(m > 1)$ better than A-optimal H_{BR} if

$$\sum_{\bar{s}^*} x_i/(N-1)^{1/m} \geq \bar{x}_s^*$$

In particular, if $\sum_{\bar{s}^*} x_i \geq \bar{x}_s^*$ then H_{RHC} is ϕ_α-better than H_{BR}. Hence, although H_{RHC} is not weakly universally optimal in the sense of Definition 10.11, it is weakly universally optimal over the subclass of strategies with *argm* $\mathbf{R}$ satisfying $\{\mathbf{R} : \mathbf{R} \in \mathcal{R}$ and $\text{tr}(\mathbf{R}) \leq \text{tr}\{\mathbf{R}(\mathbf{R}_{RHC})\}\}$, which again is a fairly large class containing some common competitors to H_{RHC}.

The following theorem states that under certain conditions H_{RHC} is a Ψ-better strategy in the sense of Definition 10.11.

Theorem 10.4: Let $g(x) = \sqrt{x}$. If $n \mid N$ then $\Psi\{\mathbf{R}(H_{RHC})\} \leq \psi\{\mathbf{R}(p, e)\}$ for any Ψ satisfying (*i*), (*ii*) and (*iii*)′ and any representative strategy satisfying

$$\sum_{s \in \mathcal{S}} P(s)/X_s \geq \frac{1}{n\bar{X}}, \text{ where } X_s = \sum_{i \in s} x_i \tag{10.3.5}$$

The proof follows by showing that under the stated conditions $\text{tr}\{\mathbf{R}(H_{RHC})\} \leq \text{tr}\{\mathbf{R}(p, e)\}$. The condition (10.3.5) is satisfied by srs and Lahiri-Midzuno scheme. Hence, H_{RHC} is better than any representative estimator used along with srs or *ppas*-sampling with respect to all the criteria satisfying (*i*), (*ii*) and (*iii*)′ which includes ϕ_m-criteria with $m \geq 1$. In particular, it is better than H_0, H_M and Hartley-Ross estimator used along with srs.

10.4 SOME OPTIMALITY CRITERIA FOR CHOOSING SAMPLING DESIGNS

We consider in this section some further optimality criteria for choosing sampling designs under superpopulation model based prediction-theoretic approach. All these lead to purposively selected sampling designs. However, these are quite different from Royall's purposive designs p^*.

(*a*) *Purposive minimax sampling designs:* Wynn (1977*a*) consider the maximum value of prediction error of $\hat{Y}_i (i \in \bar{s})$ as the criterion for choosing a sample under ξ(**X**, *v*). An optimum sample s_w is the one for which

$$\bar{g}(s_w) = \min_{s \in \mathcal{S}_n} \bar{g}(s)$$

where,
$$\bar{g}(s) = \max_{i \in \bar{s}} g_i,\ g_i = \mathcal{E}(\hat{Y}_i - Y_i)^2 / v_i .$$

A minimax purposive design p_w, where $p_w(s) = 1(0)$ for $s = s_w$ (otherwise), minimises the maximum prediction error of $\hat{Y}_i$, relative to the variance of the observation overall the non-sampled units. An upper bound to the minimax risk is given by

Theorem 10.5: For the regression model ξ(**X**, *v*),

$$\min_{s \in \mathcal{S}} g(s) \le (n + 1)/(n - q + 1) \tag{10.4.1}$$

where $q = r + 1$, number of parameters in **β** in ξ(*X*, *v*).

Note that the upper bound (10.4.1) is independent of *N*. The minimax sample s_m is obtained from an initial sample s_0 as follows:

(*i*) Choose an initial sample s_0 such that $\mathbf{X}'_{s_0} \mathbf{X}_{s_0}$ is non-singular;

(*ii*) Choose s_1 from $s'_1 = s_0 \cup j_0$ (where $g_{j0} = \bar{g}(s_0)$, $j_0 \in \bar{s}_0$) such that

$$|\mathbf{X}'_{s_1} \mathbf{X}_{s_1}| = \max_{s \subset s'_1} |\mathbf{X}'_{s_t} \mathbf{X}_{s_t}|;$$

(*iii*) Treat s_1 as the new initial sample and obtain s_2 following step (*ii*).

Proceed in this way until the sequence $|\mathbf{X}'_{s_t} \mathbf{X}_{s_t}|$ converges to some value $|\mathbf{X}'_{s_p} \mathbf{X}_{s_p}|$ (say). At this stage the bound in (10.4.1) has already been crossed.

Example 10.1: Model ξ(1, 1; 1) with $x_i = 0(1)$, $i = 1, ..., N/2(N/2 + 1, ..., N)$, *N* even. Assume *n* odd. The minimax sample selects $(n + 1)/2$ units out of $\{1, ..., N/2\}$ and $(n - 1)/2$ units from the remaining. The minimax value $\sigma^2(n + 1)/(n - 1)$ is achieved at any $i (i \in \bar{s})$ with $x_i = 1$.

Example 10.2: Consider the population in (Murthy, 1977, p. 178) with *x* as the geographical area and *y* as the area under winter paddy for 108 villages.

Suppose we fit the model ξ(1, 1; 1). Let $n = 18$ and the initial sample s_0 be a systematic sample with random start 1. The table on the following page shows the different steps of iteration in achieving the minimax sample.

Iteration	Unit		$\bar{g}(s)$	X'_sX_s
	Added	Deleted		
1	108	67	1.4523	4.4362
2	107	73	1.2848	6.1027
3	106	79	1.1880	7.4281
4	105	85	1.1270	8.3535
5	104	91	1.1143	8.9073
6	2	61	1.1056	9.3171
7	4	55	1.1026	9.6562
8	102	97	1.1022	9.9842
9	6	49	1.0988	10.2800

After this step there is no further improvement. The bound (10.4.1) (=19/17) is crossed after the 4th stage.

(b) D-optimal and G-optimal designs: Following the optimum design theory in regression problems as developed by Kiefer and Wolfwitz (1959), Wynn (1977*a*, *b*) defined *D*-optimal and *G*-optimal sampling designs as follows:

Definition 10.12: A sampling design p_D is termed a *D*-optimal design if $p_D(s) = 1(0)$ for $s = s_D$ (otherwise), where s_D is such that

$$|\mathbf{X}'_{s_D}\mathbf{X}_{s_D}| = \max_{s\in\mathcal{S}_n} |\mathbf{X}'_s\mathbf{X}_s|$$

Definition 10.13: A sample s_G is *G*-optimal if

$$\sup_{i\in\bar{s}_G} v_s(\mathbf{x}_i) = \inf_{s\in\mathcal{S}_n} \sup_{i\in\bar{s}} v_s(\mathbf{x}_i),$$

$s_G \in \mathcal{S}_n$, where $v_s(\mathbf{x}_i)$ is the variance function $\mathbf{x}'_i(\mathbf{X}'_s\mathbf{X}_s)^{-1}\mathbf{x}_i$ under $\xi(\mathbf{X}, v)$. Wynn's (1977*a*) minimax purposive sampling design is a *G*-optimal design.

A more general theory on *D*-optimal and *G*-optimal designs applicable for continuous case has been developed by Wynn (1977*b*). [Consider the model:

$$\mathscr{E}(Y_i) = f(\mathbf{x}_i)'\boldsymbol{\beta},\ \mathscr{V}(Y_i) = \sigma^2,\ \mathscr{C}(Y_i, Y_{i'}) = 0,\ i \neq i'$$

where $\mathbf{x}_i$ lies in a compact set of R_m, $f(\mathbf{x}) = (f_1(\mathbf{x}), \ldots, f_m(\mathbf{x}))'$ is a vector of linearly independent continuous functions of $\mathbf{x}$ and $\boldsymbol{\beta} = (\beta_1, \ldots, \beta_m)$. For a given s, define the moment matrix $\mathbf{M}_s$ as

$$\mathbf{Z}'_s\mathbf{Z}_s = \mathbf{M}_s$$

where, $$Z_s = [f(\mathbf{x}_k)', k \in s]$$

Also for a given s and i, define

$$v_s(\mathbf{x}_i) = f'(\mathbf{x}_i)\mathbf{M}_s^{-1} f(\mathbf{x}_i)$$

For this model,

$$\mathscr{E}(\hat{Y}_i - Y_i)^2 = \sigma^2[1 + v_s(\mathbf{x}_i)], i \in \bar{s},$$

A sample is said to be D-optimal if it maximises $\mathbf{M}_s$ and a sample is G-optimal if it maximises the $\underset{i \in \bar{s}}{\text{supremum}}\ v_s(\mathbf{x}_i)$ overall $s \in \mathscr{S}_n$. It can be proved that for a D-optimal sample

$$v_s(\mathbf{x}_i) > v_s(\mathbf{x}_j) \text{ for } i \in s,\ j \in \bar{s}]$$

❑ ❑ ❑

CHAPTER 11

Variance Estimation Under Different Superpopulation Models

11.1 INTRODUCTION

We discussed in Section 2.4 variance estimation for different estimators under fixed population set-up. In some cases we exploited superpopulation models to get an assessment of the average value of the design variance of an estimator and contrived to estimate this model-expected variance. The usual textbook formulae for unbiased estimation of sampling variance of simple (linear) estimators of, say, totals and means are inadequate for modern complex surveys which often involve estimation of non-linear functions, say, population ratio, difference of ratios, regression coefficient, correlation coefficient. There are two approaches to the estimation of variance of a non-linear estimator. One is linearisation, in which the non-linear estimator is approximated by a linear one for the purpose of variance estimation. The second is replication in which several estimates of the population parameters are derived from different comparable parts of the original sample. The variability of these estimates is used to estimate the variance of the parameter estimator.

When, however, a superpopulation model is available and is exploited to build up suitable predictors of population characteristics (mean, total, etc.) the conditionality principle (Cox and Hinkley, 1974) states that the inference should be based on the model for a given sample. Here, therefore, instead of estimating the design variance (expected p-variance) of an estimator, one is required to estimate the model-variance of a predictor and in the case of population total, the quantity $\mathscr{E}(\hat{T}-Y)^2$ (alternatively, $\mathscr{E}(\hat{T}-\mu)^2$) for a given sample. (Recently, in the context of a ratio predictor, the conditional mean square error given an ancillary statistic is suggested by Wu and Deng (1983) for the construction of confidence interval. Specifically, they considered estimation of $\mathscr{E}\{(\hat{T}_R-Y)^2 \mid \overline{x}_s\})$.

Royall and Cumberland (1968*a*) studied the robust estimation of $\mathscr{V}\{\hat{T}^*(\mathbf{X}, v)\}$ under alternative superpopulation models with different variance functions. Several authors studied the estimation of model-variance of ratio predictor, regression predictor, Horvitz-Thompson predictor under different models.

Sarndal (1982) considered the *ADU* and *ADC*-estimation of design dispersion matrix of generalised regression estimators of different linear functions of population values under $\xi(\mathbf{X}, v)$. It has been shown that under suitable conditions one of these estimators (predictors) is model-unbiased for model-dispersion matrix of the predictors.

We review these results in this chapter. In Sections 11.2 through 11.6 we consider Taylor series linearisation procedure, random group (RG), balanced repeated replication (BRR), Jackknife (JK) and Boot-strap (BS) techniques of variance estimation in brief. All these procedures (except the Boot-strap resampling) have been considered in details in the excellent treatise by Wolter (1985). Section 11.7 considers robust estimation of $\mathcal{V}\{\hat{T}^*(\mathbf{X}, v)\}$ under alternative models. Estimation of model-variance of ratio, regression and Horvitz-Thompson predictors have been considered in Sections 11.8 to 11.10. Lastly, we consider asymptotically unbiased estimation of design variance of $\hat{T}_{GR}$ under $\xi(\mathbf{X}, v)$.

11.2 TAYLOR SERIES LINEARISATION

Here, a non-linear estimator is approximated by a linear estimator and the variance of the linear approximation is estimated by using standard variance estimation methods appropriate to the specific sampling design or by any of the replication procedures discussed subsequently. This leads to a biased but typically consistent estimator of the variance (actually of the mse, since the linear approximation is a biased estimator) of the non-linear estimator.

Let $\boldsymbol{\theta} = (\theta_1, ..., \theta_p)'$ denote a p-dim. vector of population parameters with $\hat{\boldsymbol{\theta}} = (\hat{\theta}_1, \ldots, \hat{\theta}_p)'$ as the corresponding vector of estimators (often unbiased, but sometimes biased, but consistent) based on s of size $n(s)$. Often θ_i is the population total (mean) of a survey character. We want to estimate $\psi = g(\boldsymbol{\theta})$ for which a natural estimator is $\hat{\psi} = g(\hat{\boldsymbol{\theta}})$. If $g(\mathbf{t})$, $\mathbf{t} = (t_1, ..., t_p)$ possesses continuous derivatives of order two in an open interval containing θ and $\hat{\theta}$, then we may write

$$\hat{\psi} - \psi = \sum_{j=1}^{p} \frac{\partial g(\mathbf{t})}{\partial t_j}(\hat{\theta}_j - \theta_j)\Bigg]_{\mathbf{t}=\boldsymbol{\theta}} + R_{n(s)}(\hat{\boldsymbol{\theta}}, \boldsymbol{\theta}) \tag{11.2.1}$$

where $R_{n(s)}(\hat{\boldsymbol{\theta}}, \boldsymbol{\theta})$ is the remainder term, often considered to be negligible (which may not be the case always, particularly for small $n(s)$) in the Taylor series expansion. Hence,

$$\begin{aligned}\text{MSE}(\hat{\psi}) &= E\{g(\hat{\boldsymbol{\theta}}) - g(\boldsymbol{\theta})\}^2 \\ &= \sum_{i=1}^{p}\sum_{j=1}^{p} \frac{\partial g(\mathbf{t})}{\partial t_i}\frac{\partial g(\mathbf{t})}{\partial t_j}\text{Cov}(\hat{\theta}_i, \hat{\theta}_j)\Bigg]_{\mathbf{t}=\boldsymbol{\theta}}\end{aligned}$$

$$= \mathbf{h}' \sum_{\mathbf{n(s)}} h \text{ (say)} \tag{11.2.2}$$

where $\sum_{n(s)}$ is the dispersion matrix of $\hat{\theta}$. Second and higher order approximations are possible by retaining additional terms in the Taylor series. The variance estimation of $g(\hat{\boldsymbol{\theta}})$ is

$$v_{\text{Tay}}\{g(\hat{\boldsymbol{\theta}})\} = \hat{\Sigma}_{n(s)}(\mathbf{h}) \tag{11.2.3}$$

where parameters in (11.2.2) are replaced by their sample-based estimates. In general, v_{Tay} is not unbiased either for the true mse $\{g(\hat{\boldsymbol{\theta}})\}$ or for that of the approximation in (11.2.1). It is, however, consistent provided that $\hat{\boldsymbol{\theta}}$ and $\hat{\Sigma}_{n(s)}$ are consistent.

A criticism against the method is about the convergence of the Taylor series used to develop (11.2.1). For ratio estimation Koop (1972) gave a simple example where the convergence condition is violated. Again, for complex estimators, the analytic partial differentiation needed to derive the linear substitute has been found to be intractable. Woodruff and Causey (1976) describes a solution to this problem that uses a numerical procedure to obtain the necessary partial derivatives. Binder (1983) provides a general approach to the analytic derivation of variance estimators for linear Taylor series approximations for a wide class of estimators. Empirical evidences have shown, however, that the linearisation variance estimators are generally of adequate accuracy, particularly, when the sample size is large. The approximations may be unreliable in case of highly skewed populations.

11.3 RANDOM GROUP METHOD

In the random group (RG) method, first developed at the US Bureau of Census, the original sample is divided into a number k of groups at random. Let $\hat{\theta}, \hat{\theta}_1, \ldots, \hat{\theta}_k$ be the estimators obtained from the original sample and k random groups respectively, all the estimators using the same estimating procedure. The RG estimate of θ is $\bar{\hat{\theta}} = \sum_i \hat{\theta}_i / k$. If $\hat{\theta}$ is linear, $\bar{\hat{\theta}} = \hat{\theta}$. The RG estimate of $V(\bar{\hat{\theta}})$ is

$$v(\bar{\hat{\theta}}) = \sum_{i=1}^{k} (\hat{\theta}_i - \bar{\hat{\theta}})^2 \Big/ k(k-1) \tag{11.3.1}$$

To estimate $V(\hat{\theta})$, one may use either $v_1(\hat{\theta}) = v(\bar{\hat{\theta}})$ or

$$v_2(\hat{\theta}) = \sum_{i=1}^{k} (\hat{\theta}_i - \hat{\theta})^2 \Big/ k(k-1) \tag{11.3.2}$$

If a conservative estimator of $V(\hat{\theta})$ is desired, $v_2(\hat{\theta})$ is preferable to $v_1(\hat{\theta})$. If

$$E(\hat{\theta}_i) = \mu_i \, (i = 1, \ldots, k),$$

$$E\{v(\hat{\bar{\theta}})\} = V(\hat{\bar{\theta}}) + \frac{1}{k(k-1)}\left[\sum_{1}^{k}(\mu_i - \bar{\mu})^2 - 2\sum\sum_{i<j=1}^{k}\text{Cov}(\hat{\theta}_i, \hat{\theta}_j)\right] \quad (11.3.3)$$

The RG procedure has its origin in the works of Mahalanobis (1944) and Deming (1956), who advocated the use of interpenetrating or replicated samples, where k independent samples are selected using the same sampling design and the estimators of the type (11.3.1) is used to estimate the variance of the overall estimator. In RG method, the major difference is that the replicates are not formed independently.

The coefficient of variation of $v(\hat{\bar{\theta}})$ is an increasing function of kurtosis of $\hat{\theta}_1, \beta_4(\hat{\theta}_1)$ and a decreasing function of k for a wide range of complex surveys (this theoretical result is derived assuming $\hat{\theta}_i$ ($i = 1, ..., k$) are independently and identically distributed; however, when N, n are large and $n/N \simeq 0$, the result holds even for non-independent RG's). Consequently, the larger the number of groups, the higher is the precision, though computational cost will increase at the same time. The optimum value of k is a trade-off between cost and precision. The RG method is suitable in surveys using a large number of primary stage units (psu's) where many psu's are selected per stratum.

11.4 BALANCED REPEATED REPLICATIONS

The method of balanced half-sample repeated replications (BRR) has proved very useful for surveys in which two psu's are selected per stratum. Following Plackett and Burman (1946), McCarthy (1966), (1969*a,b*) introduced the concept of BRR, also known as balanced half-samples, balanced fractional samples, pseudoreplication.

Suppose that two units are selected by srswr from each of the L strata in a population for estimating $\bar{Y} = \sum_h W_h\bar{Y}_h[W_h = N_h/N, \bar{Y}_h = Y_h/N_h$, $N_h(N)$ is the stratum (population) size, $Y_h = \sum_{i=1}^{N_h} y_{hi}$, y_{hi} is the value of y on the unit (h, i)]. By selecting one unit from the sampled units in each stratum at random we can form 2^L sets (of which 2^{L-1} are distinct) of two half-samples (HS's) each such that each set forms a complete replicate. In a set (α) denote the two HS's as S_α, $S_{\alpha'}$ with the corresponding estimates $\bar{y}_{st,\alpha} = \left(\sum_h W_h y_{h1}\right)$, $\bar{y}_{st,\alpha'}\left(=\sum_h W_h y_{h2}\right)$. The customary estimate $\bar{y}_{st} = (\bar{y}_{st\,\alpha} + y_{st\,\alpha'})/2$. The αth replicate-estimate of $V(\bar{y}_{st})$,

$$v_\alpha(\bar{y}_{st}) = v_\alpha = (\bar{y}_{st\alpha} - \bar{y}_{st})^2 = (\bar{y}_{st\alpha} - \bar{y}_{st\alpha'})^2/4 \quad (11.4.1)$$

is unbiased. Now,

$$\bar{y}_{st\alpha} = \sum W_h(y_{h1}\,\delta_{h1\alpha} + y_{h2}\delta_{h2\alpha}) \tag{11.4.2}$$

where

$$\delta_{h1\alpha} = 1(0) \text{ if sampled unit } (h,\ 1) \in S_\alpha \text{ (otherwise)}$$
$$\delta_{h2\alpha} = 1 - \delta_{h1\alpha} \tag{11.4.3}$$

Also,
$$\bar{y}_{st\alpha} - \bar{y}_{st} = \frac{1}{2}\sum W_h \delta_H^\alpha d_h \tag{11.4.4}$$

where
$$\delta_h^\alpha = 2\delta_{h1\alpha^{-1}},\ d_h = y_{h1} - y_{h2} \tag{11.4.5}$$

Hence,
$$v_\alpha = \frac{1}{4}\sum W_h^2 d_h + \frac{1}{2}\sum \sum_{h<h'}^{L} W_h W_{h'} \delta_h^\alpha \delta_{h'}^\alpha,\ d_h d_{h'} \tag{11.4.6}$$

It follows that

$$\sum_{\alpha=1}^{2^L} \bar{y}_{st\alpha} \Big/ 2^L = \bar{y}_{st},\ \sum_{\alpha=1}^{2^L} v_\alpha \Big/ 2^L = v(\bar{y}_{st}),$$

the customary estimator $\sum_h W_h^2 d_h^2/4$. When L is large, computation of $v(\bar{y}_{st})$ as the average of v_α over 2^L HS's become formidable.

However, if we choose a set η of K HS's such that

$$\sum_{\alpha\in\eta} \delta_h^2 \delta_{h'}^\alpha = 0,\ h < h' = 1,\ \ldots,\ L \tag{11.4.7}$$

then
$$\bar{v}_{(K)} = \sum_{\alpha\in\eta} v_\alpha / K = v(\bar{y}_{st}).$$

Plackett and Burman (1946) developed a method for constructing $m \times m$ orthogonal matrices with entries +1, –1 where m is a multiple of 4. These can be used directly to obtain values of δ_h^α satisfying (11.4.7). The orthogonal matrix of size K where K is a multiple of 4, between L and $L + 3$, can be used dropping the last $K - L$ columns. The entires in the matrix can be substituted as $\delta_{h'}^\alpha$, each row (column) standing for a replicate (stratum). McCarthy referred to the set η as balanced. If, further, the condition

$$\sum_{\alpha\in\eta} \delta_h^\alpha = 0,\ h = 1,\ \ldots,\ L \tag{11.4.8}$$

is satisfied, then $\sum_{\alpha\in\eta} \bar{y}_{st\alpha}/K = \bar{y}_{st}$. The set of replicates satisfying (11.4.7), (11.4.8) is said to be in full orthogonal balance.

For designs with wor-sampling of psu's, v_α is positively biased. In this case a separate adjustment is necessary to account for this bias, though the bias is generally negligible.

In the non-linear case, in which the usefulness of BRR is manifested most, let $\hat{\theta}, \hat{\theta}_\alpha, \hat{\theta}_{\alpha'}$ be the estimates of θ based on the whole sample, S_α, $S_{\alpha'}$ respectively. Let

$\bar{\hat{\theta}}_\alpha = (\hat{\theta}_\alpha + \hat{\theta}_{\alpha'})/2$. We note that even for a balanced set of HS's, $\sum_{\alpha \in \eta} \hat{\theta}_\alpha / K \neq \hat{\theta}$ in general. Empirical studies by Kish and Frankel (1970), among others, however, show that $\hat{\hat{\theta}}_\alpha$ is very close to $\hat{\theta}$ in general. Writing

$$\bar{v}_{(K)}(\hat{\theta}) = \sum_{\alpha \in \eta} (\hat{\theta}_\alpha - \hat{\theta})^2 / K$$

$$\bar{v}'_{(K)}(\hat{\theta}) = \sum_{\alpha \in \eta} (\hat{\theta}_{\alpha'} - \hat{\theta})^2 / K$$

we have the alternative variance estimators

$$(i)\ \bar{v}_{(K)}(\hat{\theta})$$

$$(ii)\ \bar{v}'_{(K)}(\hat{\theta})$$

$$(iii)\ [\bar{v}_{(K)}(\hat{\theta}) + \bar{v}'_{(K)}(\hat{\theta})]/2 = \bar{\bar{v}}_{(K)}(\hat{\theta}) \qquad (11.4.9)$$

$$(iv)\ \sum_{\alpha \in \eta} (\hat{\theta}_\alpha - \hat{\theta}_{\alpha'})^2 / 4K = \bar{v}^{+}_{(K)}(\hat{\theta}).$$

(*i*), (*ii*), (*iii*) are sometimes regarded as estimators of mse $(\hat{\theta})$, while (*iv*) is regarded as estimator of $V(\hat{\theta})$.

Since (*iii*) is the average of (*i*) and (*ii*), it is at least as precise as the others and equally biased. However, (*iii*) is comparatively costlier than (*i*) (and (*ii*)) and perhaps, significantly so, when many estimators are produced. All the estimators (*i*)–(*iv*) become identical in the linear case.

Another set of variance estimators can be attained by writing $\hat{\hat{\theta}}$ in place of $\hat{\theta}$ in (11.4.9). Such estimators are unbiased for linear $\hat{\theta}$ only if the number of HS's is $T > L$. If L is a multiple of 4, $T(= L + 4)$ HS's must be used to retain the unbiasedness. (Lemeshow and Epp, 1977). The estimates using $\hat{\hat{\theta}}$ are generally not preferred to those using $\hat{\theta}$, since they give smaller and less conservating estimates of mse, as they do not include the components for bias of $\hat{\theta}$. Empirical works by McCarthy (1969*a*, *b*), Kish and Frankel (1970), Levy (1971), Frankel (1971) and others show that BRR provides satisfactory estimates of the true variance.

Two modifications of BRR have been proposed, that require fewer replicates but the corresponding estimates are less precise and equally biased as the full BRR. In one modification strata are combined into groups, not necessarily of the same size. For each replicate all strata into a group g are assigned the same value δ_α^h.The constraints (11.4.7) is imposed for pairs h, h' of strata which are not in the same group g. Thus if G groups are formed, the number of replicates required is the multiple of 4, which lies in the range G to $G + 3$. The Plackett and Burman matrices of size K may then be used to derive the values δ_α^h.

The second procedure for reducing the number of replicates, discussed by McCarthy (1966) and developed by Lee (1972, 1973) is the method of partially balanced replicated replications (PBRR). Here the strata are divided into groups and full balancing are applied to the strata within each group. If T replicates are required for L strata, $G = L/T$ groups are formed with T strata in each. A $T \times T$ orthogonal matrix is then used to ensure a full balance within each group. Lee (1972, 1973), Rust (1984) suggest methods of implementing PBRR that would minimise the loss in precision over fully balanced BRR.

Rust (1984, 1986) shows that the method of PBRR and combined strata are equivalent. However, the combined strata method has a greater flexibility in the sense that the number of strata per group varies.

For general designs in which strata sample sizes vary, BRR can be implemented by dividing the psu's in each stratum into two groups of equal sizes (assuming $n_h = 2m_h$, m_h an integer), and then using these groups as units (Kish and Frankel, 1970). In this case the BRR-variance estimator is somewhat less precise than the customary variance estimator. Valiant (1987) considers the large sample prediction properties of the BRR-separate ratio and regression estimator under a superpopulation model when n_h, is large and compares these with Jackknife and linearisation procedures.

11.5 JACKKNIFE PROCEDURES

Quenouille (1949, 1956) originally introduced Jackknife (JK) as a method of reducing the bias of an estimator. Tukey (1958) suggested the use of this technique for variance estimation. Durbin (1959) first considered its use in finite population. Extensive discussion of JK method is given in Miller (1964, 1972), Gray and Schucany (1972) and in a recent monograph by Efron (1982).

Let θ be the parameter to be estimated. An estimator $\hat{\theta}$ is obtained from the full sample. Assuming $n = mk$ (m, k integers), we partition the sample into k groups of m observations each. Let $\hat{\theta}_{(\alpha)}$ be the estimator of θ computed from the whole sample except the αth group. Define pseudo-values $\hat{\theta}_\alpha$ as

$$\hat{\theta}_\alpha = k\hat{\theta} - (k-1)\hat{\theta}_{(\alpha)} \tag{11.5.1}$$

Quenouille's estimator is

$$\bar{\hat{\theta}} = \sum_{\alpha=1}^{k} \hat{\theta}_\alpha / k \tag{11.5.2}$$

Tukey suggested that $\hat{\theta}_\alpha$ are approximately iid. The JK-estimator of variance is

$$\begin{aligned} v_1(\bar{\hat{\theta}}) &= \sum_{\alpha=1}^{k} (\hat{\theta}_\alpha - \bar{\hat{\theta}})^2 \Big/ k(k-1) \qquad (11.5.3) \\ &= \{(k-1)/k\} \sum_{\alpha=1}^{k} (\hat{\theta}_{(\alpha)} - \hat{\theta}_{(.)})^2 \end{aligned}$$

where $\hat{\theta}_{(\cdot)} = \sum_{1}^{k} \hat{\theta}_{(\alpha)}/k$. In practice $v_1(\bar{\hat{\theta}})$ is used not only to estimate variance of $\bar{\hat{\theta}}$ but also of $\hat{\theta}$. Alternatively, one may use

$$v_2(\hat{\theta}) = \sum_{\alpha=1}^{k} (\hat{\theta}_{\alpha} - \hat{\theta})^2 / k(k-1) \qquad (11.5.4)$$

which is always at least as large as $v_1(\hat{\theta})$.

The number of groups k is determined from the point of view of computational cost and the precision of the resulting estimator. The precision is maximised when each drop-out group is of size one and each unit is dropped only once. Rao (1965), Rao and Webster (1966), Chakraborty and Rao (1968), Rao and Rao (1971) in their studies on ratio estimator based on superpopulation model showed that both bias and variance of $\bar{\hat{\theta}}$ are maximised for the choice $k = n$.

Brillinger (1966) showed that both v_1 and v_2 give plausible estimates of the asymptotic variance. Shao and Wu (1989), Shao (1989) considered the efficiency and consistency of JK variance estimators. For non-linear statistic $\hat{\theta}$ that can be expressed as functions of estimated means of p variables, such as ratio, regression, correlation coefficient, Krewski and Rao (1981) established the asymptotic consistency of variance estimators from JK, the linearisation and BRR methods. In the case of two sampled psu's per stratum, Rao and Wu (1985) showed that the linearisation and JK variance estimators are asymptotically equivalent.

11.6 THE BOOTSTRAP

The Bootstrap (BS) method for finite population sampling was introduced and discussed by Gross (1980), Bickel and Freedman (1984), Chao and Lo (1985), McCarthy and Snowdon (1985), Booth, Butler and Hall (1991), among others. Rao and Wu (1988) have shown how to apply the BS in design-based survey sampling under different sampling designs including stratified cluster sampling with replacement, stratified simple random sampling without replacement, unequal probability random sampling without replacement and two-stage cluster sampling with equal probabilities and without replacement. We consider the elements of variance estimation by BS.

Suppose we have p variables $y_1, \ldots, y_p$ with $Y_{hik}(y_{hik})$, $\bar{Y}_k, \bar{Y}_{hk}, \bar{y}_{hk}$ as the value of y_k on the ith unit in the hth stratum [of size N_h, $W_h = N_h/N$] in the population (sample in size n_h), population mean, stratum population mean, stratum sample mean of y_k respectively $\left(h = 1, \ldots, L;\ \bar{Y}_k = \sum_h W_h \bar{Y}_{hk}\right)$. Suppose we want to estimate $\theta = g(\bar{Y}_1, \ldots, \bar{Y}_p) = g(\bar{\mathbf{Y}})$, a non-linear function of $\bar{\mathbf{Y}}$. This includes population ratio, regression, correlation coefficient, etc. A natural estimator of θ, whenever n_h (≥ 2)

psu's are selected wr is $g(\hat{\bar{\mathbf{Y}}}) = g(\bar{\mathbf{y}})$ where $\bar{\mathbf{y}} = (\bar{y}_1, \ldots, \bar{y}_p)$, $\bar{y}_k = \sum_h W_h \bar{y}_{hk}$ $(\bar{\mathbf{y}}_h = (\bar{y}_{h1}, \ldots, \bar{y}_{hp}))$. In this case, the random vector $\mathbf{y}_{hi} = (y_{hi1}, \ldots, y_{hip})$, $i = 1, \ldots, n_h$ are iid with $E(\mathbf{y}_{hi}) = \bar{\mathbf{Y}}_h = (\bar{Y}_{h1}, \ldots, \bar{Y}_{hp})$. The vectors $\mathbf{y}_{hi}$, $\mathbf{y}_{h'j}$ $(h \neq h')$ are independent but not necessarily identically distributed. The BS resampling procedure is as follows:

(*a*) Draw a random sample wr $\{\mathbf{y}^*_{hi}, i = 1, \ldots, n_h\}$ of size n_h from $\{\mathbf{y}_{hi}, i = 1, \ldots, n_h\}$ independently from each stratum. Calculate

$$\bar{\mathbf{y}}^*_h = \sum_i \mathbf{y}^*_{hi} / n_h \quad \bar{\mathbf{y}}^* = \sum_h W_h \bar{\mathbf{y}}^*_h \text{ and } \hat{\theta}^* = g(\bar{\mathbf{y}}^*).$$

(*b*) Repeat step (*a*) a large number B of times and calculate the corresponding $\hat{\theta}^{*1}, \ldots, \hat{\theta}^{*B}$.

(*c*) Calculate the Monte-Carlo estimate,

$$v_b(a) = \sum_1^B (\hat{\theta}^{*b} - \hat{\theta}^{*\cdot})^2 / (B-1) \tag{11.6.1}$$

where $\hat{\theta}^{*\cdot} = \sum_b \hat{\theta}^{*b} / B$. The estimator $v_b(a)$ approximates well the BS variance-estimator

$$v_b = \text{var}_*(\hat{\theta}^*) = E_*(\hat{\theta}^* - E_*(\hat{\theta}^*))^2 \tag{11.6.2}$$

where E_* denotes expectation wrt BS-resampling. The BS-estimator $E_*(\hat{\theta}^*)$ of θ is approximated by $\hat{\theta}^{*\cdot}$.

In the linear case with $p = 1$, $\theta = \bar{Y}$, $\hat{\theta}^* = \sum W_h \bar{y}^*_h = \bar{y}^*$ and v_b reduces to

$$\text{var}_*(\bar{y}^*) = \sum_h W_h^2 \sigma_h^2 / n_h \tag{11.6.3}$$

where $\sigma_h^2 = \sum_i (y_{hi} - \bar{y}_h)^2 / n_h$. Comparing (11.6.3) with the customary estimator $v(\bar{y}) = \sum_h W_h^2 s_h^2 / n_h$ where $s_h^2 = \sum_i (y_{hi} - \bar{y}_h)^2 / (n_h - 1)$, it follows that $\text{var}_*(\bar{y}^*)/v(\bar{y})$ does not converge in probability to 1 when n_h is bounded. Hence, $\text{var}_*(\bar{y}^*)$ is not a consistent estimator of $V(\bar{y})$ unless n_h is a constant for all h. Moreover, v_b in (11.6.2) is not a consistent estimator of the variance of a general non-linear estimator. To get rid of this difficulty, Rao and Wu (1988) suggested the following resampling procedure:

(*i*) Draw a random sample $\{y^*_{hi}, i = 1, \ldots, m_h\}$ w_r $(m_h \geq 1)$ from $\{y_{hi}, i = 1, \ldots, n_h\}$. Calculate,

$$\tilde{y}_{hi} = \bar{y}_h + \sqrt{m_h}(y^*_{hi} - \bar{y}_h)/\sqrt{n_h - 1}$$

$$\tilde{y}_h = \sum_i \tilde{y}_{hi}/m_h = \bar{y}_h + \sqrt{m_h}(\bar{y}^*_h - \bar{y}_h)/\sqrt{n_h - 1}$$

$$\tilde{y} = \sum_h W_h \tilde{y}_h, \tilde{\theta} = g(\tilde{y})$$

(*ii*) repeat the step (*i*) **B** times independently and calculate the corresponding estimates $\tilde{\theta}^1, \ldots \tilde{\theta}^B$. The BS-estimator $E_*(\tilde{\theta})$ of θ is approximated by $\tilde{\theta}^{\cdot} = \sum \tilde{\theta}^b / B$.

(*iii*) The BS variance estimator of $\hat{\theta}$,

$$\tilde{v}_b = E_*(\tilde{\theta} - E_*(\tilde{\theta}))^2$$

is approximated by the Monte-Carlo estimate,

$$\tilde{v}_b(a) = \sum_1^B (\tilde{\theta}^b - \tilde{\theta}^{\cdot})^2/(B-1)$$

For $m_h = n_h - 1$, $\tilde{y}_{hi} = y^*_{hi}$ and the method reduces to the usual Bootstrap.

In general, for $n_h \geq 5$, $m_h \simeq n_h - 3$. Rao and Wu made some empirical studies on the choice of m_h.

In the linear case of $p = 1$, $\theta = \bar{Y}, \tilde{v}_b$ reduces to the customary unbiased variance estimator $v(\bar{y})$.

In the non-linear case it has been shown that

$$\tilde{v}_b = v_L + 0(n^{-2})$$

where v_L is the customary linearisation variance estimator.

Through Sections 11.2 to 11.6 we have given introduction to linearisation and different replication techniques to variance estimation which are equally applicable to design-based and model-based surveys. Our emphasis in this treatise being on model-based surveys we shall in subsequent sections consider estimation of model variance of different predictors under suitable superpopulation models and design-variance of model-based predictors.

11.7 ROBUST ESTIMATION OF $\mathcal{V}(\hat{T}^*)$ UNDER $\xi(X, v)$

Consider $\xi(\mathbf{X}, v)$ and $\hat{T}^*$ as given in (4.4.5).

$$\mathcal{V}(\hat{T}^* - Y) = \sigma^2[\mathbf{x}'_{\bar{s}}\mathbf{A}^{-1}\mathbf{x}_{\bar{s}} + \mathbf{1}'_{N-n}\mathbf{V}_{\bar{s}}\mathbf{1}_{\mathbf{N}-n}] \quad (11.7.1)$$

where $\mathbf{A} = \mathbf{X}'_s\mathbf{V}_s^{-1}\mathbf{X}_s$. An estimator of σ^2 is

$$s^2 = \sum_s (Y_i - \mathbf{x}'_i\hat{\beta}^*)^2/v_i(n - r - 1) \quad (11.7.2)$$

Hence,
$$G_0 = [\mathbf{x}'_{\bar{s}}\mathbf{A}^{-1}\mathbf{x}_{\bar{s}} + \mathbf{1}'_{\mathbf{N}-n}\mathbf{V}_{\bar{s}}\mathbf{1}_{\mathbf{N}-n}]s^2 \quad (11.7.3)$$

is a model-unbiased estimator of $\mathcal{V}(\hat{T}^*)$ under $\xi(\mathbf{X}, v)$. However, if the true dispersion matrix of $\mathbf{Y}$ is not proportional to $\mathbf{V}$, (11.7.3) is not model-unbiased for

$\mathscr{V}(\hat{T}^*)$. Suppose that the true model is $\xi'(\mathbf{X}, w)$ where $\mathscr{D}(\mathbf{Y}) = \mathbf{W}$, $\mathscr{D}(\mathbf{Y}_s) = \mathbf{W}_s$, $\mathscr{D}(\mathbf{Y}_s) = \mathbf{W}_{\bar{s}}$. Under ξ',

$$\mathscr{E}'(Y_i - \bar{\mathbf{x}}_i'\hat{\boldsymbol{\beta}}^*)^2 = w_i(1 - d_i) \tag{11.7.4}$$

where,
$$d_i = 2v_i^{-1}\mathbf{x}_i'\mathbf{A}^{-1}\mathbf{x}_i' - w_i^{-1}\mathbf{x}'\mathbf{A}^{-1}\mathbf{B}\mathbf{A}^{-1}\mathbf{x}_i' \tag{11.7.5}$$

$$\mathbf{B} = \mathbf{X}_s'\mathbf{V}_s^{-1}\mathbf{W}_s\mathbf{V}_s^{-1}\mathbf{X}_s$$

Under mild restrictions d_i is of order n^{-1}. Thus $(Y_i - x_i'\hat{\beta}^*)^2$ is approximately unbiased estimator of $\mathscr{V}'(Y_i)$ under ξ'. Hence for any $\mathbf{g}'Y_s$, $\mathscr{V}'(\mathbf{g}'\mathbf{Y}_s) = \mathbf{g}'\mathbf{W}_s\mathbf{g}$ is approximately unbiasedly estimated by

$$\mathbf{g}'\dot{\mathbf{W}}_s\mathbf{g} \tag{11.7.6}$$

where $\dot{\mathbf{W}}_s = \text{Diag}((Y_i - \mathbf{x}_i'\hat{\beta}^*)^2, i \in s)$, even if w_i is not proportional to v_i, $i = 1, ..., N$.

When $\mathbf{W} = \sigma^2\mathbf{V}$, d_i reduces to

$$v_i^{-1}\mathbf{x}_i'\mathbf{A}^{-1}\mathbf{x}_i = k_i \text{ (say)} \tag{11.7.7}$$

Hence, another estimator of $\mathscr{V}(\mathbf{g}'\mathbf{Y}_s)$ is

$$\mathbf{g}'\dot{\mathbf{W}}_s^*\mathbf{g} \tag{11.7.8}$$

where $\dot{W}_s^* = \dot{W}_s$ Diag $((1 - k_i)^{-1}, i \in s)$ which is strictly unbiased under ξ but approximately so under ξ'.

Another estimator of $\mathscr{V}'(\mathbf{g}'\mathbf{Y}_s)$, which is also unbiased under ξ, is

$$\alpha\mathbf{g}\dot{\mathbf{W}}_s\mathbf{g} \tag{11.7.9}$$

where,
$$\alpha = \mathbf{g}'\mathbf{V}_s\mathbf{g}\{\mathbf{g}'\mathbf{V}_s \text{ Diag } (1 - k_i, i \in s)\, \mathbf{g}\}^{-1}$$

$$= \sum_s g_i^2 v_i \Big/ \sum_s g_i^2 v_i (1 - k_i) \tag{11.7.10}$$

Since $0 < k_i < 1$, both the adjusted estimators (11.7.8) and (11.7.9) are greater than (11.7.6). Now,

$$\hat{T}_s^*(\mathbf{X}, v) = \mathbf{1}_n'\mathbf{Y}_s + \mathbf{x}_{\bar{s}}'\hat{\boldsymbol{\beta}}^*$$

Hence,
$$\hat{T}^*(\mathbf{X}, v) - Y = \mathbf{x}_{\bar{s}}'\hat{\boldsymbol{\beta}}^* - \mathbf{1}_{N-n}'\mathbf{Y}_{\bar{s}}$$

so that
$$\mathscr{V}'(\hat{T}^*(\mathbf{X}, v) - Y) = \mathbf{g}'\mathbf{W}_s\mathbf{g} + \mathbf{1}_{N-n}'\mathbf{W}_{\bar{s}}\mathbf{1}_{N-n} \tag{11.7.11}$$

where, $$\mathbf{g}'\mathbf{Y}_s = \mathbf{x}'_{\bar{s}}\hat{\boldsymbol{\beta}}^* \tag{11.7.12}$$

There are two estimators for the first components of (11.7.11), namely, (11.7.8) and (11.7.9), which are strictly unbiased when $\mathbf{W} = \sigma^2\mathbf{V}$, and are, otherwise, approximately unbiased for any diagonal $\mathbf{W}$.

Three estimators of the second component, which are unbiased (approximately) when $\xi = \xi'$ (otherwise), are

(*a*) $\mathbf{1}'_{N-n}\mathbf{V}_{\bar{s}}\mathbf{1}_{N-n}\mathbf{1}'_n\dot{\mathbf{W}}_s\mathbf{1}_n(\mathbf{1}'_n\mathbf{V}_s\,((\text{Diag}\,(1-\mathbf{k}_i)\mathbf{1}_n)^{-1}$

$$= \left(\sum_{\bar{s}} v_i\right)\left(\sum_s (Y_i - \mathbf{x}'_i\hat{\boldsymbol{\beta}}^*)^2\right)\Big/\sum_s v_i(1-k_i)$$

(*b*) $\mathbf{1}'_{N-n}\mathbf{V}_s\mathbf{1}_{N-n}\mathbf{1}'_n\dot{\mathbf{W}}^*_s\mathbf{1}_n(\mathbf{1}'_n\mathbf{V}_s\mathbf{1}_n)^{-1}$ (11.7.13)

$$= \frac{\left(\sum_{\bar{s}} v_i\right)}{\sum_s v_i}\sum_s \frac{1}{1-k_i}(Y_i - \mathbf{x}'_i\hat{\boldsymbol{\beta}}^*)^2$$

(*c*) The weighted least square statistic

$$\mathbf{1}'_{N-n}\mathbf{V}_{\bar{s}}\mathbf{1}_{N-n}s^2 = \left(\sum_{\bar{s}} v_i\right)\sum_s (Y_{\bar{i}} - \mathbf{x}'_i\hat{\boldsymbol{\beta}}^*)^2/(n-r-1)v_i\,.$$

Each of (*a*), (*b*), (*c*) is or order N in probability while $\mathbf{g}'\dot{\mathbf{W}}^*_s\mathbf{g}$ and $\alpha\mathbf{g}'\dot{\mathbf{W}}_s\mathbf{g}$ are of order N^2/n. Thus, the contributions of each of (*a*), (*b*), (*c*) is asymptotically negligible compared with $\mathbf{g}'\dot{\mathbf{W}}^*_s\mathbf{g}$ and $\alpha\mathbf{g}'\dot{\mathbf{W}}_s\mathbf{g}$.

(*a*) and (*b*) are natural counterparts of $\alpha\mathbf{g}'\dot{\mathbf{W}}_s\mathbf{g}$ and $\mathbf{g}'\dot{\mathbf{W}}^*_s\mathbf{g}$ respectively. Hence we get two estimators of $\mathscr{V}'(\hat{T}^* - Y)$:

(*i*) $\alpha\mathbf{g}'\dot{\mathbf{W}}_s\mathbf{g} + (a)$

$$= \frac{\sum g_i^2 v_i}{\sum g_i^2 v_i(1-k_i)}\mathbf{g}'\dot{\mathbf{W}}_s\mathbf{g} + \frac{(N-n)\bar{v}_{\bar{s}}\sum_s (Y_i - x'_i\hat{\beta}^*)^2}{\sum_s v_i(1-k_i)} \tag{11.7.14}$$

$= G_1$ (say)

(*ii*) $\mathbf{g}'\dot{\mathbf{W}}^*_s\mathbf{g} + (b) = G_2$ (say) (11.7.15)

where α, $\mathbf{g}'$ are given by (11.7.10) and (11.7.12) respectively.

Example 11.1: For model where Y_i, Y'_i $(i \neq i')$ are independent with $\mathscr{E}(Y_i) = \beta$, $\mathscr{V}(Y_i) = \sigma^2$, $\hat{T}^* = N\bar{Y}_s$, $G_1 = G_2 = \{N(N-n)/n(n-1)\} \sum_s (Y_i - \bar{Y}_s)^2$.

11.8 ESTIMATION OF MODEL VARIANCE OF RATIO PREDICTOR

The ratio predictor $\hat{T}_R = \bar{Y}_s \bar{X} / \bar{x}_s$ is BLUP under $\xi(0, 1; x)$.

$$\mathscr{V}(\hat{T}_R - Y) = N\sigma^2(1-f)\bar{x}_{\bar{s}}\bar{X} / f\,\bar{x}_s \tag{11.8.1}$$

The following estimators of (11.8.1) have been suggested.

(*i*)

$$\begin{aligned} v_L &= N(1-f)\hat{\sigma}^2 \bar{x}_{\bar{s}}\bar{X} / f\,\bar{x}_s \\ &= N(1-f)\bar{x}_{\bar{s}}\bar{X} \sum_s \{d_i^2 / \bar{x}_s x_i f(n-1)\} \\ &= G_0 \text{ of } (11.7.3) \end{aligned} \tag{11.8.2}$$

obtained by using least square estimate of σ^2, where $d_i = y_i - \bar{Y}_s x_i / \bar{x}_s$. Royall and Eberherdt (1975) showed that v_i may be seriously biased if $\mathscr{V}(Y_i)$ is not proportional to x_i.

(*ii*) The design-based classical estimator of $V(\hat{T}_R)$, considered by Sukhatme and Sukhatme (1970, p. 144), Cochran, (1977, p. 155),

$$v_c = N(1-f)\sum_s d_i^2 / f(n-1) \tag{11.8.3}$$

which may be seriously biased under $\xi(0, 1; x)$ with relative bias (R.B.)

$$\frac{\mathscr{E}(v_c) - \mathscr{V}(\hat{T}_R - Y)}{\mathscr{V}(\hat{Y}_R - Y)}$$

$$\simeq \frac{\bar{x}_s^2}{\bar{x}_{\bar{s}}\bar{X}}(1 - c^2/n) - 1$$

where

$$c^2 = \left[\sum_s x_i^2 / \bar{x}_s^2 - n\right] \Big/ (n-1)$$

(*iii*) Royall and Eberherdt adjusted v_c to remove its bias under $\xi(0, 1; x)$ in unbalanced samples giving

$$v_H = v_c \frac{\bar{x}_{\bar{s}}\bar{X}}{\bar{x}_s^2}(1 - c^2/n)^{-1}$$

as an estimator of (11.8.1). v_H is approximately unbiased if $\mathscr{V}(Y_i)$ is not proportional to x_i.

(*iv*) $v_2 = (\bar{X}/\bar{x}_s)^2 v_c$

suggested by Hájek (1958), Kish (1965, p. 204), Murthy (1977, p. 373) and considered by Rao and Rao (1971), Scott and Wu (1981), Wu (1982), among others. $v_H \simeq v_2$ for large n and $N >> n$. Empirical investigations of Wu and Deng (1983) suggest superiority of v_2 over v_H, v_D and v_L.

(*v*) The Jackknife estimator

$$v_J = \{N(1-f)(n-1)\bar{X}^2/f\}\sum_s T_{(j)}^2$$

where $T_{(j)}$ is the difference between the ratios $\sum_{s-j} Y_i \Big/ \sum_{s-j} x_i$ and the mean of these ratios. Royall and Eberherdt showed that v_H is asymptotically equivalent to v_J.

(*vi*) Robust estimator

$$v_D = \frac{N(1-f)\bar{x}_{\bar{s}}\bar{X}}{(\bar{x}_s)^2 nf}\left\{\sum_s d_i^2/(1-x_i/n\bar{x}_s)\right\}$$

$$= G_2 \text{ of } (11.7.15)$$

is unbiased under $\xi(0, 1; x)$, asymptotically unbiased under more general variance model and is asymptotically equivalent to v_J.

(*vii*)
$$v_E = \{(1-f)\bar{x}_{\bar{s}}X/f\ \bar{x}_s\}\left[n\bar{x}_s - \sum_s x_i^2/n\bar{x}_s\right]^{-1}\sum_s d_i^2$$

$$= G_1 \text{ of } (11.7.14)$$

v_c is asymptotically equivalent to v_E in any sequence of samples in which $\bar{X}/\bar{x}_s \to 1$ and $c^2/n \to 0$.

Failure of the condition $\mathscr{E}(Y_i) = \beta x_i$ increases the expected value of each of these estimators and their mse. If the sample is balanced, the ratio predictor is unbiased under general polynomial regression models, but v_H, v_D are positively biased.

Robustness of v_L, v_c, v_J were studied by Royall and Eberherdt under the two alternative models (*i*) $\mathscr{E}(Y_i) = \beta x_i$, $\mathscr{V}(Y_i) = v_i$ (*ii*) $\mathscr{E}(Y_i) = h(x_i)$, $\mathscr{V}(Y_i) = v_i$, Y_i's being uncorrelated in both the models. Royall and Cumberland (1978 *b*) studied v_H empirically.

Royall and Cumberland (1981) empirically analysed the performance of v_L, v_c, v_H, v_J, v_D through the samples selected, (*a*) randomly, (*b*) in a restricted manner and, (*c*) purposively. Their study shows that the performance of the randomly selected samples is poor as compared to the other two methods of selection for all the estimators. In (*b*), samples were selected if $|\bar{x}_s - \bar{X}| < \delta$, a pre-assigned quantity. In (*c*), three non-random samples were taken from each population. Two of these were extremes of the x-distribution and the third was obtained by fitting the x-distribution of the sample to the population. $F_s(x_0)$, the proportion of units in s whose values are not greater than x_0 is compared to $F(x_0)$, the corresponding proportion in the population. The sample for which $\max_x |F_s(x) - F(x)|$ is minimised, called the 'best fit sample' is selected.

11.9 ESTIMATION OF MODEL VARIANCE OF REGRESSION PREDICTOR

Under $\xi(1, 1; 1)$, $\hat{T}^* = N[\bar{Y}_s + b(\bar{X} - \bar{x}_s)] = \hat{T}_{lr}$ (say),

where, $$b = \sum_s (x_i - \bar{x}_s)Y_i \Big/ \sum_s (x_i - \bar{x}_s)^2$$

$$\mathscr{V}(\hat{T}_{lr} - Y) = N(1-f)\sigma^2(1 + h(s))/f \qquad (11.9.1)$$

where, $$h(s) = (\bar{x}_s - \bar{X})^2/\{(1-f)g(s)\},$$

$$g(s) = \sum_s (x_i - \bar{x}_s)^2/n$$

(*i*) An unbiased estimator of (11.9.1) is the weighted least square estimator

$$v_L = N(1-f)\hat{\sigma}^2(1 + h(s))/f \qquad (11.9.2)$$

$$\sigma^2 = \sum_s r_i^2/(n-2),$$

$$r_i = Y_i - \bar{Y}_s - b(x_i - \bar{x}_s)$$

(*ii*) The design-based classical estimator of (11.9.1) is

$$v_c = v_L/(1 + h(s)) \qquad (11.9.3)$$

Since $h(s) \geq 0$, $v_L \geq v_c$. For a balanced sample, $h(s) = 0$ and $v_L = v_c$.

(*iii*) G_1 of (11.7.14) is

$$v_H = \{(1-f)/f\}^2 \sum_s r_i^2 \left(1 + \frac{\Delta(s)}{g(s)}\right)^2 \Bigg/ \left(1 - \sum_s \frac{w_i k_i}{n}\right) + (N-n)\hat{\sigma}^2$$

where, $$w_i \propto \{g(s) + \Delta(s)\},\ \Delta(s) = (x_i - \bar{x}_s)(\bar{x}_{\bar{s}} - \bar{x}_s),\ i \in s$$

$$k_i = 1 + \{(x_i - \bar{x}_s)^2/g(s)\} \qquad (11.9.4)$$

(*iv*) G_2 of (11.7.15) is

$$v_D = \{(1-f)/f\}^2(n/(n-1))$$

$$\sum_s \frac{r_i^2[\{1+\Delta(s)/g(s)\}^2+(1-f)/f]}{1-\{(x_i-x_s)^2/(n-1)g(s)\}} \tag{11.9.5}$$

(*v*) The Jackknife estimator is

$$v_J = \{(1-f)(n-1)/n\}\sum_s D_{(j)}^2 \tag{11.9.6}$$

where $D_{(j)}$ is the difference between the regression estimator based on $s-j$ and the average of these n estimators. v_D, v_H, v_J remain unbiased under $\xi(1, 1; 1)$ and is approximately so for the same with a different variance function. Under $\xi(1, 1; x) = \xi'$ (say)

$$\mathscr{V}_{\xi'}(\hat{T}_{lr}-Y) = (N\sigma^2/f)\{(2-f)\bar{X}-\bar{x}_s+(\bar{X}-\bar{x}_s)^2C(s)\} \tag{11.9.7}$$

where,
$$C(s) = \left(\sum_s x_i^3 - 2\bar{x}_s\sum_s x_i^2 + n\bar{x}_s^3\right)\Big/ng(s)^2$$

$$\mathscr{E}_{\xi'}(v_L) = \{N(1-f)\sigma^2/f\}(1+h(s))\ \{\bar{x}_s+\{\bar{x}_s-C(s)g(s)\}/(n-2)\}$$

$\mathscr{E}_{\xi'}(v_c) = \mathscr{E}_{\xi'}(v_L)$ with $h(s) = 0$. When n is large, f is small, $\mathscr{E}_{\xi'}(v_c) \simeq N\sigma^2\bar{x}_s/f$. If $g(s) \simeq g$, where $g = \sum (x_i-\bar{X})^2/N = g(\mathscr{P})$, $\mathscr{E}_{\xi'}(v_L) \simeq (N\sigma^2\bar{x}_s/f)[1+(\bar{x}_s-\bar{X})^2/g]$.

For simple random sampling, v_c is a biased estimator of the design variance of $\hat{T}_{lr}$, though the bias is negligible in large samples. In the prediction theoretic approach based on the given sample this bias is revealed whenever $\bar{x}_s \neq \bar{X}$.

11.10 ESTIMATION OF MODEL-VARIANCE OF HORVITZ-THOMPSON PREDICTOR

Estimators of $V(e_{HT})$, apart from v_{HG}, v_{YG}, v_{JA} (equation (1.4.15.2)) are

$$v_c = N\bar{X}^2\sum_s (d_i/x_i)^2/(n-1)f \tag{11.10.1}$$

where $d_i = y_i - bx_i$, $b = \sum_s (y_i/nx_i)$, usual variance estimator in ppswr sampling, considered by Durbin (1953);

The Jackknife estimator,

$$v_J = (1-f)(1-1/n)\sum_{j\in s}\{\hat{T}(j)-\hat{T}(\cdot)\}^2,$$

$\hat{T}(j)$ is the HTE of Y, based on $s - \{j\}$, $\hat{T}(\cdot)$ is the mean of $\hat{T}(j)$'s. Royall and Cumberland (1981c) showed that $v_J = (1 - f)v_c$.

As noted before, if a superpopulation model is available, conditionality principle states that inference should be based on model for a given s. We, therefore, consider estimation of $\mathscr{V}(\hat{T}_{HT} - Y)$. Under $\xi(0, 1; x^2)$,

$$\mathscr{V}(\hat{T}_{HT} - Y) = \sigma^2 X t(x_s), \text{ where}$$

$$t(x_s) = \bar{X}/f - 2\bar{x}_s + c,$$

$$c = \sum x_i^2 / X \tag{11.10.2}$$

and $\pi_i = nx_i/X$. (11.10.2) is minimised for the sample containing units with the highest values of x. Different estimators of (11.10.2) are:

(a) The weighted least square estimator

$$v_L = \hat{\sigma}^2 X t(x_s) = v_c[1 - f(2\bar{x}_s - c)\bar{X}] \tag{11.10.3}$$

where
$$\hat{\sigma}^2 = \sum_s (d_i / x_i)^2 / (n-1).$$

(b) Royall and Cumberland (1978) robust estimator of G_1 of (11.7.14) is

$$v_D = N\bar{X}^2 \sum_s (d_i / x_i)^2 / f(n-1) - 2X \sum_s (d_i^2 / x_i^2)/(n-1)$$

$$+ n\left(\sum x_i^2 \Big/ \sum_s x_i^2\right)\left(\sum_s \frac{d_i^2}{(n-1)}\right) \tag{11.10.4}$$

In studying the estimators v_{HT}, v_{YG}, v_c, v_J, v_L, v_D, Royall and Cumberland (1981c) considered pps-systematic sampling scheme due to Hartley and Rao (1962) when

$$n\pi_{ij}/\pi_i\pi_j \simeq (n - 1)(1 + b_{ij}/N),$$

$$b_{ij} = (x_i + x_j - c)/\bar{X}$$

which when substituted in v_{HT}, v_{YG} give $\overset{*}{v}_{HT}, \overset{*}{v}_{YG}$ (say). All the estimators $\overset{*}{v}_{HT}, \overset{*}{v}_{YG}$, v_c, v_J, v_L, v_D have the same leading term

$$N\bar{X}^2 \sum_s (d_i / x_i)^2 / f(n-1)$$

and are asymptotically equivalent if $n \to \infty$, $f \to 0$. However, for moderate f they have important differences. Each of v_D, v_L, $\overset{*}{v}_{YG}$ is ξ-unbiased, while

$$\mathscr{E}\, v_c = \sigma^2 N\bar{X}^2 / f$$

$$\mathscr{E}\, v_J = \sigma^2 N\overline{X}^2 /(1/f - 1)$$

$$\mathscr{E}\, \overset{*}{v}_{HT} = \sigma^2 N\overline{X}^2 / f - \sigma^2 N\overline{X}\,\overline{x}_s + \beta^2 X(\overline{x}_s - c)\,.$$

Defining the relative bias of T as $|\mathscr{E}(T) - \mathscr{V}(\hat{T}_{HT} - Y)| / \mathscr{V}(\hat{T}_{HT} - Y)$,

$$\text{R.B. }(v_c) \simeq f(2\overline{x}_s - c)/\overline{X} \to 0$$

$$\text{R.B. }(v_J) \simeq \text{R.B. }(v_c) - f \to 0$$

generally for moderate, f, while

$$\text{R.B.}(\overset{*}{v}_{HT}) \simeq (1+g)f(\overline{x}_s - c)/\overline{X}\,,\; g = (\beta/\sigma^2) \qquad (11.10.5)$$

$\overset{*}{v}_{HT}$, as well as its model-expectation can take negative values. In the limit when $\sigma^2 = 0$, $Y_i = \beta x_i$, $g = \infty$, (11.10.2) as well as v_c, v_D, V_L, $\overset{*}{v}_{YG}$ are zero for each s, while $\overset{*}{v}_{HT} = \beta^2 X(\overline{x}_s - c)$. For Sampford's scheme, with the values of π_{ij}'s given by Asoke and Sukhatme (1976) (vide Sub-section 2.2.5), approximate relative bias (R.B.) $(\overset{*}{v}_{HT})$ remains the same as in (11.10.5)

Under $\xi' = \xi(0, 1; v_i)$,

$$\mathscr{V}'(\hat{T}_{HT} - Y)(\hat{T}^*) = (N\overline{X}^2\sigma^2/f)\left\{\sum_s v_i / nx_i^2\right\}$$

$$- 2X\sigma^2\left\{\sum_s v_i / nx_i\right\} + \sigma^2 \sum v_i \qquad (11.10.6)$$

v_D is ξ'-biased for (11.10.6), but R.B. $(v_D) \to 0$ as $n \to \infty, f \to 0$. Thus v_D and since all the estimators have the same dominant term as v_D, all of the other estimators are bias-robust. The flaw in $\overset{*}{v}_{HT}$ in ξ and ξ' are similar, since the term $\beta^2 X(\overline{x}_s - c)$ remains unaffected under ξ'. Defining $g(s) = \beta^2 \big/ \sum_s v_i / nx_i^2$, $\mathscr{E}'[\overset{*}{v}_{HT}] < 0$ when $\overline{x}_s < c - h(s)$, where $h(s) \to 0$ as $g(s) \to \infty$.

11.11 ASYMPTOTICALLY UNBIASED ESTIMATION OF DESIGN-VARIANCE OF $\hat{T}_{GR}$ UNDER $\xi(X, v)$

Consider once again $\xi(\mathbf{X}, v)$. Suppose the problem is that of estimation of Q linear functions $\mathbf{F} = (F_1, \ldots, F_Q)' = \mathbf{C'Y}$ where $F_q = \mathbf{C}_q'\mathbf{X}$, $\mathbf{C}_q' = (C_{1q}, \ldots, C_{Nq})$, C a $N \times Q$ matrix $((C_{kq}))$, C_{kq} being known constants. Important functions F_q are population mean $\overline{Y}$, $(C_{kq} = 1/N \forall k)$, domain total $Y_q\{C_{kq} = 1(0)$ if $k \in (\notin)$ domain D_q), finite population regression coefficient $\mathbf{B} = (\mathbf{X'V^{-1}X})^{-1}\mathbf{XV^{-1}Y}$ where $\mathbf{B} = (B_0, \ldots, B_r)'$.

Consider the following estimates of F_q

$$\hat{T}_q = \mathbf{C}'_{qs}\mathbf{II}_s^{-1}(\mathbf{Y}_s - \hat{\mathbf{Y}}_s) + \mathbf{C}'_q\hat{\mathbf{Y}} \qquad (11.11.1)$$

where $\mathbf{C}'_{qs}$ is the row vector (C_{kq}, $k \in s$) and $\hat{Y}_k = \mathbf{x}'_k\hat{\boldsymbol{\beta}}_s$ with

$$\hat{\boldsymbol{\beta}}_s = \hat{\boldsymbol{\beta}}(V^{-1}\mathbf{II}^{-1}) = (\mathbf{X}'_s\mathbf{V}_s^{-1}\mathbf{II}_s^{-1}\mathbf{X}_s)^{-1}\mathbf{X}'_s\mathbf{V}_s^{-1}\mathbf{II}_s^{-1}\mathbf{Y}_s\,.$$

(11.11.1) is an extension of generalised regression estimator $\hat{T}_{GR}(V^{-1}II^{-1})$ of Y. Let $\hat{T} = (\hat{\mathbf{T}}_1, \ldots, \hat{T}_Q)$. Thus,

$$\hat{\mathbf{T}} = \mathbf{C}'_s\mathbf{II}_s^{-1}(\mathbf{Y}_s - \hat{\mathbf{Y}}_s) + \mathbf{C}'\hat{\mathbf{Y}}$$

where $\mathbf{C}'_s$ is the part of $\mathbf{C}$ corresponding to $k \in s$.

Consider the asymptotic framework of Note (7.1). Conceive a sequence of populations and samples with (N_ν, n_ν), $\nu = 1, 2, \ldots$ increasing so that in the limit $S_{tu} = \sum_{k\in s} x_{kt}x_{ku}w_k/\pi_k \rightarrow \sum x_{kt}x_{ku}w_k = \gamma_{tu}$ (say).

Assume also $\hat{\boldsymbol{\beta}}_s \xrightarrow{P} \boldsymbol{\beta}$. Also assume ADU-ness of S_{tu} and $\hat{\boldsymbol{\beta}}_s$. It follows that $\hat{\boldsymbol{\beta}}_s$ is ADU and ADC for $\mathbf{B}$. Now

$$\hat{\mathbf{T}} = \mathbf{G}'_s\mathbf{II}_s^{-1}\mathbf{Y}_s$$

where,

$$\mathbf{G}'_s = \mathbf{C}'_s - \mathbf{M}'_s\mathbf{H}_s^{-1}\mathbf{X}'_s\mathbf{V}_s^{-1}$$
$$\mathbf{M}'_s = \mathbf{C}'_s\mathbf{II}_s^{-1}\mathbf{X}_s - \mathbf{C}'\mathbf{X}$$
$$\mathbf{H}_s = \mathbf{X}'_s\mathbf{V}_s^{-1}\mathbf{II}_s^{-1}\mathbf{X}_s.$$

Thus $\hat{T}_q = \sum_s g_{skq}Y_k/\pi_k$, g_{skq} being $(k, q)^{\text{th}}$ component of $\mathbf{G}_s$. Under the above assumptions $\hat{\mathbf{T}}$ is approximately unbiased for $\mathbf{T}$. We are interested here in estimating the design-dispersion matrix $\mathbf{D}(\mathbf{T}) = ((\text{Cov}_p(\hat{T}_q, \hat{T}_r))$. The following two methods have been suggested by Sarndal (1982).

(*a*) Taylor expansion method:

$$\hat{\mathbf{T}} - \mathbf{F} = \mathbf{a}_s - \mathbf{M}'_s\mathbf{H}_s^{-1}\mathbf{D}_s \qquad (11.11.2)$$

where,

$$\mathbf{a}_s = \mathbf{C}_s\mathbf{II}_s^{-1}\mathbf{e}_s^* - \mathbf{C}'\mathbf{e}^*,\ \mathbf{D}_s = \mathbf{X}'_s\mathbf{V}_s^{-1}\mathbf{II}_s^{-1}\mathbf{e}_s^*,$$

$\mathbf{e}^* = (e_1, \ldots, e_N^*)$, $\mathbf{e}_s^* = (e_k^*, k \in s))_{n\times 1}$, $e_k^* = Y_k - \mathbf{x}'_kB$. The Taylor expansion of $\hat{\mathbf{T}}$ around the expected value of each of the sample sum gives

$$\hat{\mathbf{T}} - \mathbf{F} \simeq \mathbf{a}_s.$$

Thus,

$$V_p(\hat{\mathbf{T}}) \simeq E(\mathbf{a}_s\mathbf{a}'_s) \quad [\text{as } E(\mathbf{a}_s) \simeq \mathbf{0}]$$

Hence, $\text{Cov}_p(\hat{T}_q, \hat{T}_r) \simeq \text{Cov}\left(\sum_s C_{kq}\overset{*}{e}_k/\pi_k, \sum_s C_{kr}\overset{*}{e}_k/\pi_k\right).$

Thus, an estimate of $\text{Cov}_p(\hat{T}_q, \hat{T}_r)$ is approximately the Yates-Grundy estimator of covariance, viz.

$$\sum_s{}' (\pi_{kl}/\pi_k\pi_{l-1})(z_{kq}/\pi_k - z_{lq}/\pi_l)(z_{kr}/\pi_k - z_{lr}/\pi_l)$$

$$= v_T(q, r) \text{ (say)}$$

where $$z_{kq} = C_{kq}e_k,\ e_k = Y_k - \mathbf{x}_k'\hat{\boldsymbol{\beta}}_s.$$

Writing $YG_s(d_{kq}, d_{kr}) = \sum_s{}' (\pi_{kl}/\pi_k\pi_l - 1)(d_{kq}/\pi_k - d_{lq}/\pi_l)(d_{kr}/\pi_k - d_{lr}/\pi_l)$ as the YG-transformation of $d(d_{kq}, d_{lr}), k \in s$, we have $v_T(q, r) = YG_s(e_{kq}, e_{kr})$.

(*b*) *Model method.* Here an approximate estimate of $\text{Cov}_p(\hat{T}_q, \hat{T}_r)$ is

$$v_M(q, r) = YG_s(\overset{*}{z}_{kq}, \overset{*}{z}_{kr})$$

where $$\overset{*}{z}_{kq} = g_{skq}e_k.$$

Both $((v_T))$ and $((v_M))$ are approximately p-unbiased for $\mathbf{V(T)}$ whether the assumed model $\xi(\mathbf{X}, v)$ is true or false. The model is used here only to generate the form of the estimator $\mathbf{T}$ and hence, $((v_T))$ and $((v_M))$. In the limit both the methods tend to be equivalent.

It follows that under $\xi(\mathbf{X}, v)$ and under stratified random sampling

$$\mathscr{E}((v_M)) \simeq \mathscr{E}(\hat{T} - \mathbf{F})(\hat{\mathbf{T}} - \mathbf{F})'.$$

provided sample size $\mathscr{E} n_h$ is large and $N_h \gg n_h$. Thus $((v_M))$ is approximately p-unbiased for $\mathbf{V}_p(\mathbf{T})$ and under the stated conditions is approximately model-unbiased for $\mathscr{D}(\hat{\mathbf{T}})$.

Example 11.2: Model $\xi(0, 1; x)$. $\hat{\beta}_s = \tilde{Y}_s/\tilde{X}_s, \tilde{Y}_s = \left(\sum_s Y_k/\pi_k\right)\Big/\left(\sum_s 1/\pi_k\right) \hat{\hat{Y}}$ $= \bar{X}\hat{\beta}_s$ (Hájek's estimator). Two estimators of $\mathbf{V}_p(\hat{\hat{Y}})$ are (*i*) $v_T(\hat{\hat{Y}}) = YG_s(e_k^2/N^2)$ (*ii*) $v_M(\hat{\hat{Y}}) = YG_s\{(\bar{X}_{ek})^2/(\tilde{x}_s\tilde{\pi}_s)^2\}$, $e_k = Y_k - \hat{\beta}_s x_k$, $\tilde{\pi}_s = \sum_s 1/\pi_k$.

Under srs, $v_M = \hat{v}$ (say) $= (\bar{X}/\bar{x})^2 v_T$, where $v_T = \hat{v}$ (say) $= \{(1 - f)/n(n - 1)\}$ $\sum_s (Y_k - \hat{\beta}x_k)^2$, the usual estimator of mse of ratio estimator of population mean. Royall and Cumberland (1978) derived robust variance estimator $v_h = h_s\hat{v}$ where,

$$h_s = \{(\bar{x}_{\bar{s}} - \bar{X})/\bar{x}_s\}^2 \bigg/ \left\{1 - \sum_s (x_k - \bar{x}_s)^2 \Big/ n(n-1)\bar{x}_s^2\right\}$$

If n is large and $N \gg n$. $h_s \simeq (\bar{x}/\bar{x}_s)^2$ and $v_h \simeq \hat{v}$.

For the stratified random sampling, $\hat{\bar{Y}}_c = \bar{X}\hat{\beta}_c$, the combined ratio estimator,

$$\hat{\beta}_c = \bar{Y}_{st}/\bar{x}_{st}, \bar{Y}_{st} = \sum W_h \bar{Y}_h,\ W_h = N_h/N,\ \bar{Y}_h = \sum_{s_h} Y_k/n_h. \text{ Here}$$

$$v_T(\hat{\bar{Y}}_c) = \sum W_h^2(1 - f_h)\hat{v}_{sh}/n_h$$

$$v_M(\hat{\bar{Y}}_c) = (\bar{X}/\bar{x}_{st})^2 v_T(\hat{\bar{Y}})$$

where,
$$\hat{v}_{sh} = (n_h - 1)^{-1}\sum_{s_h} \{Y_k - \bar{Y}_h - \hat{\beta}_c(x_k - \bar{x}_h)\}^2,\ f_h = n_h/N_h.$$

If $N_h \gg n_h$,

$$\mathscr{E}\{v_M(\hat{\bar{Y}}_c)\} \simeq \mathscr{E}(\hat{\bar{Y}}_c - \bar{Y})^2 \simeq \left(\frac{\bar{X}}{\bar{x}_{st}}\right)^2 \mathscr{E}\{v_T(\hat{\bar{Y}})\}$$

where,
$$\mathscr{E}\{v_T(\hat{\bar{Y}}_c)\} = \sigma^2 \sum W_h^2 \bar{x}_h/n_h.$$

If the sample selects large x_k values, v_M will on an average give a shorter confidence interval for $V(\hat{\bar{Y}}_c)$ than v_T.

Example 11.3: Model $\xi(1, 1; 1)$. $\hat{\bar{Y}} = \bar{Y}_s + \hat{\beta}(\hat{X} - \bar{x}_s)$, where

$$\hat{\beta} = \left\{\sum_s (x_k - \bar{x}_s)/\pi_k\right\} \bigg/ \left\{\sum_s (x_k - \bar{x}_s)^2/\pi_k\right\}$$

Here
$$e_k = Y_k - \tilde{Y}_k - \hat{\beta}(x_k - \tilde{x}_s).$$

$$v_T(\hat{\bar{Y}}) = YG_s(e_k^2/N^2),\ v_M(\hat{\bar{Y}}) = YG_s\{g_{sk}^2 e_k^2\},$$

$$g_{sk} = \left(\sum_s 1/\pi_k\right)^{-1} + (x_k - \tilde{x}_s)(\bar{X} - \bar{x}_s) \bigg/ \left\{\sum_s (x_k - \tilde{x}_s)^2/\pi_k\right\}$$

In the special case of stratified random sampling, $\hat{\bar{Y}}$ becomes combined regression estimator and $v_T(\hat{\bar{Y}})$ agrees in form with Cochran's (1977, p. 203) estimator. For srs, $\hat{\bar{Y}}$ reduces to the traditional regression estimator with

$$v_T(\hat{\bar{Y}}) = (1-f)\sum_s e_k^2 / n(n-1),$$

$$v_M(\hat{\bar{Y}}) = (1-f)\sum_s e_k^2(1+t_s^2 w_k) \Big/ n(n-1),$$

where, $$t_s^2 = n^2(\bar{x}_s - \bar{X})^2 \Big/ \sum_s (x_k - \bar{x}_s)^2$$

and, $$w_k = (x_k - \bar{x}_s)^2 \Big/ \sum_s (x_k - \bar{x}_s)^2 .$$

Again, $$\mathscr{E}\{v_T(\hat{\bar{Y}})\} = (1-f)\sigma^2 (n-2)/n(n-1)$$

$$\mathscr{E}\{v_M(\hat{\bar{Y}})\} = \{(1-f)\sigma^2/n\}\{(n-2)/(n-1) + R_s/n\}$$

$$\mathscr{E}\{(\hat{\bar{Y}} - \bar{Y})^2\} = \{(1-f)\sigma^2/n\}\{1 + t_s^2/n\,(1-f)\}$$

where $$R_s = t_s^2\{1 - \alpha_{s4}/(n-1)\sigma_{s2}^2\},$$

$$\sigma_{sj} = n^{-1}\sum_s (x_k - \bar{x}_s)^j .$$

Hence, $$\mathscr{E}\{v_T(\hat{\bar{Y}})\} \simeq \mathscr{E}\{v_M(\hat{\bar{Y}})\} \simeq \mathscr{E}(\hat{\bar{Y}} - \bar{Y})^2 \simeq (1-f)\sigma^2/n.$$

Example 11.4: Consider a domain of study $D_q = U_h\ \mathscr{P}_{hq}$, $\mathscr{P}_{hq}$ of size N_{hq} is the part of D_q falling in stratum h, $\mathscr{P}_h$. Let s_h of size n_h be a sample from $\mathscr{P}_h$ and s_{hq} of random size n_{hq} fall in $\mathscr{P}_{hq}$. Means are $\bar{x}_h, \bar{x}_{sh}, \bar{x}_{shq}, \bar{x}_{s\cdot q}$.

Under model where Y_k's are uncorrelated with $\mathscr{E}(Y_k) = \beta_h x_k$, $\mathscr{V}(Y) = \sigma_h^2 x_k$, $k \in \mathscr{P}_h$ for the stratified simple random sampling design, the estimator of domain total, $\sum_{k \in D_q} Y_k$ is

$$T_q = \sum_1^L N_{hq}\{P_{hq}\bar{Y}_{sh_q} + \hat{\beta}_h(\bar{x}_{\mathscr{P}h_q} - P_{h_q}\bar{x}_{sh_q})\},$$

$$\hat{\beta}_h = \bar{Y}_{sh}/\bar{x}_{sh},\ \mathrm{P}_{hq} = n_{h_q}N_h / n_h N_{h_q} .$$

Residual is $$e_k = Y_k - \hat{\beta}_h x_k .$$

Kott (1990) considered estimators which are simultaneously model-unbiased for the model-variance and design-consistent for the design mean-square error. His procedure appears to be very robust in the sense that the estimators have small design bias, small model-bias under a class of model-mis-specifications and retain the property of unbiasedness if the assumed model holds. Further researches in this area are welcome.

❑ ❑ ❑

General Notations

$\mathscr{P} = \{1, ..., N\}$, a finite population

y_i = The value of y, main character of interest, on $i(= 1, \ldots, N)$ (for the sample, y_i is the value of y on the unit selected at the ith draw $(i = 1, \ldots, n)$) $\mathbf{y} = (y_1, \ldots, y_N)$

R^N = N-dimensional Eucledian Space

$Y = \sum_{i=1}^{N} y_i$, population total; $\bar{Y} = Y/N$, population mean

$S^2 = \sum_{i=1}^{N} (y_i - \bar{Y})^2/(N-1)$ -Population variance

x_i = Value of x, an 'auxiliary variable' on unit i; $X = \sum_{i=1}^{N} x_i$, population total of x; $\bar{X} = X/N$, population mean of x; $p_i = x_i/X$, size-measure of i.

$S = \{i_1, \ldots, i_n\}$ $(1 \le i_t \le N)$ = A sample (sequence) obtained in n draws $s = (j_1, \ldots, j_n)$, $1 \le j_1 < \ldots < j_n \le N$ = A sample (set) of n distinct units.

$v(S)[v(s)]$ = Number of distinct units (effective sample size) in $S[s]$

$\mathscr{S} = \{s\}$ or $\{S\}$ = Sample space

$\mathscr{S}_n = \{s(S) : v(s)[v(S)] = n\}$;

$\bar{s} = \mathscr{P} - s$

$p(s)[p(S)]$ = Probability of selecting $s(S)$

$P(s \mid i)[P(s \mid i, j)]$ = Probability of selecting an unordered sample s given that unit i (units i, j) was selected at first (two) draw(s).

$D(\mathscr{S}, p) = p$ = A sampling design

$\rho_n = \{p : p(s)[p(S)] > 0 \Rightarrow v(s)[v(S)] = n\}$ = Class of all fixed effective size(n)[FES (n) or FS (n)] designs.

$\pi_{i_1} \ldots {}_{i_k}$ = k^{th} order inclusion probability; $\pi_k(\pi_{kk'})$ = First (second) order inclusion probability

$\sum = \sum_{i=1}^{N}, \sum' = \sum_{i \neq i'=1}^{N}, \quad \sum_{s} = \sum_{i \in s}, \sum_{s}' = \sum_{i \neq i \in s}, \quad (n)_r = n(n-1)\ldots(n-r+1)$ $p_r(i_r \mid i_1, \ldots, i_{r-1})$ = Conditional probability of selecting i_r at the rth draw given that the units $i_1, \ldots, i_{r-1}$ were selected at the 1st, ..., $(r-1)^{\text{th}}$ draws respectively.

$p_r(i_r)$ = Probability of selecting i_r at the rth draw.

$p(i_1, \ldots, i_r)$ = The joint probability that $(i_1, \ldots, i_r)$ are selected at the 1st, ..., r^{th} draw respectively.

d' = Data obtained from a sequence sample = $\{(k, y_k), k \in S\}$

d = Data obtained from a set sample = $\{(k, y_k), k \in s\}$

srsw(o)r = Simple random sampling with (without) replacement

$f = n/N$, sampling fraction

$\bar{y}_S$ = Sample mean in srswr;

$\bar{y}_S'$ = Mean of the distinct units in srswr;

$\bar{y}_s$ = Sample mean in srswr

$\hat{Y}_R(e_R)$ = Ratio estimator of population total

$\hat{Y}_D, \hat{Y}_{GD}(e_D, e_{GD})$ = Difference estimator, generalised difference estimator of population total respectively

$\hat{Y}_{lr}(e_{lr})$ = Linear regression estimator of population total

$\hat{Y}_{HH}, \hat{Y}_{pps}, (e_{HH}, e_{pps})$ = *ppswr* (Hansen-Hurwitz) estimator of population total

ppswr = Probability proportional to size with replacement

E, V = Design-expectation and design-variance operators, respectively

$v(e)$ = Variance estimator of e

$v_{pps} = v(e_{pps})$

$\hat{Y}_{HT}(e_{HT})$ = Horvitz-Thompson estimator of population total

$V_{HT}(v_{HT})$, $V_{YG}(v_{YG})$ = $V(e_{HT})$, $v(e_{HT})$ due to Horvitz, Thompson and Yates, Grundy, respectively

e'_R = General ratio-type estimator for any given sampling design

e_D, V_D, v_D = Des Raj's estimator, $V(e_D)$, $v(e_D)$ respectively

v'_D = Murthy's (1957) improved estimator over v_D

e_M, v_M, v_M = Murthy's (1957) estimator, $V(e_M)$, $v(e_M)$, respectively

e_{GR} = Generalised regression estimator

e_{Ha} = Hájek's estimator (1959)

$H_1 \succ H_2$: Strategy H_1 is superior to H_2

$\mathscr{E}, \mathscr{V}, \mathscr{C}$ = Expectation, variance, covariance operators, respectively, wrt superpopulation model

Y_i = A random variable whose one particular realised value is y_i

$\mathbf{Y} = (Y_1, ..., Y_N)$

$\mu_i = \mathscr{E}(Y_i)$, $\mu = \mathscr{E}(Y)$, $\bar{\mu} = \mathscr{E}(\bar{Y}) = \mu/N$, $\bar{\mu}_s = \mathscr{E}(\bar{Y}_s), \bar{\mu}_{\bar{s}} = \mathscr{E}(\bar{Y}_{\bar{s}})$
$= \mu_{\bar{s}}/(N-n)$

$M(p, \hat{T}_s) = E\mathscr{E}(\hat{T}_s - \bar{Y})^2$, where $\hat{T}_s$ is a predictor of $\bar{Y}$

$M_1(p, \hat{T}_s) = E\mathscr{E}(\hat{T}_s - \mu)^2$, where $\hat{T}_s$ is a predictor of $\bar{Y}$

$\xi(\delta_0, \delta_1, \ldots, \delta_J)$ = J^{th} order polynomial regression model, where

$\mathscr{E}(Y_i \mid x_i) = \sum_{j=0}^{j} \delta_j \beta_j x_i^j$, $\mathscr{V}(Y_i \mid x_i) = \sigma^2 v(x_i)$, $\mathscr{C}(Y_i, Y_{i'}) = 0 (i \neq i')$, $\delta_j = 0(1)$ according as β_j is absent (present) in μ_i;

$\hat{T}^*(\xi), \hat{\hat{T}}^*(\xi)$ = Royall's (1970) predictor, BLU projection type predictor, respectively, under model ξ

$\hat{T}_g^* = \hat{T}^*(0, 1; x^g)$, $\hat{T}_1^* = \hat{T}_R$ = Ratio predictor

$\hat{T}_{ab}$ = Predictor corresponding to estimator $\hat{Y}_{ab}(e_{ab})$ (sometime $\hat{Y}_{ab}$ denotes the predictor also);

p^* = Royall (1970) purposive sampling plan

$\bar{p}(J), s_b(J)$ = Royall and Herson's (1973) balanced sampling design, balanced sample of order J, respectively

$\mathbf{1}_q = (1, \ldots, 1)'_{q \times 1}$; $\mathbf{1} = \mathbf{1}_N$

$I_q = \text{Diag}(1, \ldots, 1)_{q \times q}$; $I = I_N$

$\mathbf{J}_N$ = A $N \times N$ matrix with entry 1 at each place;

x_{kj} = Value of variable x_j on unit $k (j = 0, 1, \ldots, r$, in general)

$\mathbf{X} = ((x_{kj}))_{N \times (r+1)}$

$\mathbf{Y}_s = (Y_k; k \in s)_{n \times 1}$; $Y_{\bar{s}} = (Y_k, k \in \bar{s})_{(N-n) \times 1}$

$\mathbf{X}_s = ((x_{kj}; k \in s, j = 0, \ldots, r))_{n \times (r+1)}$

$\mathbf{X}_{\bar{s}} = ((x_{kj}; k \in \bar{s}, j = 0, \ldots, r))_{(N-n) \times (r+1)}$

$x_{js} = \sum_{k \in s} x_{kj}, \bar{x}_{js} = x_{js}/n, x_{j\bar{s}} = \sum_{k \in \bar{s}} x_{kj}, \bar{x}_{j\bar{s}} = x_{j\bar{s}}/(N-n), X_j = \sum_k x_{kj}$

$\mathbf{x}_s = (x_{0s}, \ldots, x_{rs})'$, $\bar{x}_{\bar{s}} = (\bar{x}_{0s}, \ldots, \bar{x}_{rs})'$

$\mathbf{x}_{\bar{s}} = (x_{0\bar{s}}, \ldots, x_{r\bar{s}})'$ $\bar{\mathbf{x}}_{\bar{s}} = (\bar{x}_{0\bar{s}}, \ldots, \bar{x}_{r\bar{s}})'$

$x_k = (x_{k0}, \ldots, x_{kr})'$

$D(\cdot)$ = Design-dispersion matrix of $(\cdot)$

$\mathscr{D}(\cdot)$ = Design-dispersion matrix of $(\cdot)$

$\mathscr{D}(Y) = \sigma^2 \mathbf{V}$ where $\mathbf{V} = ((v_{kk}))$, $\mathscr{C}(Y_k, Y_{k'}) = \sigma^2 v_{kk'}$

$\mathbf{V}_s = ((v_{kk};\, k, k' \in s))_{n\times n}$; $\mathbf{V}_{sr} = ((v_{kk'};\, k \in s, k' \in \bar{s}))_{n\times(N-n)}$; $\mathbf{V}_{rs} = \mathbf{V}'_{sr}$.

$\mathbf{V}_{\bar{s}} = ((v_{kk};\, kk' \in \bar{s}))_{(N-n)\times(N-n)}$

$\boldsymbol{\beta} = (\beta_0, \ldots, \beta_r)_{r\times 1}$ = Vector of regression coefficients

$\xi(\mathbf{X}, v)$ = Multiple regression model with $\mathscr{E}(\mathbf{Y}) = \mathbf{X}\beta$, $\mathscr{D}(\mathbf{Y}) = \sigma^2\mathbf{V}$ where $\mathbf{V}$ = Diag $(v_k,\, k = 1, \ldots, N)$, $v_{kk} = v_k$

$\xi(\mathbf{X}, \mathbf{V})$ = Multiple regression model with a general dispersion matrix $\sigma^2\mathbf{V}$, not necessarily diagonal and not necessarily having a constant correlation coefficient

$\xi(\mathbf{X}, \mathbf{V}, \rho)$ = Multiple regression model with a constant correlation coefficient ρ $\mathbf{II}$ = Diag$(\pi_1, \ldots, \pi_N)_{N\times N}$; $\mathbf{II}_s$ = Diag $(\pi_k;\, k \in s)_{n\times n}$

$\hat{\beta}(\mathbf{Q}) = (\mathbf{X}_s'\mathbf{Q}_s\mathbf{X}_s)^{-1}(\mathbf{X}_s'\mathbf{Q}_s\mathbf{Y}_s)$, where $\mathbf{Q}$ is an arbitrary positive semi-definite matrix, $\mathbf{Q}_s$ is a $n \times n$ submatrix of $\mathbf{Q}$ corresponding to $k \in s$

$\hat{\beta}^* = \hat{\beta}(\mathbf{V}^{-1})$

$\hat{T}_{GR}(\mathbf{Q})$ = Generalised regression predictor using $\hat{\beta}(\mathbf{Q})$ for β.

$\hat{T}^*_{GR} = \hat{T}_{GR}(\mathbf{V}^{-1})$

$\hat{\beta}_{\mathbf{II}}$ = A predictor of β arising from a II^{-1}-weighting

$\hat{Y}(p, \mathbf{Q}, \mathbf{R}) = \hat{Y}(\mathbf{Q}, \mathbf{R})$ = Wright's' (1983) (p, $\mathbf{Q}$, $\mathbf{R}$)-predictor of Y

$\hat{Y}(\mathbf{Q}, \mathbf{O}) = \hat{\hat{Y}}_Q$, $\hat{\mathbf{Y}}(\mathbf{V}^{-1}, \mathbf{II}^{-1}) = \hat{Y}^*_{GR}$

$\mathscr{C}(\mathbf{X})$ = Column space of $\mathbf{X}$

$\hat{T}^*_{Bh}$ = Brewer's (1979) predictor.

❑ ❑ ❑

Abbreviations

AISM : Annals of Institute of Statistical Mathematics
AJS : Australian Journal of Statistics
AMS : Annals of Mathematical Statistics
Am. St. : American Statistician
Ap. St. : Applied Statistics
AS : Annals of Statistics
Biom. : Biometrika
Biom. J. : Biometrika Journal
Bioms. : Biometrics
BISI : Bulletin of International Statistical Institute
CSAB : Calcutta Statistical Association Bulletin
CS(TM) : Communications in Statistics (Theory and Methods)
ISR : International Statistical Review
JASA : Journal of American Statistical Association
JISA : Journal of Indian Statistical Association
JISAS : Journal of Indian Society of Agricultural Statistics
JRSS : Journal of Royal Statistical Society
JSPI : Journal of Statistical Planning and Inference
Sc. JS : Scandinavian Journal of Statistics
SM : Survey Methodology

❑ ❑ ❑

References

Adhikary, A. (1991): On the performance of the nearest proportional to size sampling design. *CS (TM)*, **20 (12)**, pp. 1933–41.

Agarwal, C.L. (1979): A note on unequal probability sampling. *JISAS*, **31 (2)**, pp. 83–85.

Agarwal, O.P. (1959): Bayes and minimax procedures in sampling from a finite and infinite populations I. *AMS,* **30**, pp. 206–18.

Agarwal, R., Singh, P. and Singh, D. (1984). IIPS sampling schemes through grouping. *Biom. J.*, **5**, pp. 527–33.

Agarwal, S.K. and Goel, B.P. P. S. (1977): A note on unequal probability sampling. *JISAS*, **29**, pp. 98–100.

_____, Kumar, P. and Dey, A. (1982): On unequal probability sampling of two units without replacement. *JRSS* B, **44** (1), pp. 43–46.

Ajgaonkar, S.G.P. (1967): Unbiased estimator of the variance of the Narain, Horvitz and Thompson estimator, *Sankhya* A, **29**, pp. 55–60.

Amahia, G.N., Choubey, Y.P. and Rao, T.J. (1989): Efficiency of a new estimator in *pps*-sampling of multiple characteristics, *JSPI,* **15**, pp. 75–84.

Andreatta, G. and Kaufman, G.M. (1986): Estimation of finite population when sampling is without replacement and proportional to magnitude, *JASA*, **81**, pp. 657–66.

Arnold, S.F. (1979): Linear models with exchangeably distributed errors, *JASA*, **74**, pp. 194–99.

Asoke, C. and Sukhatme, B.V. (1976): On Sampford's procedure of unequal probability sampling without replacement, *JASA,* **71**, pp. 912–18.

Avadhani, M.S. and Sukhatme, B.V. (1967): Controlled sampling with varying probabilities and without replacement. *AJS*, **9**, pp. 8–15.

_____ (1968): Stratified sampling for designing controlled simple random sampling. *AJS*, **10**, pp. 1–17.

_____ (1973): Controlled sampling with unequal probabilities and without replacement. *ISR*, **41 (2)**, pp. 175–82.

Bansal, M.L. and Singh, R. (1985): An alternative estimator for multiple characteristic in *pps*-sampling, *JSPI*, **11**, pp. 313–20.

Basu, D. (1958): On sampling with and without replacement, *Sankhya*, **20**, pp. 287–94.

_____ (1969): Role of sufficiency and the likelihood principles in sample survey theory. *Sankhya*, **A, 31 (2)**, pp. 441–54.

_____ (1971): An essay on the logical foundation of survey sampling. Part I, in V.P. Godambe and D.A. Sprott eds. Foundations of Statistical Inference. Toronto, Holt. Rinehart and Winston, pp. 203–42.

_____ and Ghosh, J.K. (1967): Sufficient statistics in sampling from a finite universe. *BISI*, **42** (2), pp. 85–89.

Bayless, D. L. and Rao, J.N.K. (1970): An empirical study of the stabilities of estimators and variance estimators in unequal probability sampling ($n = 3, 4$), *JASA*, **65**, pp. 1645–67.

Bellhouse, R.D. (1984): A review of optimal designs in survey sampling. *Can. J. Stat.*, **12**, pp. 57–65.

Bellhouse, D.R. and Joshi, V.M. (1984): On the admissibility of regression estimator. *JRSS* B, **46**, pp. 268–69.

Berg, S. (1974): A note on the RHC method, *Sc. JS*, **57**, pp.108–14.

Bethlehem, S.C. and Schuerhoff, M.H. (1984): Second order inclusion-probability in sequential sampling without replacement with unequal probabilities, *Boim.*, **71**, pp. 642–44.

Bhattacharyya, S. (1993): Some studies on estimation of mean and variance in finite population sampling, Ph.D. Thesis submitted to the Indian Statistical Institute, Calcutta.

Bickel, P.J. and Freedman, D.A. (1984): Asymptotic normality and the bootstrap in stratified sampling, *AS*, **12**, pp. 470–82.

Bickel, P.J. and Lehmann, E.L. (1981): A minimax property of the sample mean in finite populations, *AS*, **9**, pp. 1119–22.

Binder, D.A. (1983): On the variance of asymptotically normal estimators from complex surveys, *ISR*, **51**, pp. 279–92.

Biyani, S.H. (1980): On inadmissibility of the Yates-Grundy variance estimator in unequal probability sampling. *JASA*, **75**, pp. 709–12.

Blackwell, D. and Girshick, M.A. (1954): Theory of games and statistical decisions. Wiley, N.Y.

Bobbington, A.C. (1975): A simple method of drawing a sample without replacement, *Ap. St.*, **24**, 136.

Bolfarine, H. and Rodrigues, J. (1988): On the simple projection predictor in finite population, *AJS*, **30 (3),** pp. 338–41.

Booth, J.G., Butler, R.W. and Hall, P.G. (1991): Bootstrap methods for finite population. Australian National University Tech. Rep. SMS–96–91, Canberra.

Breckling, J.U., Chambers, R.L., Dorfman, A.H., Tam, S.M. and Walsh, A.M. (1990): Maximum likelihood inference from sample survey data; Australian National University, Tech. Rep. No. SMS–025–90.

Brewer, K.R.W. (1963): A model of systematic sampling with unequal probabilities. *AJS,* **5**, pp. 5–13.

_____ (1963b): Ratio estimation in finite populations: some results deducible from the assumptions of an underlying stochastic process. *AJS*, **5**, pp. 93–105.

_____ (1967): A note on Fellegi's method of sampling without replacement with probability proportional to size, *JASA*, **62**, pp. 79–85.

_____ (1975): A simple procedure for sampling πps without replacement. *AJS*, **17 (3),** pp. 166–72.

_____ (1979): A class of robust sampling designs for large scale surveys, *JASA*, **74**, pp. 311–15.

Brewer, K.R.W. and Hanif, M. (1983): Sampling with unequal probabilities. Lecture notes in Statistics, No. 15, Springer Verlag, New York.

_____, M and Tam, S. (1988): How nearly can model-based prediction and design-based estimation be reconciled? *JASA*, **83**, pp. 128–32.

_____, Samiuddin, M. and Asad, H. (1991): Small sample performance of five unequal probability sampling estimators. (Mss)

_____ and Undy, G.C. (1962): Samples of two units drawn with unequal probability without replacement, *AJS*, **4**, pp. 89–100.

Brillinger, D.R. (1966): The application of the jackknife to the analysis of sample surveys. Commentary, **8**, pp. 74–80.

Callebaut, D.K. (1965): Generalisation of Cauchy-Schwartz inequality. *J. Math. Analysis & Applications*, **12**, pp. 491–94.

Carrol, J.L. and Hartley, H.O. (1964): The symmetric method of unequal probability sampling without replacement. *Bioms* (abstract only), **20**, pp. 908–9.

Cassel, C.M., Sarndal, C.E. and Wretman, J.H. (1976): Some results on generalised difference estimator and generalised regression estimator for finite population. *Biom.*, **63,** pp. 614–20.

_____ (1977): Foundations of inference in survey sampling, John Wiley & Sons.

_____ (1979): Prediction theory for finite population when model-based and design-based principles are combined. *Sc. JS*, **6**, pp. 97–106.

Chakravorty, M.C. (1963): On the use of incidence matrices of designs in sampling for finite universe, *JISA*, **1**, pp.78–85.

Chakravorty, R.P. and Rao, J.N.K. (1968): The bias and stability of the jackknife variance estimators in ratio estimation. *JASA*, **63**, p.748.

Chao, M.T. (1982): A general purpose unequal probability sampling plan. *Biom.*, **69**, pp. 653–56.

_____ and Lo, S.H. (1985): A bootstrap method for finite population. *Sankhya*, A **47**, pp. 399–405.

Chaudhuri, A. (1969): Minimax solutions of some problems in sampling from a finite population. *CSAB*, **18**, pp. 1–24.

_____ (1971): Some sampling schemes to use Horvitz-Thompson estimator in estimating a finite population total, *CSAB*, **20**, pp. 37–66.

_____ (1974): On some properties of the sampling scheme due to Midzuno. *CSAB*, **23**, pp. 1–20.

_____ (1981): Non-negative unbiased variance estimators, in Current topics in survey sampling, ed. Krewski, D., Plateck, R. and Rao, J.N.K., pp. 317–28, Academic Press, New York.

_____ and Arnab, R. (1979): On the relative efficiency of sampling strategies under a superpopulation model. *Sankhya, C*, **41 (1)**, pp. 40–43.

_____ and Vos, J.W.E. (1988): Unified theory and strategies of survey sampling, North Holland.

Cheng, C.S. and Li, K.C. (1983): A minimax approach to sample surveys. *AS*, **11**, pp. 552–63.

Cheng, C.S. and Li, K.C. (1987): Optimality criteria in survey sampling. *Biom.*, **74**, pp. 337–45.

Chikkagoudar, M.S. (1967): On *PPP*-sampling with and without replacement *AJS,* **9 (3),** pp. 109–18.

_____ (1967b): Sampling with preliminary random stratification. *AJS*, **9**, pp. 57–70.

_____ (1969): Inverse sampling without replacement. *AJS*, **11**, pp. 155–65.

Choudhry, G.H. (1979): Selecting a sample of size *n* with *ppswr* from a finite population, *SM*, **5**, (1), 79–95.

_____ and Singh, M.P. (1979): Sampling with unequal probabilities and without replacement — a rejective method, *SM,* pp. 162–77.

Chromy, J.R. (1979): Sequential sample selection methods. Unpublished.

_____ (1981): Variance estimation for a sequential sample selection procedure in Current topics in survey sampling, ed. Krewski, D. Plateck, R. and Rao, J.N.K., pp. 329–47, Academic Press, New York.

Cochran, W.G. (1939): The use of the analysis of variance in enumeration by sampling. *JASA*, **34**, pp. 492–510.

_____ (1946): Relative accuracy of systematic and stratified random samples for a certain class of populations. *AMS,* **17**, pp. 164–77.

_____ (1977): Sampling techniques. 3rd ed. New York, Wiley.

Cox, D.R. and Hinkley, D.V. (1974): Theoretical Statistics. London, Chapman and Hall.

Cramer, H. (1946): Mathematical methods of statistics. Princeton University Press.

Dahiya, R.C. Korear, R.M. and Lew, R.A. (1977): On inverse sampling. *Sankhya*, **C 39,** pp. 162–69.

Daleneus, T. (1962): Recent advances in sample survey theory and methods, *AMS*, **33**, pp. 325–49.

Das, A.C. (1951): Two phase sampling and sampling with varying probabilities. *BISI*, **33**, pp. 105–12.

_____ (1962): On MVU estimates of population parameters of a finite population based on varying probability samples. *CSAB*, **11**, pp. 39–45.

Das, M.N. and Mahanty, G.S. (1973): On PPS-sampling without replacement ensuring selection probabilities exactly proportional to size. *AJS*, **15**, pp. 87–94.

Deming, W.E. (1956): On simplification of sampling designs through replications with equal probabilities and without replacement and without stages. *JASA*, **51**, pp. 24–53.

Deng, L.Y. and Chhikara, R.S. (1991): On asymptotically design-unbiased estimators of a finite population mean. *Biom.*, **78**, pp. 189–95.

Deshpande, M.N. (1978): A new sampling procedure with varying probabilities. *JISAS*, **30**, pp. 110 –24.

_____ (1982): A new fixed size sampling procedure with unequal probability of selection. *JISAS*, **34**, pp. 71–75.

_____ (1984): A new method of sampling with unequal probabilities. *JISAS*, **36 (2),** pp. 23–36.

_____ (1984b): A note on Rao-Hartley and Cochran's method: *JISAS*, **36 (3),** pp. 114–16.

_____ and Ajgaonkar, S.G.P. (1977): On multitrial sampling methods. *Biom.*, **64**, pp. 422–24.

_____ (1987): A generalisation of Midzuno sampling procedure. *AJS*, **29**, pp. 188–92.

Des Raj (1956): A note on the determination of the optimum probabilities in stratified samples without replacement. *Sankhya*, **17**, pp. 197–200.

_____ (1966): On a method of sampling with unequal probabilities, *Ganita*, **17**.

_____ and Khamis, H.S. (1958): Some remarks on sampling with replacement. *AMS*, **29**, pp. 350–57.

Dodds, D.J. and Fryer, J.G. (1971): Some families of selection procedures of sampling with probability proportional to size. *JRSS*, **B, 33**, pp. 263–74.

Durbin, J. (1953): Some results in sampling theory when the units are selected with unequal probabilities. *JRSS*, **B, 15**, 262–9.

_____ (1967): Estimation of sampling error in multistage surveys. *Ap. St.,* **16,** pp. 152–64.

Efron, B. (1982): The jackknife, the bootstrap and other resampling plans. *SIAM*, monograph no. 38, Philadelphia.

Ericson, W.A. (1974): A note on the non-existence of minimum variance unbiased estimator for a finite population total. *Sankhya*, **C, 36,** pp. 181–87.

Fellegi, I.P. (1963): Sampling with varying probabilities without replacement, rotating and non-rotating samples. *JASA*, **58**, pp. 183–201.

Foreman, E.K. and Brewer, K.R.W. (1971): The efficient use of supplementary information in standard sampling procedures. *JRSS*, **B, 33**, pp. 391–400.

Frankel, M.R. (1971): Inference from survey samples. Ann. Arbor: Instt. for Social Research, University of Michigan.

Fuller, W.A. (1970): Sampling with random strata boundaries. *JRSS*, **B, 32,** pp. 209–26.

_____ (Mss): A procedure for selecting non-replacement unequal probabilities samples.

_____ and Isaki, C.T. (1981): Survey designs under superpopulation models in Current topics in survey sampling, ed. Krewski D. Plateck, R. and Rao, J.N.K; pp. 199–226, Academic Press, New York.

Gabler, S. (1981): A comparison of Sampford's procedure versus unequal probability sampling with replacement: *Biom.*, **68**, pp. 725–27.

_____ (1984): On unequal probability sampling: sufficient conditions for the superiority of sampling without replacement. *Biom.*, **71,** pp. 171–75.

_____ (1985): Horvitz-Thompson strategy versus RHC strategy. *Biom..* **72,** pp. 111–13.

_____ (1987): The nearest proportional to size sampling design. *CS (TM),* **16(4)**, pp. 1117–31.

_____ (1990): Minimax solutions in sampling from finite populations. Springer Verlag, New York.

Godambe, V.P. (1955): A unified theory of sampling from finite population. *JRSS,* **B, 17,** pp. 269–75.

_____(1960): An admissible estimate for any sampling design. *Sankhya*, **22,** pp. 285–88.

_____ (1966): A new approach to sampling from finite population, I, II, *JRSS,* **B, 28,** pp. 310–28.

_____ (1969): Admissibility and Bayes estimation in sampling finite population V, *AMS,* **40**, pp. 672–76.

_____ (1982): Estimation in survey sampling: robustness and optimality. *JASA,* **77,** pp. 393–403.

_____ and Joshi, V.M. (1965): Admissibility and Bayes estimation in sampling from finite population. I, *AMS,* **36,** pp. 1707–22.

Godambe, V.P. and Thompson, M.E. (1977): Robust — near optimal estimation in survey practice. *BISI,* **47(3),** pp. 129-46.

Goodman, R. and Kish, L. (1950): Controlled selection — a technique in probability sampling. *JASA,* **45,** pp. 350–72.

Gray, H.L. and Schucany, W.R. (1972): The generalised jackknife statistics. Marcel Drekker, New York.

Gross, S. (1980): Median estimation in sample surveys. Proc. Sec. Survey Res. Meth. Am. Statist. Assoc., pp.181–84.

Grundy, P.M. (1954): A method of sampling with probability exactly proportional to size. *JRSS,* **B, 16,** pp. 236–38.

Gupta, V.K., Nigam, A.K. and Kumar, P. (1982): On a family of sampling schemes with inclusion-probability proportional to size. *Biom.*, **69,** pp. 191–96.

Hájek, J. (1958): Some contributions to the theory of probability sampling. *BISI,* **36(3),** pp. 127–34.

_____ (1959): Optimum strategies and other problems in probability sampling. *Časopis Pěst. Mat,* **84,** pp. 387–123.

Hájek, J. (1964): Asymptotic theory of rejective sampling with varying probability from a finite population, *AMS,* **35,** pp. 1491–1523.

_____ (1971): "Comments on an essay on the logical foundation of survey sampling, part one, by D. Basu" in Foundation of statistical inference ed. Godambe, V.P. and Sprott, D.A., pp. 203–42, Holt, Rinehart and Winston, Toronto.

Hanif, M. and Brewer, K.R.W. (1980): Sampling with unequal probabilities without replacement: a review, *ISR,* **48,** pp. 317–35.

_____, Mukhopadhyay, P. and Bhattacharyya, S. (1993): On estimating the variance of the Horvitz-Thompson estimator, *Pak. J. Stat.*, **9(2)A,** pp. 123–36.

Hansen, M.H., Madow, W.G. and Tepping, B.J. (1983): An evaluation of model-dependent and probability sampling inferences in sample surveys. *JASA,* **78,** pp. 776–93.

Hanurav, T.V. (1962): An existence theorem in sample survey. *Sankhya,* **A, 24,** pp. 327–30.

_____ (1962b): On Horvitz and Thompson estimator. *Sankhya,* **A, 24,** pp. 429–36.

_____ (1962c): Some sampling schemes in probability sampling. *Sankhya,* **A, 24,** pp. 421–28.

_____ (1966): Some aspects of unified sampling theory, *Sankhya,* **A, 28,** pp. 175–203.

_____ (1967): Optimum utilisation of auxiliary information: πps-sampling of two units from a stratum. *JRSS,* **B, 23,** pp. 374–91.

Hartley, H.O. (1966): Systematic sampling with unequal probabilities and without replacement. *JASA,* **61,** pp. 739–48.

_____ and Chakravorty, R.P. (1967): Evaluation of approximate variance formulas in sampling with unequal probabilities and without replacement. *Sankhya,* **B, 29,** pp. 201–08.

Hartley, H.O. and Rao, J.N.K. (1962): Sampling with unequal probabilities and without replacement. *AMS,* **33,** pp. 350–74.

_____ (1968): A new estimation theory for sample surveys. *Biom.,* **55,** pp. 547– 57.

_____ and Rao, J.N.K. (1969): A new estimation theory for sample surveys, II: in New developments in survey sampling, ed. Johnson, N.L. and Smith, H., Jr., pp. 147–69. Wiley Interscience, New York.

_____ and Ross, A. (1954): Unbiased ratio estimates. *Nature,* **174,** pp. 270–71.

Hege, V.S. (1965): Sampling designs which admit uniformly minimum variance unbiased estimates. *CSAB,* **14,** pp. 160–62.

Herson, J. (1976): An investigation of relative efficiency of least square prediction by conventional probability sampling plan. *JASA,* **71,** pp. 700–3.

Herzel, A. (1986): Sampling without replacement with unequal probabilities sample designs with pre-assigned joint inclusion-probabilities of any order. *Metron XLIV* **(l),** pp. 49–68.

Hodges, J.L. Jr. and Lehmann, E.L. (1982): Minimax estimation in simple random sampling, in Statistics and Probability Essays in honour of C.R. Rao, ed. Kalianpúr, G., Krishnaiah, P.R. Ghosh, J.K., pp. 325–27, North Holland.

Horvitz, D.G. and Thompson, D.J. (1952): A generalisation of sampling without replacement from a finite universe. *JASA*, **64,** pp. 175–95.

Isaki, C.T. and Fuller, W.A. (1982): Survey designs under the regression super-population model. *JASA,* **77,** pp. 89–96.

Jessen, R.J. (1942): Statistical investigation of a sample survey for obtaining firm facts. Iowa Agr. Expt. Sta. Res. Bull., pp. 304.

_____ (1969): Some methods of probability non-replacement sampling. *JASA,* **64,** pp. 174–93.

Joshi, V.M. (1965): Admissibility and Bayes estimation in sampling finite populations II, *AMS,* **36,** pp. 1229.

_____ (1965b): Admissibility and Bayes estimation in sampling finite populations III, *AMS,* **36,** pp. 1730–42.

_____ (1966): Admissibility and Bayes estimation in sampling finite populations IV, *AMS*, **36, 37,** pp. 1658–70.

_____ (1969): Admissibility of the estimates of the mean of a finite population in New developments in survey sampling, ed. Johnson N.L. and Smith H. Jr, pp. 188–212, Wiley Interscience, New York.

_____ (1970): Note on the admissibility of the Sen-Yates-Grundy variance estimator and Murthy's estimator and its variance estimator for samples of size two, *Sankhya,* **A, 32,** pp. 431–38.

_____ (1979): Joint admissibility of the sample means as estimators of the means of finite populations *AS,* **7,** pp. 995–1002.

Kempthrone, O. (1969): Some remarks on statistical inference in finite sampling in New developments in survey sampling, ed. Johnson, N.L. and Smith, H., Jr., pp. 671–95. Wiley Interscience, New York.

Kharshikar, A.V. (1980): On almost sure non-negativity of the Horvitz-Thompson and Sen-Yates-Grundy estimator of the variance. *JISAS,* **18,** pp. 91–98.

Kiefer, J. (1959): Optimum experimental designs. *JRSS,* **B 21,** pp. 272–319.

_____ (1970): Optimum experimental designs. Proc. Intl. Congress Math. 3, Nice, Gauthiers Villars, Paris, pp. 249–54.

_____ (1975): Construction and optimality of generalised Youden design in a survey of statistical designs and linear models, ed. Srivastava J.N., pp. 333–54, North Holland, Amsterdam.

_____ and Wolfwitz, J. (1959): Optimum designs in regression problems. *JRSS,* **B 21**, pp. 271–94.

_____ and Wynn, H.P. (1981): Optimum balanced block and Latin square designs for correlated observations. *AMS,* **29,** pp. 675–99.

Kish, L. (1965): Survey Sampling. John Wiley and Sons.

_____ and Frankel, M.R. (1970): Balanced repeated replication for standard errors. *JASA,* **65,** pp. 1071–94.

Konijn, H.S. (1973): Statistical theory of Sample Survey Design and Analysis. North-Holland.

Koop, J.C. (1972): On the derivation of expected value and variance of ratios without the use of infinite series expansions, *Metrika,* **19,** pp. 156–70.

Kott, P. S. (1986): Some asymptotic results for the systematic and stratified sampling of a finite population. *Biom.,* **73,** pp. 485–91.

_____ (1986b): When a mean of ratios is the best linear unbiased estimator under a model. *Am. St.,* **40(3),** pp. 202–4.

_____ (1988): Model based finite population correction for the Horvitz, Thompson estimator. *Biom.,* **75,** pp. 797–99.

_____ (1990): Estimating the conditional variance of a design consistent estimator. *JSPI,* **24,** pp. 287–96.

Krewski, D. and Rao, J.N.K. (1981): Inference from stratified samples: properties of the linearisation, jackknife and balanced repeated replication methods. *AS,* **9,** pp. 1010–19.

Kumar, N. (1991): Controlled sampling for unbiased regression estimator (Mss).

Kumar, P. (1987): On sampling three units using varying probabilities. Statistics, **18,** pp. 373–77.

_____ and Agarwal, S.K. (1982): On a simple procedure of inclusion probabilities proportional to size (IPPS) sampling. *Biom., J.,* **24,** pp. 801–16.

_____, Gupta, V.K. and Agarwal, S.K. (1985): On variance estimation in unequal probability sampling. **AJS**, **27(2),** pp. 195–201.

Kunth, D. (1973): Art of computer programming 2, 2nd edn. Reading, Mass: Addison Wesley.

Lahiri, D.B. (1951): A method of sample selection providing unbiased ratio estimates. *BISI,* **3(2),** pp. 133–40.

Lanke, J. (1974): On non-negative variance estimators in survey sampling. *Sankhya,* **C, 30,** pp. 33–42.

_____ (1975): Some contributions to the theory of survey sampling. *AV-Centralin i Lund.*

Lee, K.H. (1972): Partially balanced designs for half-sample replications method of variance estimation. *JASA,* **67,** pp. 324–34.

_____ (1973): Using partially balanced designs for the half-sample replication method of variance estimation. *JASA,* **68,** pp. 612–14.

Lemeshow, S. and Epp, R. (1977): Properties of the balanced half-samples and the jackknife variance estimation techniques in the linear case. *CS(TM),* **6(13),** pp. 1259–74.

Lenz, S.T. (1989): Non-negative variance estimation of the Horvitz-Thompson estimator for samples of fixed size when sampling without replacement, Mannheim.

Levy, P.S. (1971): A comparison of balanced half-sample replication and jackknife estimators of the variance of the ratio estimates in complex sampling with two primary units per stratum (Mss). National Center for Health Statistics, Rockville, MD.

Li, K.C. (1983): Consistent, asymptotically efficient sampling strategies (Mss).

Liao, H. and Sedransk, J. (1975): Sequential sampling for comparison of domain means. *Biom.,* **62,** pp. 691–93.

Little, R.J. (1983): Estimating a finite population mean from unequal probability sampling. *JASA,* **78,** pp. 596–604.

Liu, T.P. (1984): The existence of Sen-Yates-Grundy form of a dispersion. *AJS,* **29,** pp. 97–101.

Madow, W.G. (1949): On the theory of systematic sampling II, *AMS,* **20,** pp. 333–54.

Mahalanobis, P.C. (1944): On large scale sample surveys. Philosophical transaction of the Royal Society, **B, 231,** pp. 329–451.

_____ (1946): Recent developments in statistical sampling in the Indian Statistical Institute. *JRSS,* **A, 109,** pp. 325–78.

McCarthy, P. J. (1946): Replication. An approach to the analysis of data from complex surveys in Vital and Health Statistics, Sr. 2, No. 14, US Deptt. of Health, Education and Welfare, Washington, US Govt, printing office.

_____ (1969): Pseudoreplication: Half samples. *ISR,* **37,** pp. 239–64.

_____ (1969b): Pseudoreplication: Further evaluation and application of balanced half-sample techniques. Vital and Health Statistics, Sr. 2, No. 31, National Center for Health Statistics, Public Health Service, Washington, DC.

_____ and Snowdon, C.B. (1985): The bootstrap and finite population sampling in Vital and Health Statistics, Sr. 2, No. 95. Public Health Service Publn. 85-1369, US Govt, printing office, Washington.

Mickey, M.R. (1959): Some finite population unbiased ratio and regression estimators. *JASA,* **54,** pp. 594–612.

Midzuno, H. (1950): An outline of the theory of sampling systems. *AISM,* **1,** pp. 149–51.

_____ (1952): On the sampling system with probability proportional to sum of sizes. *AISM,* **3,** pp. 99–107.

Miller, R.G. Jr. (1964): A trustworthy jackknife. *AMS,* **35,** pp. 1594–1605.

_____ (1974): The jackknife—a review. *Biom.,* **61,** pp. 1–15.

Montenari, G.E. (1979): Post-sample efficient QR prediction in large sample surveys. ISR, 55(2), 191–202.

Mukherjee, R. and Sengupta, S. (1989): Optimal estimation of a finite population total under a general correlated model. *Biom.,* **76,** pp. 789–94.

Mukhopadhyay, P. (1972): A sampling scheme to realise a pre-assigned set of inclusion-probabilities of first two orders. *CSAB,* **21,** pp. 87–122.

_____ (1974): πps-sampling schemes to base HTE. *CSAB,* **23,** pp. 21–44.

_____ (1975): An optimum sampling design to base HT-method of estimating a finite poulation total, *Metrika,* **22,** pp. 119–27.

_____ (1977): Further studies in sampling theory. Unpublished Ph. D. thesis submitted to the University of Calcutta.

_____ (1977b): Robust estimation of finite population total under certain linear regression models. *Sankhya,* **C, 39,** pp. 71–87.

_____ (1985): Estimation under linear regression models. *Metrika,* **32,** pp. 339–49.

_____ (1985b): Asymptotic properties of some BLU-estimators. Orissa Math. Soc. *Jour.* **4,** pp. 105–13.

_____ (1991): Optimal estimation of a finite population mean under exchangeable general linear models. *JISA,* **29,** pp. 79–88.

_____ (1991): Varying probability without replacement sampling designs a review. IS1 Tech. Rep. No. ASC/91/3.

_____ and Bagchi, T. (1991): Generalising Midzuno sampling strategies (Mss).

_____ (1992): On comparing sampling strategies. *JISAS,* **44(1),** pp.138–48.

Mukhopadhyay, P. and Bhattacharyya, S. (1975): On an optimal property of Hájek's estimator (Mss).

_____ (1990): On sampling designs estimating a finite population variance (Mss).

_____ (1991): On a comparison between Midzuno's strategy and *PPSWR*-strategy. *Statistics,* **22,** pp. 291–94.

_____ (1994): Prediction under balanced samples. *JSPI,* **39,** pp. 85–93.

Mukhopadhyay, P. and Vijayan, K. (1990): On non-negative unbiased estimation of quadratic forms in finite population sampling. University of Western Australia Tech. Rep. No. 41.

_____ (1992): On controlled sampling designs (to appear in JSPI).

Murthy, M.N. (1957): Ordered and unordered estimators in sampling without replacement. *Sankhya,* **18,** pp. 379–90.

_____ (1963): Generalised unbiased estimation in sampling from finite populations. *Sankhya,* **B, 25,** pp. 245–62.

_____ (1977): Sampling theory and methods. 2nd edn. Stat. Pub. Soc. Calcutta.

Narain, R.D. (1951): On sampling with varying probabilities without replacement. *JISAS,* **B,** pp. 169–75.

Neyman, J. (1934): On the two different aspects of representative method of stratified sampling and the method of purposive selection. *JRSS,* **97,** pp. 558–606.

Nigam, A.K., Kumar, P. and Gupta, V.K. (1984): Some methods of inclusion probability proportional to size sampling. *JRSS,* **B, 40,** pp. 564–71.

Ohlsson, E.: A generalisation of Rao-Hartley-Cochran sampling procedure. Deptt. of Math. Royall Instt. of Technology, Stockholm Tech. Rep. no. 17.

Padmawar, V.R. (1981): A note on comparison of certain sampling strategies. *JRSS,* **B, 43,** pp. 321–26.

_____ (1982): Optimal strategies under superpopulation models. Unpublished Ph.D. thesis submitted to the Indian Statistical Institute, Calcutta.

_____ and Mukhopadhyay, P. (1984): On estimation of symmetric functions of a finite population. *Metrika,* **31,** pp. 89–97.

Pascual, J.N. (1961): Unbiased ratio estimators in stratified sampling. *JASA,* **56,** pp. 70–82.

Patel, H.C. and Dharmadhikari, S.K. (1978): Admissibility of Murthy's and Midzuno's estimator within the class of linear unbiased estimators of finite population total, *Sankhya,* **C 40,** p. 21.

Pathak, P.K. (1961): On the evaluation of moments of distinct units in a sample. *Sankhya,* **A 23**, pp.409–14.

_____ (1962): On simple random sampling with replacement. *Sankhya,* **A, 24,** pp. 287–301.

_____ (1969): Asymptotic efficiency of Des Raj's strategy I, II. *Sankhya,* **A, 29,** pp. 283–298, 299–304.

Pereira, C.A.B. and Rodrigues, J. (1983): Robust linear prediction in finite population. *ISR,* **51,** pp. 293–300.

Pfeffermann, D. (1984): A note on large sample properties of balanced samples. *JRSS,* **B, 46,** pp. 38–41.

Plackett, R.G. and Burman, P.J. (1946): The design of optimum factorial experiments. *Biom.,* **33,** pp. 305–25.

Prasad, N.G.N. and Srivenkataraman, T.A. (1980): A modification to Horvitz Thompson estimator under Midzuno-sampling scheme. *Biom.,* **67,** pp. 709–11.

Quenouille, M.H. (1949): Approximate tests of correlation in time series. *JRSS* **B, 11,** pp. 68–84.

_____ (1956): Notes on bias in estimation. *Biom.,* **43,** pp. 353–60.

Raghavarao, D. (1971): Construction and combinatorial problems in design of experiments, John Wiley and Sons.

Ramakrishnan, M.K. (1971): Generalisation of Yates and Grundy estimates. Res. Rep. 92/MKR/l, Manchester, Sheffild.

_____ (1975): A generalisation of Yates-Grundy variance estimator, *Sankhya* **C, 37,** pp. 204–06.

Rao, C.R. (1971) : Some aspects of statistical inference in problems of sampling from finite populations, in foundations of statistical inference ed. Godambe, V.P. and Sprott D.A., Toronto, Holt, Rinehart and Winston, pp. 177–202.

Rao, J.N.K.(1961): On the estimate of variance in unequal probability sampling. *AISM,* **13 (1),** pp. 57–60.

_____ (1963): On two systems of unequal probability sampling without replacement, *AISM,* **15,** pp. 61–12.

_____ (1963b): On three procedures of unequal probability sampling without replacement. *JASA,* **57,** pp. 202–15.

_____ (1965): On two simple properties of unequal probability sampling without replacement. *JISAS,* **3,** 173–80.

_____ (1965b): A note on the estimation of ratios by Quenouille's method. *Biom.,* **52,** pp. 647–49.

_____ (1966): Alternative estimators in *pps* sampling for multiple characteristics, *Sankhya,* **A 28,** pp. 47–60.

_____ (1966b): On the relative efficiency of some estimators in *pps* sampling for multiple characteristics; *Sankhya,* **A 28,** pp. 61–70.

_____ (1969): Ratio and regression estimators, in New development in survey sampling, ed., Johnson N.L. and Smith H., Jr, Wiley Interscience, N.Y.

_____ (1979): On deriving mean square error and their non-negative unbiased estimators, *JISAS,* **17,** pp. 125–36.

_____ and Bayless, D.L. (1969): An empirical study of the stabilities of estimators and variance estimators in unequal probability sampling of two units per stratum, *JASA,* **64,** pp. 540–49.

Rao, J.N.K. and Bayless, D.L. (1969): An empirical study of the stabilities of estimators and variance estimators in unequal probability sampling of two units per stratum. *JASA,* **64,** pp. 540–59.

_____, Hartley, H.O. and Cochran, W.G. (1962): On a simple procedure of unequal probability sampling without replacement. *JRSS,* **B 24,** pp. 482–91.

_____ and Nigam, A.K.(1989): Controlled sampling with probability proportional to aggregate size. Tech. Repp. 133, Laboratory to Res. and Stat. and Prob., Carleton University, Ottawa.

_____ and Nigam, A.K. (1990): Optimum controlled sampling designs. *Biom.,* **77,** pp. 807–14.

_____ and Vijayan, K. (1977): On estimating the variance in sampling with probability proportional to aggregate size. *JASA,* **72,** 579–84.

_____ and Webster, J.T. (1966): On two methods of bias reduction in the estimation of ratios. *Biom.,* **53,** pp. 571–77.

_____ and Wu, C.F.J. (1985): Inference from stratified samples: second order analysis of three methods for non-linear statistic, *JASA,* **80,** pp. 620–30.

_____ (1988): Resampling inference with complex survey data. *JASA,* **83,** pp. 231–41.

Rao, P.S.R.S. and Rao, J.N.K. (1971): Small sample results for ratio estimation. *Biom.,* **58,** pp. 625–30.

Rao, T.J. (1967): On the choice of strategy for ratio method of estimation, *JRSS,* **B 29,** pp. 392–97.

_____ (1984): On Brewer's class of robust sampling designs for large scale surveys. *Metrika,* **31,** pp. 319–22.

Robinson, P. M. (1982): On the convergence of the Horvitz-Thompson estimator. *AJS,* **24,** pp. 234–838.

_____ and Sarndal, C.E. (1983): Asymptotic properties of the generalised regression estimator in probability sampling. *Sankhya,* **B, 45,** pp. 240–48.

_____ and Tsui, K.W.(1979): Optimal asymptotically design unbiased estimation of population total. Tech. Rep. 634, Dept. of Stat., University of Wisconsin, Madison.

Rodrigues, J. (1984): Robust estimation and finite populations. Prob. Math. *Statist.,* **4,** pp. 197–207.

_____ , Bolfarine H. and Rogatko, A. (1985): A general theory of prediction in finite population. *ISR,* **53 (2),** pp. 239–54.

Roy, J. and Chakraborty, I.M. (1960): Estimating the mean of finite population, *AMS,* **31,** pp. 392–8.

Royall, R.M. (1968): An old approach to finite population sample theory. *JASA,* **63,** pp. 1269–79.

_____ (1970): On finite population sampling theory under certain linear regression models. *Biom.,* **57,** pp. 377–87.

_____ (1970b): Finite population sampling — on labels in estimation. *AMS,* **41,** pp. 1774–79.

_____ (1971): Linear regression models in finite population sampling theory in Foundation of statistical inference, ed. Godambe, V.R. and Sprott, D.R. pp. 259–77, Toronto Holt, Rinehart and Winston.

_____ (1976): The linear least square prediction approach to two stage sampling. *JASA,* **71,** pp. 657–64.

_____ (1973b): Robust estimation in finite populations II. stratification on a size variable. *JASA,* **68,** pp. 890–93.

_____ and Cumberland, W.G. (1978): Variance estimation in finite population sampling. *JASA,* **73,** pp. 351–58.

_____ (1978b): An empirical study of prediction theory in finite population sampling: simple random sampling and the ratio estimator in Survey sampling and measurement, ed., Namboodiri N.K. pp. 293–314, Academic Press, N.Y.

_____ (1981): An empirical study of the ratio estimator and estimator of its variance. *JASA,* **76,** pp. 66–77.

_____ (1981b): The finite population regression estimator and estimator of its variance an empirical study. *JASA,* **76,** pp. 924–30.

_____ (1981c): Prediction models and unequal probability sampling. *JRSS,* **B, 49,** pp. 353–67.

_____ and Eberherdt, J. (1975): Variance estimates for the ratio estimator. *Sankhya,* **C, 37,** pp. 43–52.

_____ and Herson, J. (1973): Robust estimation in finite populations I. *JASA,* **68,** pp. 880–89.

_____ and Pfeffermann, D. (1982). Balanced samples and robust Bayesian inference in finite population sampling. *Biom.,* **69,** pp. 2401–9.

Rust, K. (1984): Techniques for estimating variances for sample surveys. Ph.D. thesis, University of Michigan.

_____ (1986): Efficient replicated variance estimation, in Proc. Sec. Survey Res. Meth., Washington, D.C.: Am Stat. Assoc.

Samiuddin, M. and Asad, H. (1981): A simple procedure for unequal probability sampling, *Biom.,* **68,** pp. 728–31.

Sampford, M.R. (1967): On sampling without replacement with unequal probability of selection. *Biom.,* **54,** pp. 499–513.

Sankaranarayanan, K. (1969): An IPPS sampling scheme using Lahiri's method of selection. *JISAS,* **21 (2),** pp. 59–66.

_____ (1980): A note on the admissibility of some non-negative quadratic estimators. *JRSS,* **B, 42,** 387–89.

Sarndal, C.E. (1976): On uniformly minimum variance estimation in finite population sampling. *AS,* **4,** pp. 993–97.

_____ (1980): Two model based inference argument in survey sampling *AJS,* **22,** pp. 314 48.

_____ (1980b): A two way classification of regression estimation strategies in probability sampling. Canadian J. Stat., **8(2),** pp. 165–77.

_____ (1980c): On π-inverse weighting versus best linear weighting in probability sampling, *Biom.,* **67(3),** pp. 639–50.

_____ (1982): Implications of survey designs for estimation of linear functions. *JSP,* **7,** pp. 155–70.

Saxena, P.R., Singh, P. and Srivastava, A.K. (1986): An unequal probability sampling scheme. *Biom.,* **73,** pp. 761–63.

Schultz, D.M. (1990): Ordered random selection with probabilities proportional to size. South African Journal, 24, pp. 35–44.

Scott, A.J. (1975): On admissibility and uniform admissibility in finite population sampling. *AS,* **3,** pp. 489–91.

_____, Brewer, K.R.W. and Ho, E.W.H. (1978): Finite population sampling and robust estimation. *JASA*, **73,** pp. 359–61.

_____ and Smith, T.M.F. (1974): Linear superpopulation models in survey sampling. *Sankhya,* **C, 36 (3),** pp.143–46.

_____ (1975): Minimax designs for sample surveys. *Biom.,* **62,** pp. 353–57.

Scott, A.J. and Wu, C.F.J. (1981): On the asymptotic distribution of ratio and regression estimators. *JASA,* **76,** pp. 98–102.

Sekkappan, R.M. and Thompson, M.E. (1975): On a class of uniformly admissible estimators in finite populations, *AS,* **3,** 492–99.

Sen, A.R. (1952): Present status of probability sampling and its use in the estimation of a characteristic, (abstract), *Econometrika,* **20,** p. 103.

_____ (1953): On the estimate of variance in sampling with varying probabilities. *JISAS,* **5,** pp. 119–27.

Sengupta, S. (1980): On the admissibility of the generalised Des Raj estimator for *ppswr*–sampling of size two. *CSAB,* **29,** pp. 35–40.

_____ (1981): A sampling scheme to realise inclusion-probability exactly proportional to size. *JISAS,* **33(2),** pp.19–23.

Seth, G.R. (1966): On estimators of variance of estimate of population total in varying probabilities. *JISAS,* **18(2),** pp. 52–56.

Shah, K.R. and Sinha, B.K. (1991): Theory of optimal designs. Springer Verlag. New York.

Shao, J. (1989): The efficiency and consistency of approximations to the jackknife variance estimators. *JASA,* **84,** pp. 114–19.

_____ and Wu, C.F.J. (1989): A general theory of jackknife variance estimation. *AS,* **3,** pp. 1176–97.

Singh, P. (1978): The selection of samples of two units with inclusion-probability proportional to size. *Biom.,* **65,** pp. 450–54.

_____ (1978): A sampling scheme with inclusion-probability proportional to size, *Sankhya,* **C, 40,** pp. 122–28.

_____ and Garg, J.N. (1979): On balanced random sampling. *Sankhya,* **C, 41 (1),** pp. 60–66.

_____ and Singh, R. (1978): A sampling scheme with varying probabilities without replacement. *JISAS,* **C, 30 (2),** pp. 65–70.

_____ and Srivastava, A.K. (1980); Sampling schemes providing unbiased regression estimators, *Biom.,* **67,** pp. 205–9.

Singh, R. and Kishore, L. (1975): On Rao, Hartley and Cochran's method of sampling, *Sankhya,* **C, 37,** pp. 88–94.

_____ and Lal, M. (1978): On the construction of random groups in RHC scheme, *Sankhya,* **C, 40,** pp. 129–35.

Sinha, B.K. (1973): On sampling schemes to realise pre-assigned sets of inclusion-probabilities of first two orders, *CSAB,* **22,** pp. 69–110.

Skinner, C.J. (1983): Multivariate prediction from selected samples. *Biom.,* **70,** pp. 289–92.

Solomon, H. and Zacks, S. (1970). Optimal design of sampling from finite population: A critical review and indication of new research areas, *JASA,* **65,** pp. 653–77.

Srivastava, A.K. and Singh, D. (1981): A sampling procedure with inclusion probability proportional to size. *Biom.*, **68,** pp. 732–34.

Srivastava, J.N. and Salah, F. (1985): Need of *t*–design in sampling theory. *Utilitias Mathematica,* **25,** pp. 5–17.

Strenger, H. (1977): Sequential sampling for finite populations. *Sankhya,* **C. 39,** pp. 10–26.

_____ (1979): A minimax approach to randomisation and estimation in survey sampling, *AMS,* **17,** pp. 395–99.

Stevens, W.L. (1958): Sampling without replacement with *pps.* JRSS, **B, 20,** pp. 393–97.

Sukhatme, B.V. and Avadhani, M.S. (1965): Controlled selection—a technique in random sampling. *AISM,* **17**, pp. 15–28.

_____ and Sukhatme, B.V. (1970): Sampling theory of survey with applications, 2nd edn., Ames: Iowa State University Press.

Sunter, A.B. (1977): List sequential sampling with equal or unequal probabilities without replacement, **Ap. St.**, **26,** pp. 261–68.

_____ (1986): Solutions to the problems of unequal probability sampling without replacement, *ISR,* **54,** pp. 33–50.

Sutradhar, B.C. (1984): Durbin's sampling scheme versus unequal probability sampling scheme. *Sankhya,* **C, 46,** pp. 370 71.

Tallis, G.M. (1978): Note on robust estimation in finite populations. *Sankhya,* **C, 40,** pp. 136–38.

Tam, S. (1984): Optimal estimation in survey sampling under a regression super-population model. *Biom.,* **71,** pp. 645–77.

_____ (1986): Characterisation of best model-based predictors in survey sampling. *Biom.*, **73,** pp. 232–35.

_____ (1988): Asymptotic design unbiased predictors in survey sampling. *Biom.* **75,** pp. 175–77.

_____ (1988b): Some results on robust estimation in finite population sampling. *JASA,* **83,** pp. 242–48.

Tin, M. (1965): Comparison of some ratio estimators, *JASA,* **60,** pp. 294–307.

Tukey, J.W. (1958): Bias and confidence in not-quite large samples. Abstract. *AMS,* **29,** pp. 614.

Upadhyay, L.N., Singh, H.P. and Vos, J.W.E. (1985): On the estimation of ratio, product and mean using auxiliary information in sample surveys, *Statistica Netherlandica,* **39,** pp. 5–13.

Valliant, R. (1987): Some prediction properties of balanced half samples variance estimators in single stage samples. *JRSS,* **B, 49,** pp. 68–81.

Vijayan, K. (1966): Some contributions to sampling theory. Ph.D. thesis submitted to Indian Statistical Institute, Calcutta.

Vijayan, K. (1975): On estimating the variance in unequal probability sampling. *JASA,* **70,** pp. 713–16.

_____ , Mukhopadhyay, P. and Bhattacharyya, S. (1995): On non-negative unbiased estimation of quadratic forms in finite population sampling. *AJS,* **37,** pp. 169–178.

Vos, J.W.E. (1974): Sampling with unequal probabilities, *Statistica Netherlandica,* **28,** pp. 11–49 and 67–101.

Williams, W.H. (1962): On two methods of unbiased estimation with auxiliary variables. *JASA,* **57,** pp. 184–86.

Wolter, K.M. (1985): Introduction to variance estimation, Springer Verlag, New York.

Woodruff, R.S. and Causey, B.D. (1976): Computerised method approximating the variance of a complicated estimate. *JASA,* **71,** pp. 315–21.

Wright, R.L. (1983): Sampling designs with multivariate auxiliary information. *JASA,* **78,** pp. 879–84.

Wright, T. (1990); Probability proportional to size sampling using ranks. *CS(TM),* **19(1),** pp. 347–62.

Wu, C.F.J. (1982): Estimation of the variance of the ratio estimator. *Biom.*, **69,** pp. 188–89.

_____ and Deng, L.Y. (1983): Estimation of the variance of the ratio estimator: An empirical study, in Scientific inference, data analysis and robustness, ed. Box, G.E.P, Leonard, T. and Wu, *C.F.J,* pp. 245–47.

Wynn, H.P. (1977): Minimax purposive designs. *JASA,* **72,** pp. 471–77.

_____ (1977b): Optimum designs for finite population sampling in Statistical decision theory and related topics 2. ed. Gupta, S.S. and Moore, D., pp. 471–78, Academic Press, New York.

Yates, F. and Grundy, P.M. (1953): Selection without replacement within strata with probability proportional to sizes. *JRSS,* **B 15,** pp. 253–61.

Zacks, S. (1969): Bayes sequential designs of fixed size samples from finite populations. *JASA,* **64,** 1342–49.

Zyskind, G. (1967): On canonical forms, non-negative covariance matrices and best and simple least square estimators in linear models. *AMS,* **38,** pp. 1092–1109.

❑ ❑ ❑

Index

AUTHOR INDEX

E

F

G

H

I

J

K

L

M

T

U

V

W

Y

Z

SUBJECT INDEX

A

B

C

D

❑ ❑ ❑